*PHOTOPOLYMERIZATION OF
SURFACE COATINGS*

	6.1.1.1	Photochemical curing mechanism	265
	6.1.1.2	Alternative binders and sensitizers	268
6.2	Types of photoresists	274
6.2.1	Positive working photoresists	275
	6.2.1.1	Quinone diazides	275
	6.2.1.2	Acidic polymers	276
6.2.2	Negative working photoresists	276
	6.2.2.1	Photocycloaddition reactions	277
	6.2.2.2	Nitrene reactions	277
	6.2.2.3	Free-radical addition reactions	278
	6.2.2.4	Ring-opening crosslinking	278
6.3	Applications of photoresists	279
6.3.1	Printing plates	279
	6.3.1.1	Relief or raised-image plates	279
	6.3.1.2	Photolithography, planographic plates	286
	6.3.1.3	Photogravure	296
6.3.2	Photoengraving	296
6.3.3	Silkscreen printing	297
	6.3.3.1	Indirect photostencil	298
	6.3.3.2	Direct photostencil	298
	6.3.3.3	The direct/indirect photostencil	298
6.3.4	Printed circuits	300
6.3.5	Collotype printing	300
6.3.6	Proofing systems	300
6.4	References	300

7 Potential Hazards of ultraviolet systems 305

7.1	General areas of hazards	305
7.1.1	Lamps and machinery	305
7.1.2	Inks and coatings	306
	7.1.2.1	Possible hazards of acrylic ester monomers	308
7.2	Testing for physiological hazards	309
7.2.1	Skin tests	309
	7.2.1.1	Draize skin test	309
	7.2.1.2	Repeat insult skin test	311
7.2.2	Eye tests	312
	7.2.2.1	Draize eye test	312
	7.2.2.2	Eye test – FHSLA (1964)	314
7.2.3	Sensitization tests	314
	7.2.3.1	Draize/Landsteiner test	314

4.4.8	Typical formulations	201
4.4.8.1	UV clear coating varnish	202
4.4.8.2	UV silk screen ink	203
4.4.8.3	UV paste inks	203
4.4.8.4	Photoemulsion coating for screen application	204
4.5	References	205

5 Ink technology and the application of radiation curing 209

5.1	The printing processes	209
5.1.1	Roller coating	209
5.1.2	Letterpress	210
5.1.3	Lithography	211
5.1.4	Gravure	212
5.1.5	Screen printing	213
5.2	Printing inks	213
5.3	Solvent-based methods of ink drying	214
5.3.1	Absorption	215
5.3.2	Oxidation and polymerization	217
5.3.3	Evaporation drying	217
5.3.4	Precipitation	218
5.4	Solventless methods of ink drying	218
5.4.1	Thermally catalysed inks	219
5.4.2	Water-based inks	219
5.4.3	Solventless oil-based inks plus overcoat	220
5.4.4	Radiation curing inks	221
5.4.4.1	Thermal-effect radiation	221
5.4.4.2	Free radical-forming radiation	227
5.4.4.3	Ultraviolet coatings	231
5.4.4.4	Ultraviolet printing inks	244
5.4.4.5	Ultraviolet inks in the field	247
5.4.4.6	Ultraviolet gravure application	251
5.4.4.7	Ultraviolet flexographic ink application	252
5.4.4.8	Ultraviolet screen printing application	254
5.4.4.9	Advantages and disadvantages of UV curing	257
5.5	References	259

6 Photoresist technology 263

6.1	Definition, history and general application	263
6.1.1	The dichromate/polyvinyl alcohol system	265

4.1.3 Alternative photo-induced crosslinking reactions to vinyl
addition polymerization 141
    4.1.3.1 Photolysis of azido groups 142
    4.1.3.2 Photo-sensitization by cinnamoyl and related groups . 142

4.2 Resins 144
  4.2.1 Epoxy resins 145
  4.2.2 Unsaturated polyester systems 149
    4.2.2.1 Preparation of acrylic functional polyesters 152
  4.2.3 Polyurethans 153
    4.2.3.1 Preparation of a urethan 154
  4.2.4 Polyethers 156
    4.2.4.1 Preparation 156
  4.2.5 Thiol/ene system 157
    4.2.5.1 Thermoplastic polymers 159
    4.2.5.2 Thermosetting polymers 160

4.3 Diluents 161
  4.3.1 Wetting agents 162
  4.3.2 Low-viscosity resins 162
  4.3.3 Monomers 162
    4.3.3.1 Vinyls 164
    4.3.3.2 Acrylics 164
    4.3.3.3 Allylic monomers 168
  4.3.4 Plasticizing diluents 169
    4.3.4.1 Primary plasticizers 170
    4.3.4.2 Secondary plasticizers 170

4.4 Formulation 170
  4.4.1 Surface defects and levelling of film 170
  4.4.2 Mechanical properties 172
    4.4.2.1 Hardness 173
    4.4.2.2 Toughness 174
    4.4.2.3 Impact resistance and flexibility 174
  4.4.3 Chemical resistance 175
  4.4.4 Gloss 175
  4.4.5 Adhesion to various substrates 176
  4.4.6 Pigmentation of photopolymerizable systems 178
  4.4.7 Cure response of a system 180
    4.4.7.1 Lamp 182
    4.4.7.2 Photo-initiator/sensitizer and film thickness
         relationship 182
    4.4.7.3 Atmosphere 183
    4.4.7.4 Substrate 186
    4.4.7.5 Temperature 186
    4.4.7.6 Theoretical treatise 187

3.4.2.4	Acetophenone derivatives	83
3.4.3	Aromatic ketone/amine combinations	85
3.4.3.1	Benzophenone	85
3.4.3.2	Michler's ketone	88
3.4.4	α-Acyloxime esters	90
3.4.5	Thioxanthone and derivatives	91
3.4.6	Quinones	99
3.4.7	Dye photo-sensitization	102
3.4.7.1	In the absence of oxygen	102
3.4.7.2	With oxygen present	103
3.4.7.3	Photoconductive effect	104
3.4.8	Pigment photo-sensitization	110
3.4.9	Organic peroxides	117
3.4.10	Organic sulphur compounds	119
3.4.11	Metal compounds and ions	123
3.4.11.1	Uranyl salts	123
3.4.11.2	Gold salts	123
3.4.11.3	Cobalt salts	123
3.4.11.4	Metal halides	124
3.4.11.5	Metal carbonyl compounds	124
3.4.11.6	Photolysis of dichromates	124
3.4.12	Organic phosphorus-containing compounds	125
3.4.13	Chlorosilanes	126
3.4.14	Azo compounds	126
3.5	Air inhibition of photopolymerization	127
3.6	References	130

4 Photopolymerizable film-forming materials 137

4.1	Photopolymerization	137
4.1.1	Photopolymerization of compositions possessing vinyl unsaturation	138
4.1.1.1	Mono- and multifunctional unsaturated monomers	138
4.1.1.2	Polymers with unsaturation	138
4.1.2	Photo-crosslinking and photopolymerization of saturated polymers	139
4.1.2.1	Vinyl polymerization of monomers with simultaneous chain-transfer reaction with saturated polymer	139
4.1.2.2	Purpose-modified saturated polymers	139
4.1.2.3	Photopolymerization and crosslinking induced by photo-sensitizers	141

2.3.3		Parabolic reflectors	48
2.3.4		Shielding	48
2.4	Lamps and control units		50
2.5	The cooling system		51
2.6	Installation of UV drying systems		51
2.7	Irradiation intensity		52
2.8	Future lamp systems		56
2.8.1		Impulse drying	58
	2.8.1.1	Spectral output	59
	2.8.1.2	Kinetics of impulse drying	61
2.9	References		65

3 Photo-initiators and photo-sensitizers 67

3.1	Definitions 67
3.2	General mechanisms for photo-initiator/sensitizer action . . . 68
3.2.1	Fragmentation 68
3.2.2	Hydrogen abstraction 68
3.2.3	Ionic initiation 69
3.2.3.1	Electron donor/acceptor complexes 69
3.2.3.2	Exciplex formation 70
3.2.4	Photo-crosslinking reactions 71
3.2.5	Triplet energy transfer reactions 71
3.3	Aromatic carbonyl compounds 72
3.3.1	The Norrish reaction 73
3.3.1.1	Norrish type I reaction 74
3.3.1.2	Norrish type II reaction 74
3.3.1.3	Norrish type III reaction 74
3.4	Classes of photo-initiators/sensitizers 74
3.4.1	Photo-ionic polymerizing compounds 74
3.4.1.1	Aryldiazonium compounds 76
3.4.1.2	Diaryliodonium compounds 76
3.4.1.3	Triarylsulphonium compounds 78
3.4.1.4	Triarylselenonium compounds 79
3.4.2	Benzoin/acetophenone and derivatives 79
3.4.2.1	Benzoin 80
3.4.2.2	Benzoin alkyl ethers 80
3.4.2.3	Benzil ketals 82

Contents

1 Molecular structure and photochemistry 1

  1.1 Molecular structure 1
      1.1.1 Atoms: the molecular building blocks 1
      1.1.2 Bonding of atoms to form molecules 8

  1.2 Photochemistry 12
      1.2.1 The absorption process 16
      1.2.2 Fates of excited states 21
      1.2.3 Quantum yield 26
      1.2.4 The cage effect 27

  1.3 Kinetics and thermodynamics of photochemical reactions . . 33
      1.3.1 Kinetics 33
          1.3.1.1 Initiation reaction 34
          1.3.1.2 Propagation and chain transfer reactions . . . 34
          1.3.1.3 Termination reactions 35
          1.3.1.4 Inhibition and retardation reactions 36
          1.3.1.5 Auto-acceleration 36
      1.3.2 Thermodynamics 37

  1.4 References 39

2 Ultraviolet lamp systems 41

  2.1 The electromagnetic spectrum 41

  2.2 Lamps 41
      2.2.1 Mercury arc lamps 42
      2.2.2 Metal halide lamps 46

  2.3 Reflector units and shielding 46
      2.3.1 Non-focusing reflectors 47
      2.3.2 Elliptical reflectors 47

Acknowledgements

Wherever possible, specific acknowledgements have been made to copyright holders. In addition, the author thanks the following organizations whose material has provided a basis for parts of the sections itemized.

American Chemical Society	§1.2.4
Consultox Laboratories	Ch. 7
Federation of Societies for Coatings Technology	§1.2.4, §4.4.7.3
General Electric Company and American Chemical Society	§3.4.1
Arthur Holden & Sons Ltd	§5.4.4.3
Mrs R. M. Kosar	Ch. 6
MacNair-Dorland	§5.4.4.1
E. T. Marler Ltd	§5.4.4.8
New England Association of Chemistry Teachers	§3.4.8
Oil & Colour Chemists' Association	§2.8.1.2, Ch. 3, Ch. 4
Paint Research Association	§3.4.7.3, Ch. 4
Royal Photographic Society of Great Britain and Kodak Ltd.	Ch. 6
Royal Society of Chemistry	§1.2.4, §1.3
Society of Manufacturing Engineers	§2.8.1.1, §5.4.4.1
Technical Association of the Graphic Arts	§4.4.7.6
Technology Marketing Corporation	Ch. 2, Ch. 4, Ch. 5
Toxicol Laboratories Ltd	Ch. 7
Ward Blenkinsop & Co Ltd	§3.4.4, §3.4.5
John Wiley and Sons	Ch. 6

Preface

The use of photopolymerizable coatings is steadily gaining ground in many industries. Interest in and development of these systems probably received its biggest impetus from environmental legislation, such as the Los Angeles '66 Rule, which sought to reduce solvent pollution of the atmosphere.

Other benefits however, including greatly reduced energy requirements and the potential of vastly reduced cure-time also made them an attractive proposition. The parallel development of high-solid and water-reducible coatings along with low-bake formulations represent alternative routes to the goals of pollution reduction and energy conservation. No doubt all of these exciting new technologies will find their most appropriate areas of application in the near future but none offer, in the longer term, the same total package of benefits as obviously as the subject of this book.

The latter comprises almost 100 per cent reactivity of constituents, far lower curing energy, very high cure speed, greatly reduced floor-space requirements in a factory coating situation, and immediate recoatability, in contrast to current conventional systems.

This book is basically a review of the literature over more than two decades assembled in such a manner as to give an adequate understanding of the theory and present practice of the subject. The early chapters assume some knowledge of the theory of atomic and molecular structure and are intended as a revision summary guide. These lead on to a treatment of basic photochemical theory followed by the main exposition dealing with resin and initiator systems known to date and their usage in practical surface coating products.

The subject is in a much more mature state now compared with only a few years ago and the technology is being used widely in the paper and board, plastics, and metal-decorating industries. Expansion to other fields of application will surely come with advances in resin and initiator design coupled with the skilled art of formulation.

Thanks are due to Steven Gorham and Alistair Charleston for constructive comments on the manuscript and Colin Brook for skillfully making most of the drawings.

Last but certainly not least, thanks go to Valerie Morris for her patience in deciphering my handwriting and typing the work.

Gosport, Hants. 1980

C. G. Roffey
BSc., Ph.D., C.Chem., M.R.S.C,
Industrial Chemist

To my parents,
George Edward and Kathleen Constance Roffey
and also to
Wendy Ann Webb

Copyright © 1982 by John Wiley & Sons Ltd.

All rights reserved.

No part of this book may be reproduced by any means, nor transmitted, nor translated into a machine language without the written permission of the publisher.

British Library Cataloguing in Publication Data:

Roffey, C. G.
　Photopolymerization of surface coatings.
　1. Coating processes
　2. Polymers and polymerization
　I. Title
　667′.9　　TP156.P6

ISBN 0 471 10063 3

Library of Congress Cataloging in Publication Data:

Roffey, C. G.
　Photopolymerization of surface coatings.
　"A Wiley–Interscience publication."
　Includes index.
　1. Polymers and polymerization.
　2. Photochemistry.
　3. Plastic coating.
　I. Title.
　QD381.8.R63　667′.9　81-12916
　ISBN 0 471 10063 3

Filmset in 'Monophoto' Times Roman
by Speedlith Photo Litho Ltd., Manchester.
Printed by Page Bros., (Norwich) Ltd.,

Photopolymerization of Surface Coatings

C. G. Roffey
Industrial Chemist

A Wiley–Interscience Publication

JOHN WILEY & SONS

Chichester · New York · Brisbane · Toronto · Singapore

	7.2.3.2	Occluded patch test	314
	7.2.3.3	Magnusson and Kligman maximization test	315
7.2.4	Toxicity		315
	7.2.4.1	Acute toxicity – LD_{50} determination	315
	7.2.4.2	Gross toxicity	317
7.2.5	Carcinogenic and mutagenic substances		318
	7.2.5.1	Bioassays	318
	7.2.5.2	Cell transformation test	318
	7.2.5.3	Mutagenicity tests	319
	7.2.5.4	Carcinogenicity tests	326
7.2.6	Inhalation tests		328

7.3 Some chemical aspects of photobiology 336
   7.3.1 Eye effects 336
   7.3.2 Skin effects 336

7.4 References 337

Index 339

1
Molecular structure and photochemistry

1.1 MOLECULAR STRUCTURE

The ground state of an atom or molecule is the state where all electrons fill the orbitals available in a progression of increasing energy. An orbital is a region in space where there is maximum probability of finding a given electron. The atom or molecule has only a single ground state of lowest energy, but may have, on absorption of a quantum of radiation, several different possibilities for higher-energy excited states. An electronically excited state is a state where one of several electrons occupies a higher energy orbital, after leaving a vacancy in the lower orbitals. These excited higher energy states have the same chemical 'structure' as the ground state but a different reactivity from it. These are therefore isomers of the ground state.[1] It is these excited states that are responsible for photochemical reactivity. A ground-state electron can only absorb light, not emit it, whereas an excited-state electron can either emit light by dropping downwards in energy to the ground state or a lower excited state, or it may absorb light by promotion upwards in energy to a higher excited state.

The structures and formation of atoms and molecules and the absorption and emission of light, with the terms used to define them (spectroscopic nomenclature) are outlined in this section. This will enable the reader to obtain the basic principles of the language used by spectroscopists and photochemists.

1.1.1 Atoms: The Molecular Building Blocks

Atoms can be considered in an oversimplified view as consisting of a central nucleus surrounded by one or more orbital electrons, according to the early Rutherford theory.[2]

This is shown schematically in Figure 1.

The electrons were considered as revolving around a positively charged nucleus. The coulombic force of attraction was regarded as balancing the centripetal force of acceleration. This view was, however, unsatisfactory as the classical electromagnetic theory of radiation required that an accelerating electric charge had to emit radiation continuously, therefore losing energy. As a consequence, the electrons would then spiral and drop into the nucleus. Classical mechanics therefore presented an unstable view of the atom.

Niels Bohr in 1913 improved upon this theory.[3] He postulated the atom (in the simplest case hydrogen) as consisting of a central nucleus with a circulating

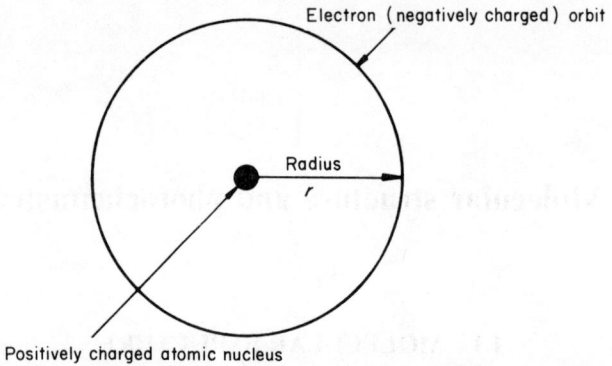

Figure 1 Rutherford model of the atom.

electron that could occupy any one of a number of orbits. Each of these orbits was regarded as being in a fixed position. The discrete orbits or orbitals in three dimensions, of increasing energy as the radius increases, that are permitted, were postulated by Bohr as those for which the angular momentum mvr is an integral multiple of $h/2\pi$, that is,

$$mvr = nh/2\pi, \quad \text{where } n = 1, 2, 3, \ldots$$

m = mass of the electron
v = tangential velocity
r = radius of the orbit
h = Planck's constant

The integer n was defined as the *principal quantum number*.

The Bohr picture of the atom is depicted in Figure 2.

If a hydrogen electron fell from an outer orbit to an inner one, it was regarded as losing energy, emitting energy in the form of an energy quantum. This energy quantum was postulated as being only one of a fixed size and wavelength that was just adequate to move it by the proper amount. The characteristic lines in the hydrogen-atom spectrum are explained by this property of absorbing or emitting only certain wavelengths of radiation. The definite quantity of energy corresponding to a particular frequency of radiation that will be emitted or absorbed during a transition of an electron from one energy level to another in an atom is given by the equation

$$\Delta E = h\nu.$$

Where ΔE is the difference in energy between two electronic energy levels, E_1 and E_2 ($E_2 > E_1$), ν is the radiation frequency absorbed by the atom, and h is Planck's constant.

This theory was inadequate as it did not explain why the orbits were fixed in the positions that they occupied. The Bohr orbits were chosen to fit experimental results found for the absorption and emission of the observed wavelengths of light concerned.

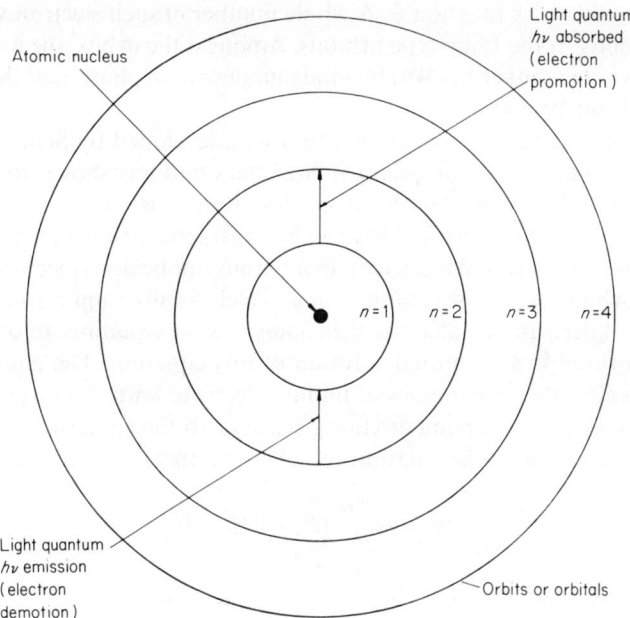

Figure 2   Bohr picture of the atom.

Heisenberg[4] introduced another concept. He considered the task of measuring the position of a particle. If a microscope were able to make an electron visible, then light or similar radiation would need to be shone on to that electron in order to see it. As an electron is minute, a single photon of light incident upon it would cause it to move and change its position. This very performance of measuring its position would cause the latter to be changed.

Heisenberg thus demonstrated that there is no means of devising a method of ascertaining the position of a subatomic particle without being quite uncertain as to its exact motion. The converse is also true, that there is no means of determining a particle's exact motion without forfeiting the certainty as to its exact position. To calculate both exactly (position and momentum) at the same time is not possible.

Wave or quantum mechanics is founded upon the *Heisenberg uncertainty principle*. The consequence of this indeterminacy principle is that the models of the atom such as that of Rutherford where electrons are fixed in orbits at a given radius from a nucleus, and therefore of known position and momentum, are not valid. The position of the electron in atomic and molecular entities must be based upon probabilities. A wave function depicted in terms of ψ is used to accomplish this, and was developed by Schrödinger.

Schrödinger in 1926[5] pictured the atom in terms of the de Broglie theory of the wave nature of particles. If the electron is regarded as a wave, then he postulated that the electron did not circle round the nucleus in the manner a planet circles the sun, but comprised a wave that curved all round the nucleus in three dimensions representing a probability of an electron's presence within the volume (orbital) contained by the wave function ψ. A whole number of such electron waves were found to exactly fit the Bohr-type orbitals. Amongst the orbits, the waves would not fit in a whole number but would amalgamate 'out of phase' and these type of orbits could not be stable.

A mathematical description of the atom was developed by Schrödinger and known as *wave mechanics* or *quantum mechanics* and has shown to be a more adequate approach to viewing the atom than that of Bohr.

Electrons have a dual nature. They can be considered to have the properties of both particles and waves. An electron in an atom can be described as occupying an atomic orbital or by a wave function ψ which results from a solution of the mathematical derivation called the Schrödinger wave equation. In other words, an atomic orbital is a permitted solution of this equation. The equation is an attempt to calculate probabilities of finding electrons with given values of their co-ordinates without inferring anything relevant to their physical nature.

In three dimensions this equation may be written as

$$\nabla^2 \psi + \frac{8\pi^2 m}{h^2}(E - V)\psi = 0$$

$\nabla^2 \psi$ is the quantity

$$\frac{d^2\psi}{dx^2} + \frac{d^2\psi}{dy^2} + \frac{d^2\psi}{dz^2}$$

E is the total energy (kinetic energy + potential energy, V), m is the electron mass, and h is Planck's constant.

An orbital theoretically extends to infinity. ψ decreases fast with distance and practical representation of an orbital space is depicted by a volume element representing 90 per cent probability of an electron being present. This constitutes the basis of the diagrams for s, p, d, and f orbitals shown later.

In a molecule, electrons are said to occupy molecular orbitals[6] and the wave function describing this molecular orbital can be regarded as the linear combination of atomic orbitals (see section 1.1.2).

An orbital is defined quantitatively by its wave function ψ. The probability of finding the electron at a point defined by the co-ordinates x, y, z with the nucleus being at the origin is $\psi^2(x, y, z)$. The probability of finding the electron somewhere in space is a certainty and therefore, the integrated value of $\psi^2(x, y, z)$ over all space is normalized to equal unity, i.e.

$$\int_0^\infty \psi^2(x, y, z)\, d\tau = 1 \qquad \text{where } d\tau \text{ is the volume element.}$$

The variation of ψ with the distance from the nucleus for an s orbital is shown[7] in Figures 3 and 4. Figure 4 depicts the probability as a density (shaded areas).

The volume of space where there is maximum probability of finding an electron has been previously defined as an orbital. Each electron in an orbital can be described by four quantum numbers. These are *principal*, *azimuthal*, *magnetic*, and *spin* quantum numbers.

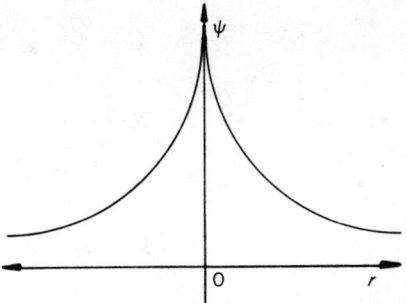

Figure 3 Variation of ψ with nuclear distance for an s orbital.[7] (Reproduced by permission of the Royal Society of Chemistry.)

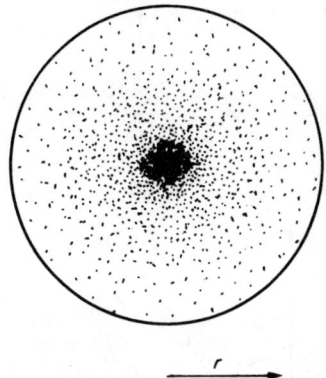

Figure 4 Probability density plot of wavefunction ψ and nuclear distance (r) for an s orbital.[7] (Reproduced by permission of the Royal Society of Chemistry.)

For an orbital, each main level has a number of sub-levels equal to the principal quantum number n. For the first shell of electrons, $n = 1$ and there is therefore only one value for the subsidiary azimuthal quantum number $l = 0$, and an s orbital only exists. When $n = 2$, $l = 0$ and 1, giving two orbitals, so that both s and p orbitals occur. Similarly, when $n = 3$, $l = 0$, 1, and 2 and s, p, and d orbitals occur. This *azimuthal* quantum number describes the shape occupied by the electron cloud which represents the greater or lesser probability of an electron being at any point near the nucleus at any given time.

l may have the values 0, 1, 2, 3,..., $(n - 1)$.

For $l = 0$, a *spherical* orbital results and is known as an s orbital.

When $l = 1$, the orbital is a *dumbell* shape and is called a p orbital. Three of these occur at right angles to each other in three dimensions, designated p_x, p_y, p_z.

When $l = 2$, a double dumbell shaped orbital results and is known as a d orbital. There are five of these designated $d_{xy}, d_{yz}, d_{xz}, d_{x^2-y^2}$, and d_{z^2}.

For $l = 3$, a more complicated f orbital occurs, totalling seven.

The shapes of these orbitals from the Schrödinger wave equation[8-9] are shown in Figure 5.

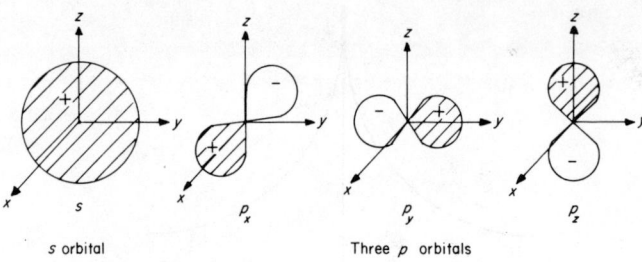
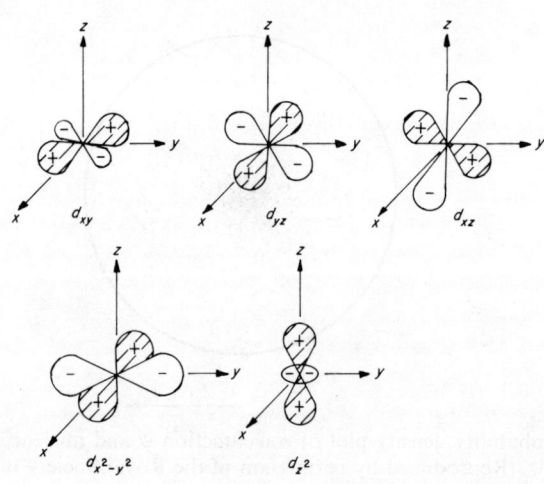
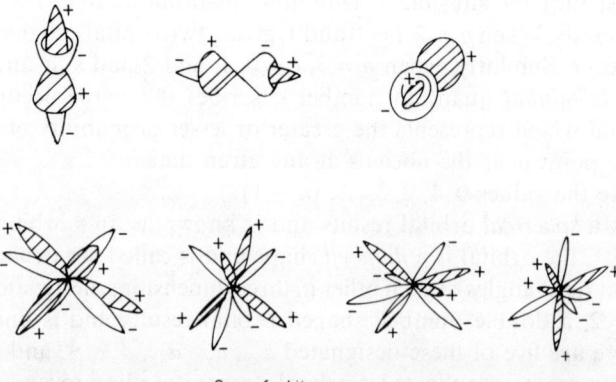

Figure 5 Shapes of *s*, *p*, *d*, and *f* orbitals.[9] (Reproduced by permission of John Wiley and Sons Ltd.)

Every sub-level is further subdivided, these subdivisions being denoted by a *magnetic* quantum number m and *spin* quantum number s. The way in which the lines in the atomic spectrum split, where electronic transitions between levels are involved, under the influence of a magnetic field determine the magnetic quantum number m. m can only have all the values from $\pm l$ and 0, where l is the *azimuthal* quantum number. There are $2l + 1$ values for the magnetic quantum number. The direction of spin of an electron on its axis may be regarded as the *spin* quantum number, s. Thus it has the values of $\pm \frac{1}{2}$ corresponding to clockwise and anticlockwise spin.

Four quantum numbers are therefore needed to define an electron's energy level in an atom. Several rules operate with respect to electrons in orbitals. The Pauli[9] exclusion principle states that no two electrons in one atom can have all four quantum numbers the same.

Hund's rules[10] are two generalizations concerning electronic occupation of orbitals. Firstly, the number of unpaired electrons or the number of singly occupied orbitals will be a maximum owing to the coulombic effect. Secondly, the electrons will have parallel spins, that is, the same spin quantum number. This is known as the spin correlation effect. These are important in the concept of singlet and triplet energy states for photo-initiators and sensitizers, described later.

The theory concerning atomic orbitals outlined, assumes that the atomic energy levels are determined solely by the quantum states of the valence electrons. In fact all the electrons and even the nucleus should be considered. Therefore, instead of the quantum number which gives the angular momentum of the single electron, a *resultant* angular momentum quantum number L should be used.

To avoid writing the number for L, the values $L = 0, 1, 2, 3$, etc., are represented by the capital letters S, P, D, F, etc., following the convention for individual orbital notation that electrons with $l = 0, 1, 2, 3$ should be called s, p, d, f electrons.

Russell–Saunders coupling[11] is used where the individual orbital moments l_1, l_2, etc., couple to give the overall orbital momentum L.

The spin moments s_1, s_2, etc., couple to give S. L and S then couple to give the overall total angular momentum J.

J can take values $L + S, L + S - 1, L + S - 2, \ldots, L - S$ if $L > S$, and there are $(2S + 1)$ values of J. $(2S + 1)$ is known as the *multiplicity* of the state. An atomic term symbol is used to describe these resultant moments. The term symbol is –

$$n^{2S+1}L_j \quad \text{where } 2S + 1 = M, \text{ the multiplicity.}$$

Certain transitions only are allowed, others forbidden. For light emission from transitions between energy states, any difference in the principal quantum number is permitted; the resultant angular momentum quantum number L changes only by $\Delta L = \pm 1$. The total angular quantum number $J = L + S$ (S is the resultant spin quantum number) allows transitions for $\Delta J = \pm 1, 0$, except that the transition $J = 0 \to J = 0$ is forbidden. This coupling of L and S is known as Russell–Saunders coupling and is applicable to the lighter elements. It breaks down significantly for heavier atoms such as mercury, Hg, where some transitions forbidden by the selection rules become allowed (e.g., the 254 nm line which is very important for ultraviolet curing systems).

1.1.2 Bonding of Atoms to Form Molecules

The individual atomic orbitals may be considered to blend into new molecular orbitals, both bonding and anti-bonding. The former bonding orbitals are of lower energy than the sum of the individual atoms, giving an energetically more favourable state which bonds. Chemical bonds are formed from the overlap of the electronic orbitals of individual atoms along the axis between two nuclei. End-to-end overlap forms a bond that is known as the sigma (σ) type. Lateral orbital overlap can give a pi (π) bond. The latter can result in the highly reactive double or triple bond formation which can be of great significance in ultraviolet radiation curing systems.

The Schrödinger wave equation can also be applied to interpret molecular orbitals as for atomic orbitals. The concept of symmetry relevant to the sign of the wave function is also very important. The linear combination of atomic orbitals (LCAO) is the simple description of the molecular orbital wave function by the addition of atomic orbital wave functions. The overlap of two atomic orbitals ψ_a (atom A) and ψ_b (atom B) results in bonding if the wave functions have the same sign in the same region of space.

An s orbital is spherically symmetrical. For an isolated atom, the wave function sign is arbitrary since the same ψ^2 value, which correlates with the probability of finding an electron is obtained from ψ^+ and ψ^- (which are defined later). With respect to its axis a p orbital is symmetric and anti-symmetric with respect to a plane. This is the reason why p orbitals participate in orientation-dependent bonds such as double and triple types. Typical sign representations are shown for s and p orbitals in Figure 6, along with their respective quantum numbers.

For efficient combination of two atomic orbitals they must have similar energy, overlap, and have similar symmetry.[10] Under these conditions, each atomic orbital splits into two of different energy giving a bonding molecular orbital ψ^+ associated with a decrease in energy and an anti-bonding orbital, ψ^- associated with an increase in energy. In molecular orbital terms ψ^+ and ψ^- are sometimes written as ψ_g (g for *gerade*, 'even' and *u* for *ungerade*, 'uneven'). The symmetry of the orbital about its centre is referred to by g and u. When the orbital is reflected about its centre and the sign of ψ is unchanged then it is *gerade*. If the converse is true, then it is *ungerade*.

Some examples of the type of orbital overlap that may occur and the shapes of the molecular orbitals generated are shown[6,9] in Figure 7.

The combination of s orbitals to give σ molecular orbitals depicted in Figure 7 can be represented by the wave functions shown in Figure 8.

The approximate order of energy for filling molecular orbitals with electrons is:

$$\sigma 1s < \sigma^* 1s < \sigma 2s < \sigma^* 2s < \sigma 2p_x \begin{matrix} \pi 2p_y \\ \pi 2p_z \end{matrix} \begin{matrix} \pi^* 2p_y, \sigma^* 2p_x \\ \pi^* 2p_z \end{matrix}$$

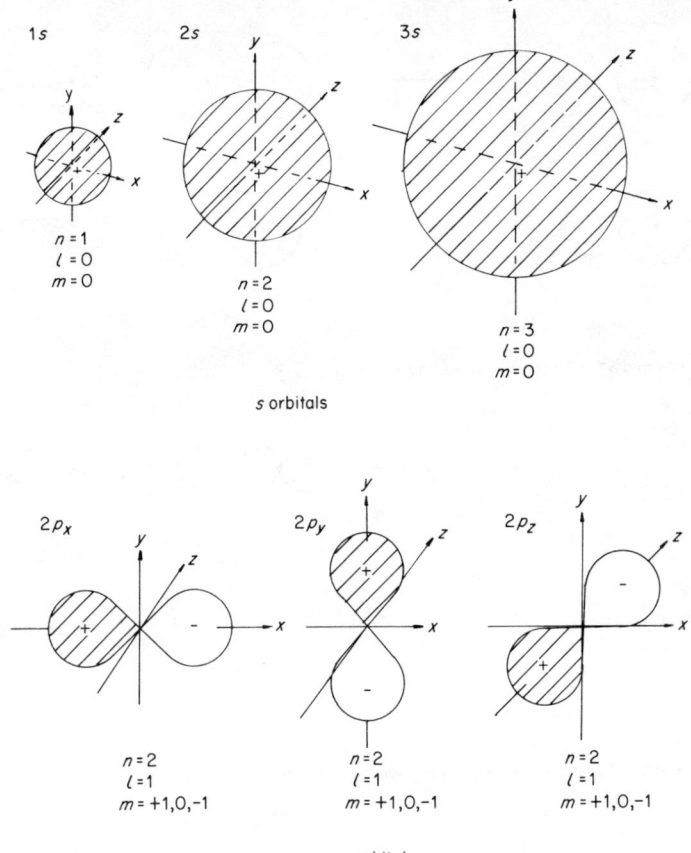

Figure 6 Typical shapes and sign representations for some s and p orbitals with appropriate quantum numbers.

For molecular orbitals the situation is a little more complicated regarding spectroscopic nomenclature than for polyelectronic atoms, as the quantization of rotational (E_{rot}) and vibrational energy (E_{vib}) has to be considered as well as electronic energy ($E_{elect.}$).

The Born–Oppenheimer approximation,[11,12] which emphasizes the great differences between E_{elect}, E_{vib}, E_{rot}, implies that they can each be treated separately, and therefore simplifies matters.

For diatomic and linear molecules, l and L for atoms are replaced by λ and Λ for individual orbitals and the whole molecule respectively.

λ represents the quantization of angular momentum (in $h/2\pi$ units) with respect to the line joining the nuclei (e.g. for a diatomic molecule).

Orbitals with $\lambda = 0, 1, 2, 3$ are the $\sigma, \pi, \delta, \phi$ orbitals, whereas states with $\Lambda = 0, 1, 2, 3$ are $\Sigma, \Pi, \Delta, \Phi$ states. Here the term symbol is basically

$$^{2S+1}\Lambda$$

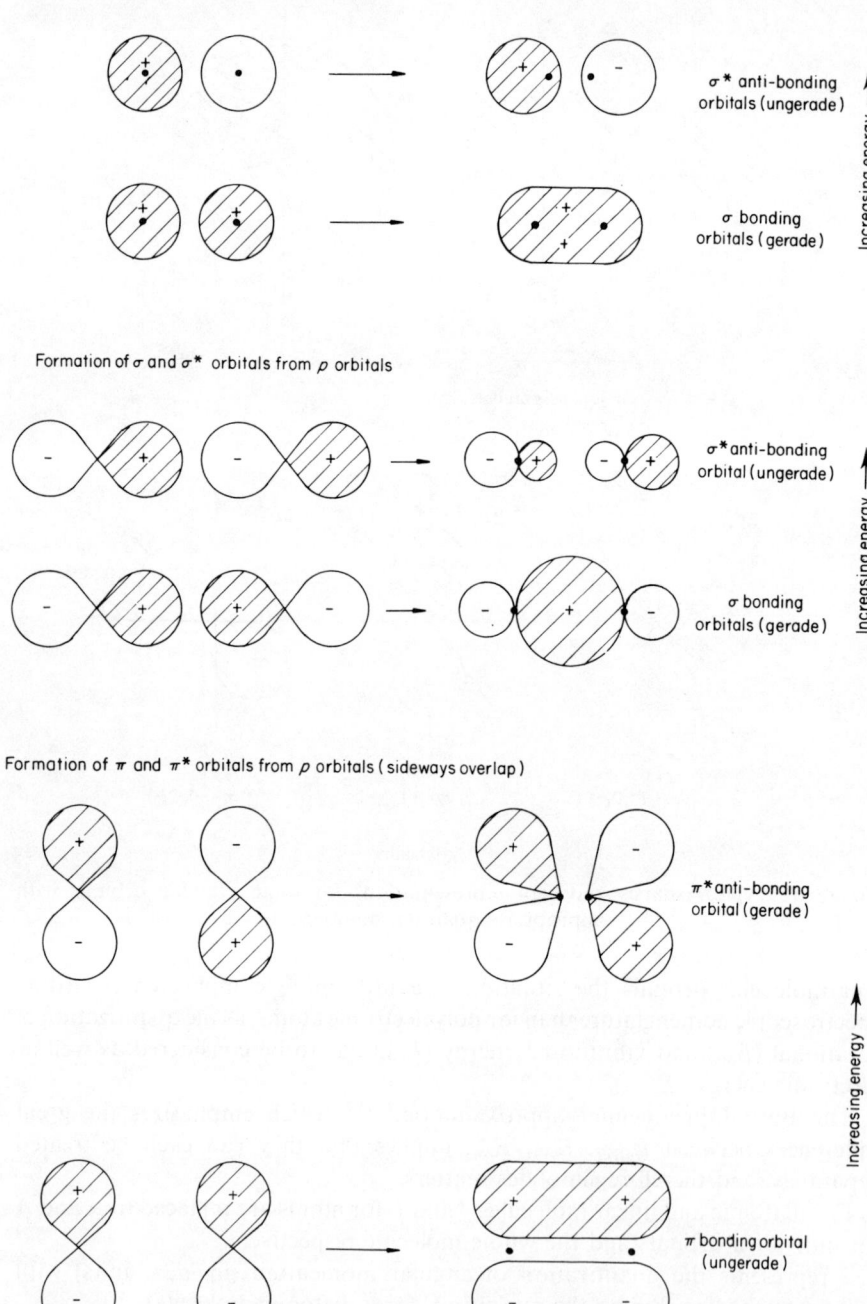

Figure 7　Some types of orbital overlap and shapes of molecular orbitals generated.

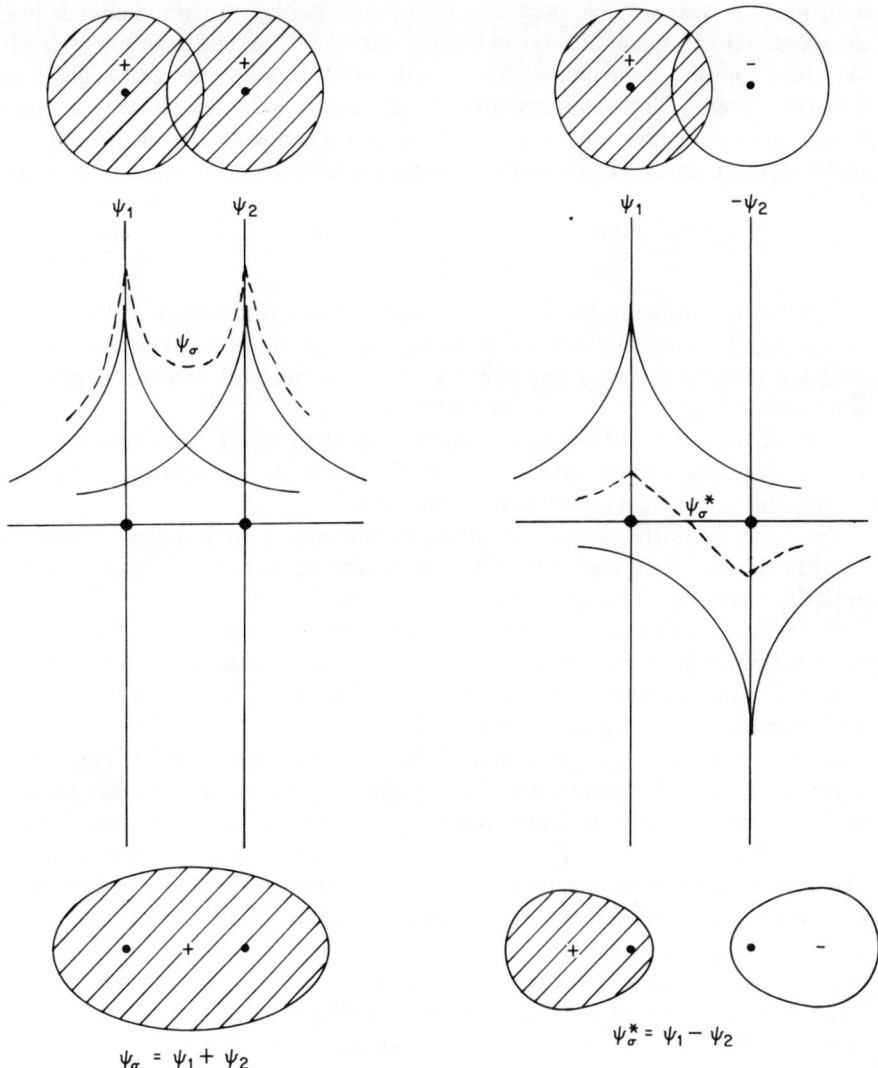

Figure 8 σ molecular orbital wave functions.

but extra pieces of information may be added. One of these is the total angular momentum, and one of several possible coupling schemes must be used to derive it.

The internuclear field does not greatly affect the spin angular momentum and the same notation as for atoms is often used for molecules. The analogue of Russell–Saunders coupling in atoms (summation of spin and orbital angular momentum), gives the resultant Ω, the analogue of J.[15] The molecular term symbol then becomes $^{2S+1}\Lambda_\Omega$.

Also, some further description of the symmetry properties of the wave function may be possible. A wave can possess some or all of the molecular symmetry. For a centro-symmetric molecule the wave function may either remain unchanged or change sign (but not magnitude) on inversion through the centre symmetry. Such wave functions are again called even or odd, gerade or ungerade, g or u, given as subscripts after Λ. The wave function for a Σ state ($\Lambda = 0$) may remain the same or change sign on reflection ($+$ or $-$) by a plane of symmetry passing through the line of atomic centres. These two possibilities are represented as superscripts after Λ.

A typical symbol would then be of the form for the Σ ($\Lambda = 0$) state

$$^1\Sigma_g^+$$

The theory outlined enables the excitation of photo-initiators to be readily interpreted. The carbonyl group is probably the most widespread chemical group used for these substances and the transitions found and orbital shapes are described in Figure 9 for a typical carbonyl compound.[11-13]

Hybridization[14] of the orbitals of an individual atom can give other types of atomic bonding orbital combinations. This can be demonstrated by the combination of the energy of s and p orbitals.

Two sp hybrid orbitals of identical shape and energy are produced. These sp hybrid orbitals, which each have one lobe much larger than the other, point in opposite directions. This is shown in Figure 10.

Hybridization allows more effective overlapping of orbitals to be feasible. As the two electron pairs present are then as far apart as possible, repulsion between them is minimized (Hund's rule) and sp hybrid orbitals form stronger bonds than those formed by overlap of s or p orbitals.

Many other forms of hybridization are available such as sp^2 trigonal, sp^3 tetrahedral, etc. sp^3 hybridization is of particular interest as it occurs for the carbon atom, oxygen atom, and nitrogen atom in many organic compounds used in photochemical systems. This can be shown for methane CH_4 which has only two unpaired $2p$ electrons in the ground state. This is in accordance with Hund's rule, which requires that in a given energy level such as the $2p$ level, the maximum number of unpaired electrons exists. The carbon atom in the excited state can form four bonds, but the s orbital and three p orbitals are not individually used to form them. The alternative process of bond hybridization occurs and four sp^3 equivalent hybrid orbitals are created, as shown in Fig. 11.

A regular tetrahedron satisfies the condition for minimum repulsion between the four orbitals if they point to the corners to give bond angles of 109° 28′ (see Figure 12).

1.2 PHOTOCHEMISTRY

Photochemistry is the study of the chemical reactions induced in a system by electromagnetic radiation excluding ionizing radiation such as electron beam, x-rays, γ-rays, etc.

These reactions depend on four stages:

1. The initial photo-act in which a molecule (M) absorbs a quantum of radiation ($h\nu$) with the consequent formation of an electronically excited state of the molecule (M*).

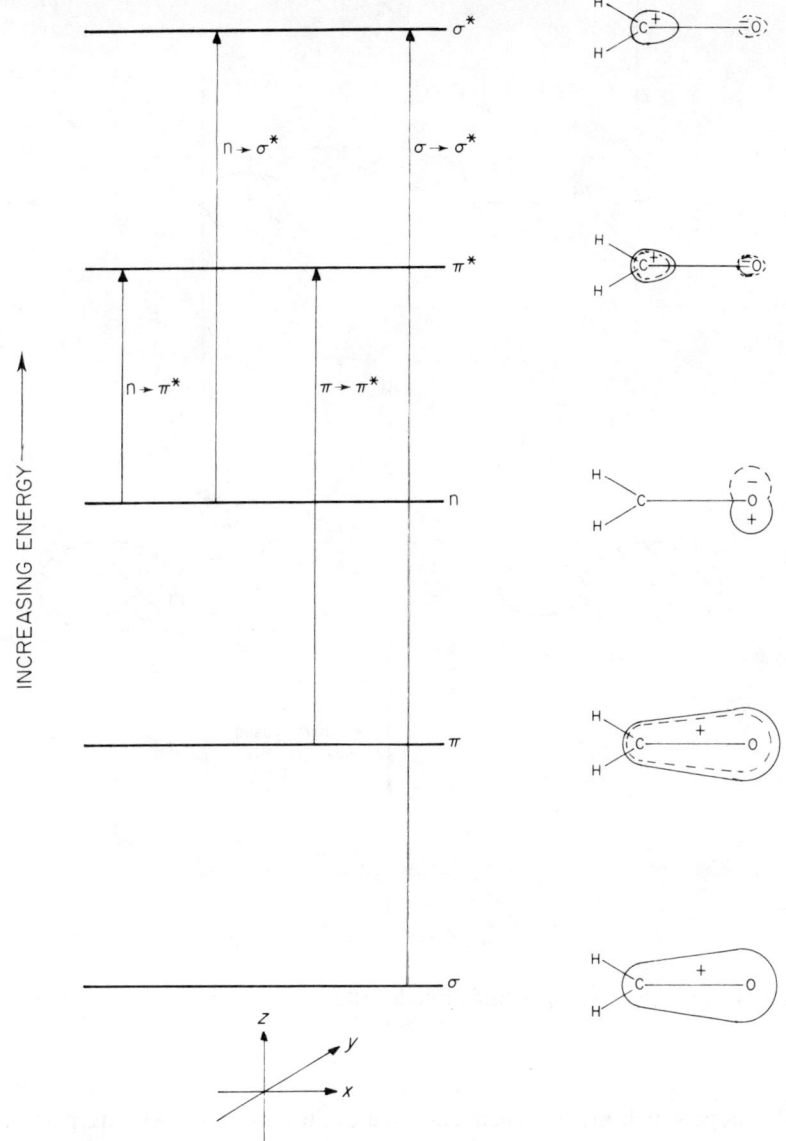

Figure 9 Transitions found and orbital shapes for a typical carbonyl compound. (Reproduced by permission of The Johns Hopkins University Press from M. Kasha, Light and Life (Eds. W. B. McElroy and B. Gloss). © 1961 The Johns Hopkins University Press.)

2. The subsequent fate of this M $\xrightarrow{h\nu}$ M* electronically excited molecule which may include dissociation, predissociation fluorescence or phosphorescence, or collisional deactivation to the ground state.
3. The dark or thermal reactions of the dissociation products.
4. Possible molecular rearrangements of the electronically excited state.

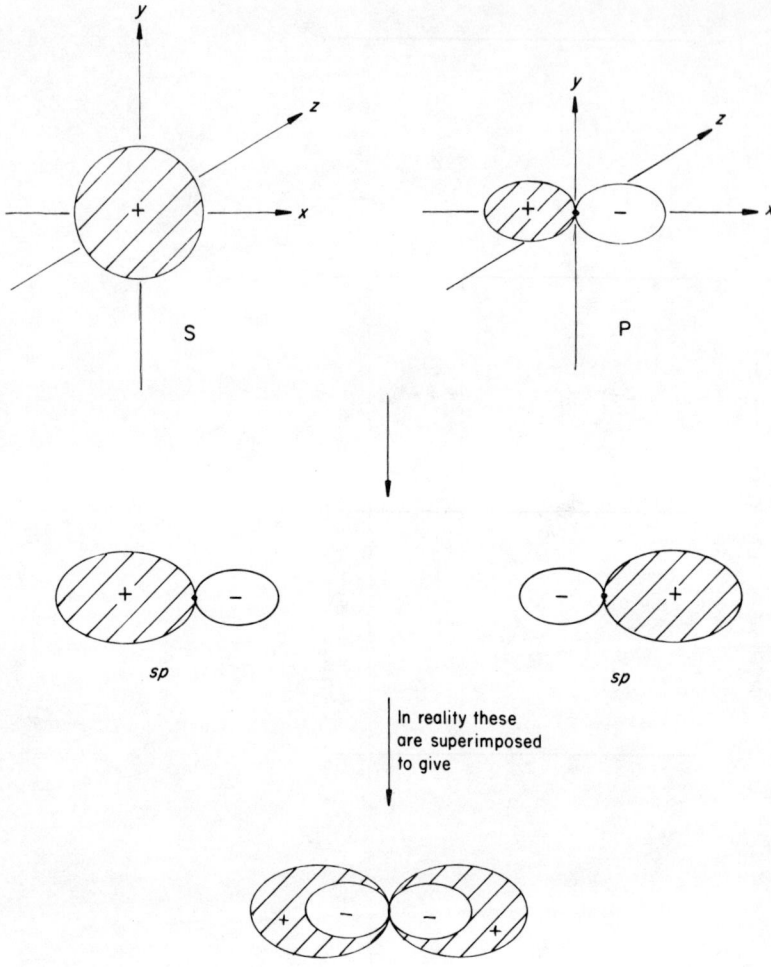

Figure 10 Formation of *sp* hybrid orbitals.[9] (Reproduced by permission of John Wiley and Sons Ltd.)

All steps which involve chemically the excited molecule M* are part of the *primary* photochemical process. Reactions such as 3 (on page 13) are the *secondary* processes.

The Stark–Einstein law[15] states that initially each molecule which forms an electronically excited state by exposure to radiation absorbs one quantum of the radiation causing the primary process. An electronic transition will only occur when there is the energy difference between two levels, ΔE, in the molecule to satisfy the Bohr frequency condition

$$v = \frac{\Delta E}{h}$$

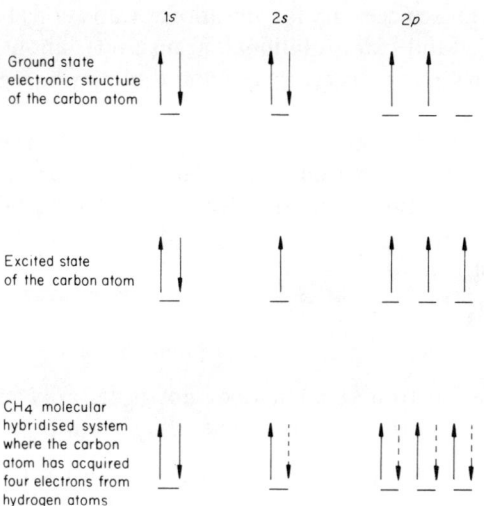

Figure 11 sp^3 hybridization in the methane molecule.

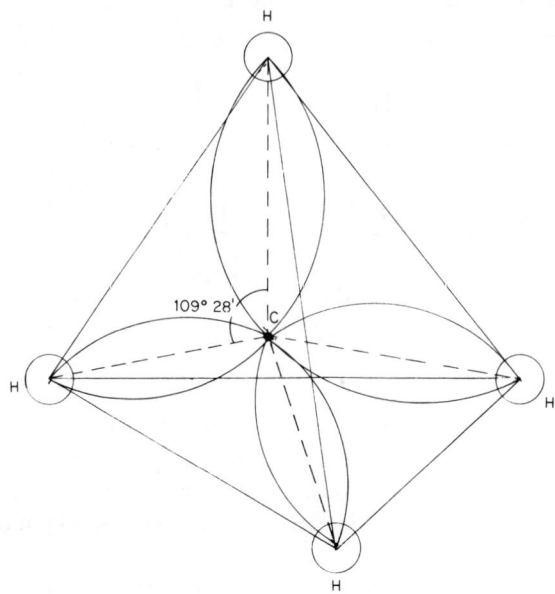

Figure 12 Stereochemistry of the methane molecule with hybridization.

where v is the frequency of the radiation absorbed by the molecule and h is Planck's constant. Photons of radiation in the visible and ultraviolet regions of the spectrum have sufficient energy to cause such an electronic transition.

The energy of a molecule consists of electronic, vibrational, rotational, and possibly translational contributions. These are all quantized, the quanta necessary to cause electronic energy transitions being of the order of

100 kcal mole^{-1}, those necessary for vibrational transitions being of the order of 5–10 kcal mole^{-1}, and for rotational transitions about 0.01 kcal mole^{-1}. Translational levels are so closely spaced that they may be regarded as forming a continuum.

The result of light on a chemical reaction can appear to be merely the speeding up of the reaction which also occurs in the dark.[7] Despite the apparent similarity, 'dark' and 'light' reactions must be distinguished. The light reaction is the reaction of the excited molecule M*, while the dark reaction proceeds through the ground state molecule M.

1.2.1 The Absorption Process

On excitation an electron can but does not necessarily reverse its spin, the resultant multiplicity (M) of the molecule being given by

$$M = (2 \times \text{resultant electron spin}) + 1$$

In Figure 13 it is shown that the multiplicity of the ground state is 1, and those of the two possible excited states, 1 and 3, designated respectively the *singlet* excited state S and *triplet* excited state T.

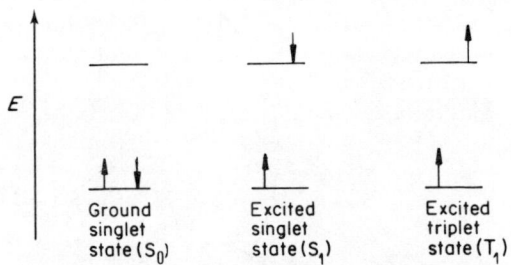

Figure 13   Multiplicity diagram.

Note that in T_1, electrons with parallel spins will be as far apart as possible to minimize the energy in accordance with Hund's Rule.

Although the triplet state has a lower energy, the preferred transition is the $S_0 \rightarrow S_1$. A selection rule operates which states that changes caused by the absorption or emission of radiation involving a change of multiplicity are of zero probability.

There are three main types of orbital involved in an electronic transition (typical examples have been given for a carbonyl compound in Figure 9).

1. *π orbital*. The electron originates from a bonding π orbital and is promoted to a π anti-bonding orbital π^* of higher energy. Both states correspond to a singlet configuration and the transition is written $^1(\pi \rightarrow \pi^*)$. Removal of an electron from the bonding orbital in the ground state will weaken the bond to an extent depending on whether the π orbital is localized or not.
2. *σ orbital*. The electron originates on a bonding σ orbital being promoted into a σ^* orbital. As σ orbitals are generally of lower energy than π orbitals, this will

require a larger quantum than the $\pi \to \pi^*$ transition. Removal of the electron from the bonding σ orbital may result in a considerable weakening of the bond in question. Transitions of this type usually occur in saturated organic molecules and are observed at wavelengths below 200 nm.
3. *Non-bonding orbital.* The electron originates in a non-bonding orbital n and is promoted to a σ^* or π^* orbital. The removal of this electron from the non-bonding orbital will have less effect on the bonding. Since non-bonding orbitals have relatively high energies, lower-energy quanta are required for this type of transition.

As the general order of orbital energies is $\sigma < \pi < n < \pi^* < \sigma^*$ for non-conjugated systems, it is found that:

1. $n \to \pi^*$ transitions lie at longest wavelengths (least energy)
2. $\pi \to \pi^*$ transitions lie at intermediate wavelengths
3. $\sigma \to \sigma^*$ transitions lie at shortest wavelengths (highest energy)

Organic photochemistry[16-18] is chiefly concerned with four types of electronic transitions, the singlet and triplet $n \to \pi^*$ (n here represents non-bonding orbitals) and the singlet and triplet $\pi \to \pi^*$. The molecule in an excited state is a different species from that of the ground state. The excited state is expected to be considerably more reactive because of its high energy content and peculiar electronic distribution. The essence of photochemistry is that activation to produce such an excited state is supplied by absorption of light.

Absorption of radiation is governed by two laws, those of Lambert and Beer.[19] Lambert's law may be expressed as $I_t = I_0 10^{-kd}$, where I_t is the intensity of radiation transmitted by the sample, I_0 is the intensity of incident radiation, k is the extinction coefficient, and d is the thickness of the sample in centimetres.

Beer's law states that $I_t = I_0 10^{-bc}$, where b is a constant, and c is the concentration of the solution, the thickness being kept constant.

These laws can be combined: $I_t = I_0 10^{-Ecd}$, where d is expressed in cm, c in mole litre^{-1}, and E is the molar absorption coefficient in l mole^{-1} cm^{-1}. This equation is applicable only to monochromatic (single wavelength) light.

Light emission may be either spontaneous or stimulated.[7] An excited atom or molecule left to itself will eventually lose its excess energy by spontaneous emission, characterized by an intrinsic lifetime or half-life.

At a time t after the commencement of emission with an intensity I_0, $I_t = I_0 e^{-kt}$, where k is the emission rate constant. The mean radiative lifetime is $1/k$, which is the time after which $I_t = (1/e)I_0$ and should not be confused with the radioactive decay 'half-life' where $I_t = I_0/2$.

The variation of potential energy with internuclear distance for two atoms or molecules can be illustrated by 'Morse' curves,[20] a typical example being of the form shown in Figure 14.

There is a long-range attractive force (curve A) and a short-range repulsive force (curve B). The latter only operates when the atoms or molecules come close together. Curve C, the usual form of a Morse curve, is the summation of A and B.

The gradual lowering of energy as two atomic nuclei are brought together from infinity (zero energy) is depicted by the fall to a minimum in the curve. This

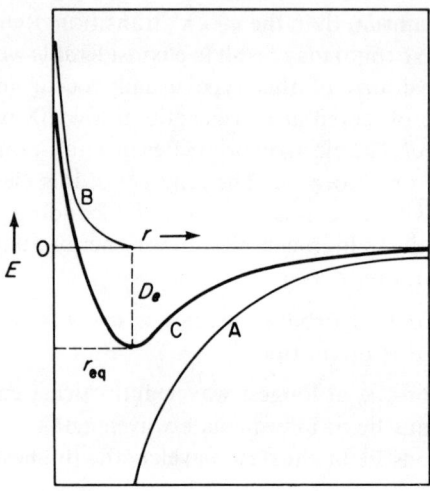

Figure 14 A typical Morse curve, depicting potential energy variation with internuclear distance.[9] (Reproduced by permission of John Wiley and Sons Ltd.)

potential energy minimum occurs at an equilibrium internuclear separation distance, r_{eq}. At this distance the molecular energy is D_e below the zero and is known as the electronic dissociation energy of the molecule. It does not represent the actual bond dissociation energy as all molecules possess zero-point energy ($\frac{1}{2}hv_0$) because of vibrational motion.

The part of the curve represented, rising with decreasing r depicts the short-range effect of nuclear repulsion where the potential energy increases rapidly as the internuclear distance decreases. These are coulombic forces due to repulsive forces on the nuclei between positive charges.

Molecular absorption of light may lead to four common transitions depictable by Morse curves. These are shown[21] in Figures 15–18 where the results of the absorption of a quantum of radiation large enough to cause electronic excitation in a molecule can be discussed in terms of potential energy curves for the state concerned.

Figure 15 shows an electronic transition from a stable ground state to a stable excited state. Vibrational energy levels are depicted by parallel horizontal lines within an electronic Morse curve. Rotational energy levels occur between vibrational levels but are not depicted as their energy is relatively so small. The resulting spectrum is composed of discontinuous bands with a fine structure of closely packed lines.

Figure 16 shows an electronic transition from a stable ground state to an unstable state (anti-bonding orbital) that immediately undergoes dissociation and gives continuous absorption spectra without bands.

Dissociation will also occur in the transition depicted in Figure 17 as the energy level attained is above the binding energy of the excited state, and again continuous absorption spectra is found without bands.

Figure 18 shows the phenomenon of predissociation where the initial transition from one stable state to another is 'crossed' or 'intersected' by the

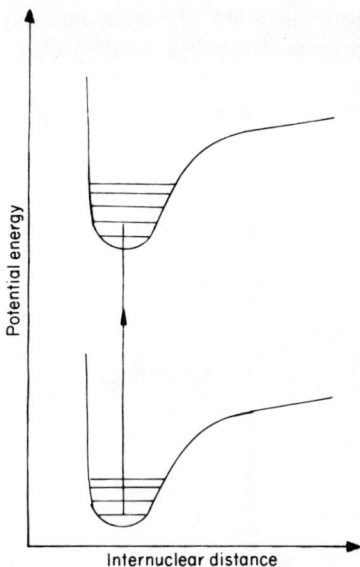

Figure 15 Morse curve depicting an electronic transition from a stable ground state to a stable excited state.

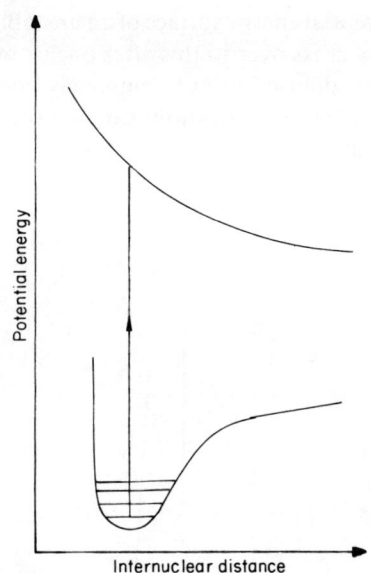

Figure 16 Morse curve depicting an electronic transition from a stable ground state to an unstable state (anti-bonding orbital).

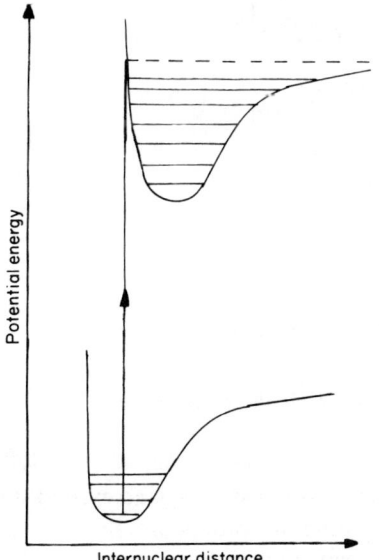

Figure 17 Morse curve depicting an electronic transition above the binding energy of the excited state leading to dissociation.

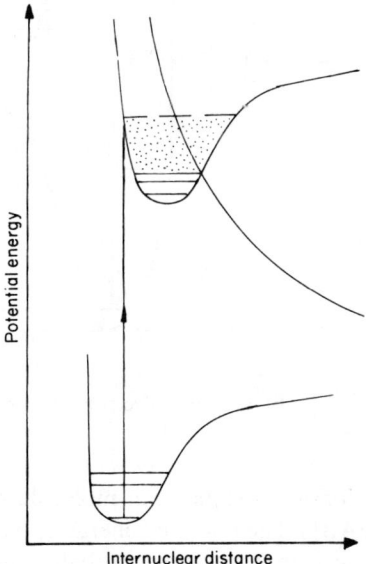

Figure 18 Morse curve showing the phenomenon of predissociation.

potential energy surface of an unstable (anti-bonding) state. The excited molecule may cross over to this after one or more vibrations. The subsequent spectra are quite diffuse but not completely continuous.

Electronic transitions can be discussed in somewhat more detail by reference to Figure 19.

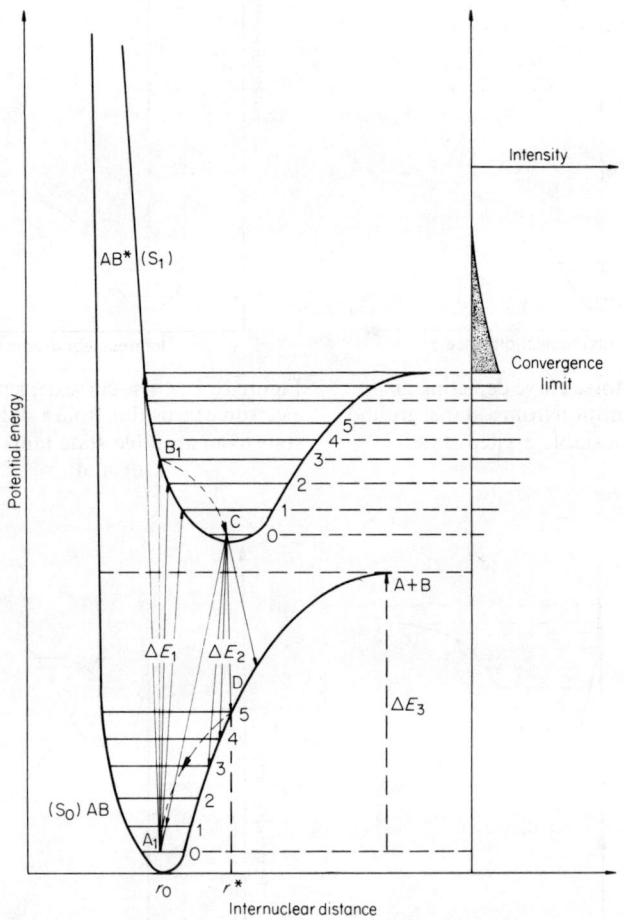

Figure 19 Electronic transitions for a simple diatomic molecule.

Consider the diatomic molecule AB, having ground state S_0 and excited state $S_1(AB^*)$. The potential energy curves representing this process are shown. It can be seen that the most probable (most intense) transition is represented by a vertical line A_1B_1. This illustrates the Franck Condon principle. The Franck Condon principle states that an electronic transition is more probable when the nuclei are in their extreme positions in an oscillation and that the time required for the electronic transitions is so small that the internuclear distance remains unchanged. As electronic transitions are so fast ($\sim 10^{-15}$–10^{-16} s) in comparison to nuclear motion ($\sim 10^{-12}$ s) that immediately after the transition the nuclei have practically the same relative position and velocities as they did just before the transition, the absorption is shown taking place from one end of the line representing a vibrational level since these are the classical rest points in the vibration of a bond. Consequently electronic transitions have a greater probability of occurring when a bond is contracted or expanded except when the molecule is in its zero-point vibrational state, where the most probable position is one corresponding to the equilibrium state. The mean bond length represented by r_0 for the ground state has increased in the excited state to r^*. This is due to the weakening of the bond when an electron is removed from the low energy orbitals which are bonding or non-bonding, whereas after the transition an anti-bonding orbital is occupied.

The molecule AB in the ground state S_0 can be dissociated or photolysed if it absorbs a photon with an energy equal to or greater than that required to carry the molecule to or beyond the 'convergence limit' (see Figure 19) above which a continuum of radiation can be observed.

The length of the horizontal lines at the right below that limit represents the probability of transition from the ground electronic state to a particular vibrational level in the excited electronic state.

The absorption of photons as described is known as the primary process. What happens as a consequence is known as a secondary process. The fate of these excited states is described below.

1.2.2 Fates of Excited States

A typical transition in a molecule will be to an upper vibrational level of the excited state. Thus a photon with an energy of ΔE_1, will raise the molecule to the third level B_1 of that state. There it will vibrate, ultimately lose energy to surrounding molecules and fall to C. It can now emit a photon with somewhat less energy ΔE_2 and fall to D. This results in *fluorescence*. After losing vibrational energy the molecule will return to A_1. The fluorescence will be at a longer wavelength than that of the absorbed radiation. Fluorescence occurs within 10^{-9}–10^{-6} s after absorption.

A typical example of this is given by iodine vapour, which has a region of band absorption leading to a continuum at shorter wavelengths. Radiation absorbed in the continuous region does not cause fluorescence but at low pressures in the discontinuous region does cause fluorescence. This can be explained by assuming that absorption in the continuum leads to dissociation of the molecule whilst the energy of the quantum corresponding to absorption in the discontinuous region is not sufficient to cause dissociation and the excited molecule loses its energy by fluorescence.

If the pressure is raised, absorption in the discontinuous region ceases to cause fluorescence. This illustrates the process of *quenching* whereby an excited molecule may be deactivated by collision with an unexcited molecule, the quencher. In the case of the iodine molecule, the internal excitation energy is converted into translational energy of the two particles, this being an inelastic collision of the second kind. If kinetic energy is converted to excited energy, then it is termed an inelastic collision of the first kind.

Fluorescence can be depicted simply as [21] in Figures 20–21.

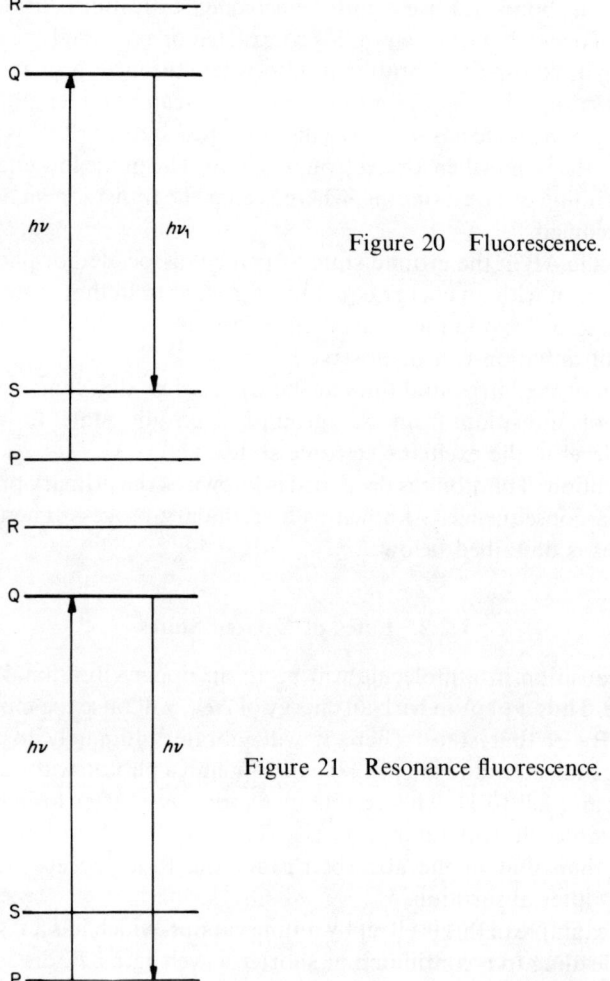

Figure 20 Fluorescence.

Figure 21 Resonance fluorescence.

In Figure 20 fluorescence is shown as resulting from a transition from levels P to Q on absorption of a quantum of radiation. Fluorescence is depicted as a re-emission of a photon from the first excited singlet state Q back to the ground state. The radiation emitted here is of longer wavelength than that of the absorption (less energy).

In contrast, Figure 21 shows *resonance fluorescence* where the radiation emitted (QP) is of the same wavelength as the absorptive radiation (PQ) as a consequence of re-emission back to the initial state.

Another consequence of the absorption of a quantum of irradiation is *phosphorescence*. Phosphorescence may also be simplistically described in a diagram such as those (on page 22), involving a metastable state (see Figure 22).

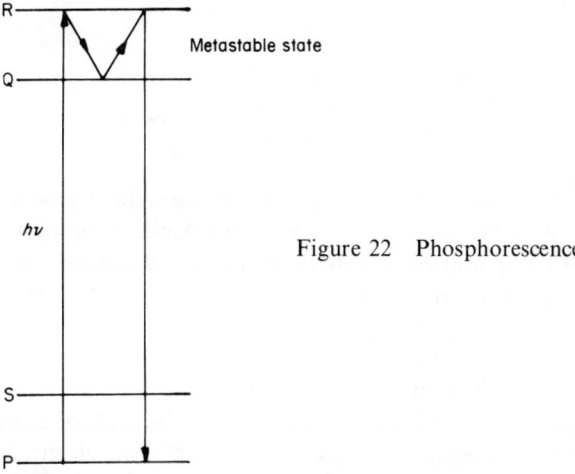

Figure 22 Phosphorescence.

Absorption initially occurs to the excited state R. An electron transition then occurs to a neighbouring level Q which may be *metastable*. A metastable level Q means that the likelihood of a transition from Q to lower levels is remote because of low quantum mechanical transition probabilities. Therefore, in order for the molecule to re-emit a light quantum it must jump back upwards to level R, involving a considerable time delay for the molecule to obtain the energy necessary for the jump by molecular collisions with the surrounding medium. This is the reason for the afterglow characteristic of phosphorescence.

Phosphorescence can therefore be described as:

1. Absorption of a photon to excite an electron from the ground state to the excited singlet state.
2. Intersystem crossing, i.e., the crossing of the electron to the triplet state.
3. Emission of a photon from this state to the ground state.

The lifetime of the phosphorescent state is therefore much longer than fluorescence.

Chemical quenching may occur by electron transfer, abstraction, or addition.

Sensitized fluorescence is the term used when a transfer of electronic energy occurs between excited atoms and atoms of a different species in the ground state.

An example of this is the irradiation of a mixture of mercury and thallium vapours with light of the mercury 253.7 nm line, to which thallium is transparent. The observed sensitized fluorescence is from the excited states of thallium which lie near or below the 4.86 eV of the $^3P_1$ mercury line. The excited mercury atoms lose energy (decay) by either fluorescence or by inelastic collisions with the other element which in turn becomes electronically excited and emits this energy by fluorescence.

Secondary processes in which the acceptor molecule may dissociate are illustrated by photo-sensitizer reactions. These have been studied using mercury as the photo-sensitizer and hydrogen as the acceptor molecule.

Thomas and Gwenn[22] showed that the following reactions satisfactorily accounted for their results.

1. $Hg(6^1S_0) + h\nu(254\,nm) \to Hg(6^3P_1)$
2. $Hg(6^3P_1) + H_2 \to Hg(6^1S_0) + 2H + KE$
3. $Hg(6^3P_1) + H_2 \to Hg(6^1S_0) + H_2 + KE$

where KE is the excess kinetic energy from the quenching process.

Reaction 2 above is a collision of the second kind, the excitation energy of mercury atoms being utilized to supply the energy of dissociation of hydrogen and the translational energy of the products.

Reaction 3 represents a process whereby Hg (6^3P_1) atoms lose their excitation energy at a rate proportional to the hydrogen pressure but without leading to removal of hydrogen from the system.

A further aspect of internal energy conversion is the occurrence of a transition to the triplet state known as *intersystem crossing*. Normal electronic excitations are from singlet to singlet state and transitions from ground-state singlet to excited-state triplet are rare. In small molecules these are forbidden by selection rules. However, if the molecule is sufficiently large these rules may not hold and such transitions may occur.

The first triplet state is of lower energy than the first excited singlet state. Decay of the triplet state may proceed by emission to the singlet ground state known as β-phosphorescence by a transition back to the initial excited state by thermal excitation. The latter process may be followed by emission to the singlet ground state, internal conversion, or solvent quenching. When intersystem crossing from the first triplet to the first excited singlet results in emission to the ground singlet it is known as α-*phosphorescence* or delayed fluorescence and is temperature dependent. As the transition is forbidden the lifetime of a triplet state is considerably longer than those of allowed singlet states and may be of the order of 10^{-4} s for the liquid phase and several seconds for the glassy state.

All the possible singlet–singlet and singlet–triplet transitions are summarized in Figure 23, showing a typical Jablonski diagram.[14]

The most important consequence of the absorption of radiation by a molecule is the dissociation of that molecule. This can occur either if the molecule has enough energy in its upper excited state to cause dissociation or if the excited state is completely repulsive (anti-bonding). This process is photodissociation and is usually caused by absorption in the continuous region of the molecular spectrum. The molecule dissociates within the period of one vibration ($\sim 10^{-13}$ s) to form two atoms, radicals, or stable molecules.

In the first excited state the molecule may be stable with respect to dissociation as far as that state is concerned. However, it is possible for dissociation to occur by a crossing to another excited state, the dissociation limit having been exceeded for this state. Also it is possible for the crossing to occur to a repulsive state.

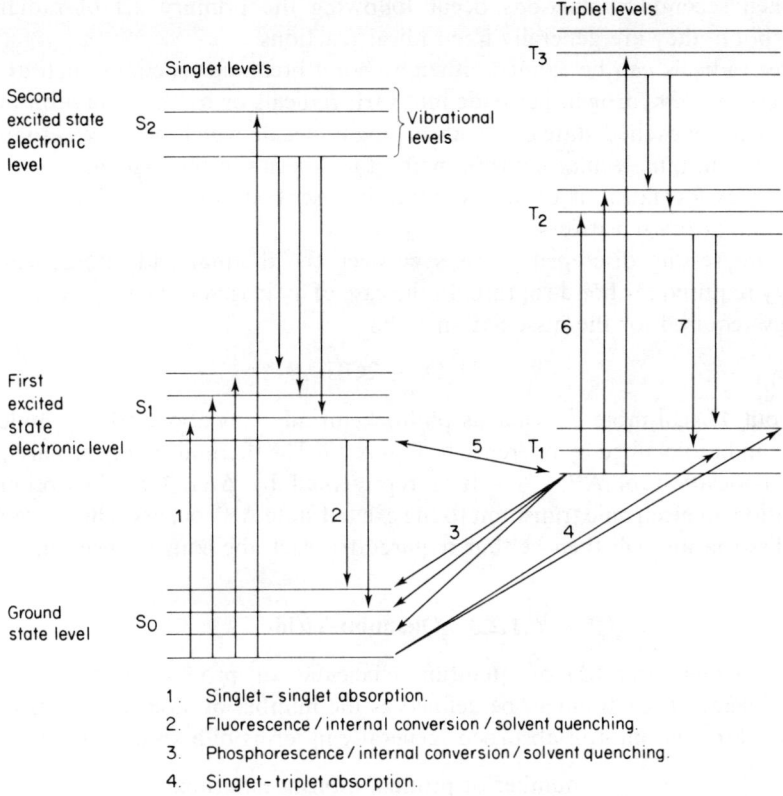

1. Singlet–singlet absorption.
2. Fluorescence / internal conversion / solvent quenching.
3. Phosphorescence / internal conversion / solvent quenching.
4. Singlet–triplet absorption.
5. Excited singlet–triplet transition. Intersystem cossing.
6. Triplet–triplet absorption.
7. Internal conversion.

Figure 23 Jablonski diagram.

Dissociation arising as a result of these crossings is termed *predissociation*. The iodine molecule is an example of this.

With polyatomic molecules the bond broken may not be the same as that responsible for radiation absorption, since the energy of the absorbing bond is greater than the observed quantum. The molecule on receipt of the quantum does not immediately dissociate but after the time of a few vibrations passes into a state where the weaker bond is being stretched and consequently breaks.

An example of this is the carbonyl bond of formaldehyde which absorbs the radiation. An energy transfer takes place causing the carbon–hydrogen bond to break.

$$HCHO \xrightarrow{h\nu} HCHO^*$$
$$HCHO^* \rightarrow H\cdot + \cdot CHO$$

Since a molecule undergoing predissociation will dissociate within the time for one rotation, the rotational fine structure will tend to be lost in the gas-phase spectrum, giving a blurred appearance.

When secondary reactions occur following the primary act of radiation absorption, they are generally free-radical reactions.

Free radicals can be formed either by bond breaking reactions such as the dissociation of hydrogen-peroxide into OH· radicals or by the formation of an electronically excited state of the absorbing molecule which produces atoms or molecules in bimolecular collision with other species in the system.

Exciplex formation is of this type and is described in Chapter 3 concerning photo-initiators/sensitizers.

An interesting discrepancy arises between the thermal and photochemical energy required for bond rupture. In the case of hydrogen peroxide, the thermal energy required for the dissociation

$$H_2O_2 \rightarrow 2OH·$$

is about 47 kcal mole^{-1}, whereas photochemically it is about 80 kcal mole^{-1}. This can be explained by reference to Figure 19. The thermal energy required for the dissociation of $AB \rightarrow A + B$ is represented by ΔE_3. On absorption of radiation an electronic transition to the excited state AB^* occurs. The energy for the dissociation will then be that required to reach the convergence limit.

1.2.3 Quantum Yield

The *quantum yield* (ϕ) or quantum efficiency for product formation of a photochemical reaction can be defined as the number of molecules of product formed for each photon absorbed; generally in terms of a specific wavelength:

$$\phi = \frac{\text{number of product molecules formed}}{\text{number of photons absorbed}}$$

Similarly, it may be defined in terms of the quantum yield of disappearance of the reactant:

$$\phi = \frac{\text{number of reactant molecules which disappear}}{\text{number of photons absorbed}}$$

When $\phi < 1$ no chain process occurs, but if $\phi \geqslant 1$ then a chain reaction can occur. For $\phi = 1$ a photochemical reaction occurs for every absorbed quantum.

It is of paramount importance to know the quantum yield so that the mechanism and course of a photochemical reaction can be understood.

Quantum yield values depend on their method of measurement. Generally the primary photochemical process is of interest as the maximum is unity, a theoretical value resulting from one molecule of a compound disappearing for one quantum absorbed. When the value is less than unity this indicates that recombination (cage reactions) or deactivation are occurring. For overall quantum yields where the value is greater than unity (and in some cases equal to unity) a chain mechanism is indicated resulting from the secondary reactions of the liberated primary free radical species. A good indication of a chain reaction occurring is a lack of proportionality between the reaction velocity and the incident light intensity.

Apart from those reactions obeying the Einstein equivalence law exactly, where one light-absorbing molecule should react for each quantum of radiation absorbed, the quantum yield will usually be a function of the concentrations of the substances (reactants and products) present, the temperature of the system, the wavelength of the light employed, and frequently with changes in other variables. Further complications can occur of the relation between the quantum yield and the variable.[20] For instance, the temperature coefficient often has a dependence upon the wavelength of the incident radiation.

1.2.4 The Cage Effect

Extremes of viscosities are met in surface coatings systems. Some coatings are water-thin, whereas others are stiff pastes (lithographic inks) and these rheology variations will affect the photochemistry of the system because of the 'cage effect'.

When a molecule is dissociated in the liquid phase into two radicals (or atoms), these can undergo *primary recombination* in the solvent cage in which they are formed. The solvent is able to remove the excess kinetic energy by collision and thermalize the particles within about one or a few molecular diameters. Primary recombination is, however, considered by Noyes[23] to occur at less than a molecular diameter, the recombination taking place in a period of the order of a vibration ($\sim 10^{-13}$ s) and certainly less than the time between diffusive displacements ($\sim 10^{-11}$ s).

After separation of about one molecular diameter and as the activation energy for radical recombination is low, the original radical pair can recombine by diffusion, as this is the rate-controlling factor, and hence the term *diffusive secondary recombination* has been applied to this process. Finally, a radical pair if escaping the above two processes may recombine with radicals from another dissociation or undergo secondary reactions with other species. The combination of the first two processes, that is where original partners recombine from the same dissociation, is generally known as the cage effect or geminate recombination. The cage effect was postulated by Franck and Rabinowitch in 1934.[24]

As well as recombination in the solvent cage a dissipation effect can occur in which the electronically excited atom, molecule or radical may interact with the solvent to convert the activation energy of the atom, molecule or radical into vibrational energy of the solvent molecule or to produce a chemical change in the colliding molecule. Both the effects will have quantitative efficiencies which will depend on the nature of the dissociation products and of the solvent e.g., its viscosity, molecular mass, strength of hydrogen bonding, etc., as well as on the frequency of the light absorbed. The greater the frequency of the absorbed radiation the less the recombination effect will be since the greater excess kinetic energy will enable the dissociation particles to escape through the walls of the solvent cage and thus prevent multiple collisions with their former partners. Similarly, if the dissociation products are heavier than the molecules of the solvent their excess kinetic energy allows them to break through the solvent cage wall and thus escape recombination in accordance with the law of conservation of momentum. For dissociation products lighter than the solvent molecules, whatever their kinetic energy they may be stopped or reflected back by the first collision with a solvent molecule.

If after escaping primary recombination, secondary recombination does not take place within about 10^{-9} s, fragments will almost certainly have diffused so far apart that there is negligible chance they will ever re-encounter each other again, but combination with fragments from other dissociations can occur. If a scavenger is introduced into the system it can only compete if it is in sufficient concentration that it can usually catch a radical in the short time for secondary recombination to take place. Secondary recombination cannot be treated with a conventional rate constant.

Noyes[23] considered the following processes other than primary recombination to predict the effect of a scavenger in the system:

$$R_0 + R_0 \rightarrow R_2 \qquad (1)$$

$$R_0 + S \xrightarrow{k_2} P \qquad (2)$$

$$R_0 + R \cdot \xrightarrow{k_3} R_2 \qquad (3)$$

The subscript 0, refers to a radical from a specific dissociation. Reaction (1) is secondary recombination and (3) is combination with a radical from another dissociation. Both of these reactions reduce the efficiency with which dissociation initiates reaction (2) with scavenger S.

Noyes found theoretically[25-26] that as the scavenger concentration increases, the yield of P should increase linearly as the square root of scavenger, where P is a stable product.

If the scavenger reacts with a radical at almost every encounter, then a medium of pure scavenger should compete with primary recombination.

The dependence of the quantum yield on concentration of very reactive scavenger published in Noyes's paper is shown in Figure 24.

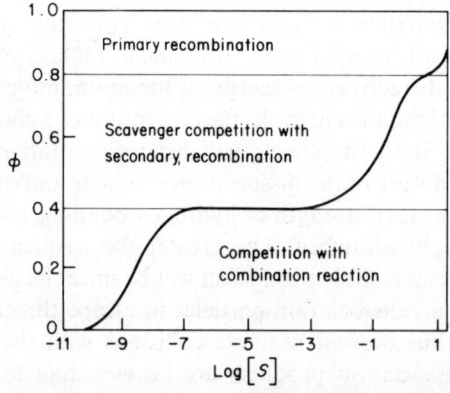

Figure 24 Quantum yield dependence on concentration of very reactive scavenger. (Reprinted with permission from R. M. Noyes, *J. Am. Chem. Soc.*, 77, 2042–2045. Copyright 1955 American Chemical Society.)

There are many examples demonstrating the cage effect, some of which will be discussed below.

Acetyl peroxide decomposes photochemically to produce methane, ethane, carbon dioxide, and methyl acetate. These products can be explained in terms of the following mechanism, where RH is any hydrogen donor, radicals in round brackets are cage radicals, and other radicals are separated from one another by solvent molecules.

$$CH_3-\underset{\underset{O}{\|}}{C}-O-O-\underset{\underset{O}{\|}}{C}-CH_3 \begin{array}{c} \xrightarrow{h\nu} \\ \xrightarrow{h\nu} \end{array} \begin{array}{l} 2CH_3-CO_2\cdot \rightarrow 2CH_3\cdot + 2CO_2 \quad (1) \\ \text{(gas phase)} \\ (CH_3-CO_2\cdot \;\; \cdot O_2C-CH_3) \quad (2) \\ \text{(solution)} \end{array}$$

$$(2CH_3-CO_2\cdot) \rightarrow 2CH_3-CO_2\cdot \quad (3)$$

$$CH_3-CO_2\cdot \rightarrow CH_3\cdot + CO_2 \quad (4)$$

$$CH_3\cdot + RH \rightarrow CH_4 + R\cdot \quad (5)$$

$$2CH_3\cdot \rightarrow C_2H_6 \quad (6)$$

$$(2CH_3-CO_2\cdot) \rightarrow CH_3-\underset{\underset{O}{\|}}{C}-O-CH_3 + CO_2 \quad (7)$$

$$(2CH_3-CO_2\cdot) \rightarrow C_2H_6 + 2CO_2 \quad (8)$$

Acetyl peroxide dissociates to form acetate radicals, reaction (1). In the gas phase, these acetate radicals lose carbon-dioxide by β-scission faster than they undergo any other reaction. However, in solution, the pair of acetate radicals from a given peroxide molecule are held in close proximity by a wall of solvent molecules, reaction (2). This solvent 'cage' holds the original pair of radicals together for about 10^{-10} s. Hence the solvent cage provides a locus for reactions between nascent radical pairs which would not be observed in the gas phase.

The decomposition of acetyl peroxide in iso-octane at 65 °C[27] gives a yield of carbon dioxide per mole of peroxide decomposed of 1.81 for a 1.3×10^{-3} M solution of peroxide. This ratio is not changed if iodine, quinone, or styrene is added as a radical scavenger. This ratio would be 2.0 if (1) were the only reaction which occurred. The missing carbon dioxide implies that a 20 per cent yield of methyl acetate is produced; since the yield of ester is not affected by scavengers, it must be concluded that the ester is formed entirely in the cage, reaction (7). The yield of methane is large (0.82 mole/mole of CO_2 formed), but is reduced to almost zero by 7 M styrene monomer. Thus, methane is largely produced from scavengeable free radicals outside the cage. It must arise from the reaction of free methyl radicals with the solvent, reaction (5). The yield of ethane is small (0.02 mole/mole of CO_2 formed), but is not decreased by the addition of styrene. Therefore, most of the ethane must be formed in the cage, reaction (8). In conclusion, methyl acetate and ethane are formed in cage reactions, but methane is formed from the reaction of free methyl radicals with the solvent.

Barrett, Mansell, and Ratcliffe[28] have carried out studies on cage effects in the photolysis of hydrogen peroxide in alcohol/water mixtures.

Radiation of 253.7 nm in dilute aqueous solution causes hydrogen peroxide to dissociate as:

$$H_2O_2 \xrightarrow{h\nu} (OH\cdot + OH\cdot) \text{ solvent cage} \to 2OH\cdot \qquad (1)$$

The quantum yield of (1) is 0.5 with respect to the decomposition of H_2O_2.[29] The overall quantum yield is twice this value owing to the additional reactions:

$$OH\cdot + H_2O_2 \to H_2O + HO_2\cdot \qquad (2)$$

$$2HO_2\cdot \to H_2O_2 + O_2 \qquad (3)$$

A chain reaction has been found to occur in the presence of alcohols as scavengers.

The following scheme was proposed, excluding reactions (2) and (3):

$$OH\cdot + RCH_2OH \to H_2O + R\dot{C}HOH$$

$$R\dot{C}HOH + H_2O_2 \to RCHO + H_2O + OH\cdot$$

Barrett, Mansell, and Ratcliffe published a diagram for the overall quantum yield variation, ϕ_T, and primary quantum yield, ϕ_p, with mole fraction of ethanol (see Figure 25).

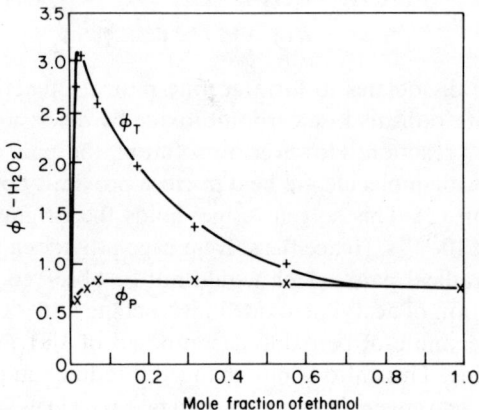

Figure 25 The variation of ϕ_T and ϕ_p for hydrogen peroxide decomposition in aqueous ethanol.[25] (Reproduced by permission of the Royal Society of Chemistry.)

ϕ_p was measured in the presence of allyl alcohol, which competes with the alcohol present for the OH· radicals. ϕ_T was found to increase with ϕ_p, presumably owing to the presence of alcohol molecules in the solvent-cage region which prevents hydroxyl radical pairs from recombining. As the alcohol concentration increases, ϕ_p increases but at higher alcohol concentrations the values of ϕ_p level off and those of ϕ_T go through a maximum and finally in the pure alcohols have values equal to ϕ_p.

At these higher alcohol concentrations (>2 M) the chain decomposition is progressively replaced by a process which may be interpreted as a solvent cage effect. The presence of more than one alcohol molecule in the solvent cage could lead to two alcohol radicals being produced in close proximity. These alcohol radicals may either diffuse from the cage and cause a chain decomposition of the hydrogen peroxide or they may interact within the original solvent cage. This interaction may be either dimerization to give a glycol or disproportionation. Ethanol and 2-propanol gave only acetaldehyde and acetone as solution products, respectively. Their quantum yields were equal to ϕ_T at all concentrations of the alcohols. The radicals, $CH_3\dot{C}HOH$ and $(CH_3)_2\dot{C}OH$ are therefore presumed to be undergoing disproportionation reactions in the solvent cage at high alcohol concentrations.

In the case of the methanol/water system is was possible to distinguish between the two reaction paths. At low methanol concentrations, formaldehyde from the chain process was the major product, with ethylene glycol occurring to the extent of 15 per cent of ϕ_T. Glycol would be formed in the termination step of the chain reaction by the dimerization of the hydroxymethyl radicals. At high methanol concentration the quantum yield for formaldehyde decreased such that in pure methanol solution the quantum yield for formaldehyde had dropped to only 5 per cent of ϕ_T. In all reactions ϕ_T also represented the sum of glycol and formaldehyde quantum yields. It is known that hydroxymethyl radicals dimerize in preference to disproportionation,[30] and this being so the results for the methanol/water system at high methanol concentrations indicate that this dimerization is taking place within the solvent cage of the original hydrogen peroxide molecule. If any of the hydroxymethyl radicals did escape the original cage a chain decomposition would be initiated.

Barrett and Roffey[31] have extended this work concerning cage reactions to another peroxide, the peroxodisulphate ion.

The overall quantum yields, ϕ_T, for peroxodisulphate decomposition in water and aqueous alcohol have been investigated by them at 253.7 nm.

In water this was found to be 0.58 in agreement with the original value found by L. J. Heidt, the proposed mechanism being,

$$S_2O_8^{2-} \underset{}{\overset{h\nu}{\rightleftharpoons}} (2SO_4^-)_{cage} \rightarrow 2SO_4^-{}_{bulk}$$

$$SO_4^- + H_2O \rightarrow HSO_4^- + OH\cdot$$

$$2OH\cdot \rightarrow H_2O + \tfrac{1}{2}O_2$$

In the presence of methanol, ethanol, and iso-propanol, a chain mechanism was found to be induced, which for methanol can be postulated as

$$SO_4^- + CH_3OH \rightarrow HSO_4^- + \cdot CH_2OH$$

followed by

$$\cdot CH_2OH + S_2O_8^{2-} \rightarrow HCHO + SO_4^- + HSO_4^-$$

An initial increase in ϕ_T has been shown to occur for all three alcohols at low concentration, thereafter decreasing to a limiting ϕ_T value at high alcohol concentration. This tended to approach the value for the primary quantum yield, ϕ_p, which was measured for methanol using allyl alcohol as a scavenger (see Figure 26).

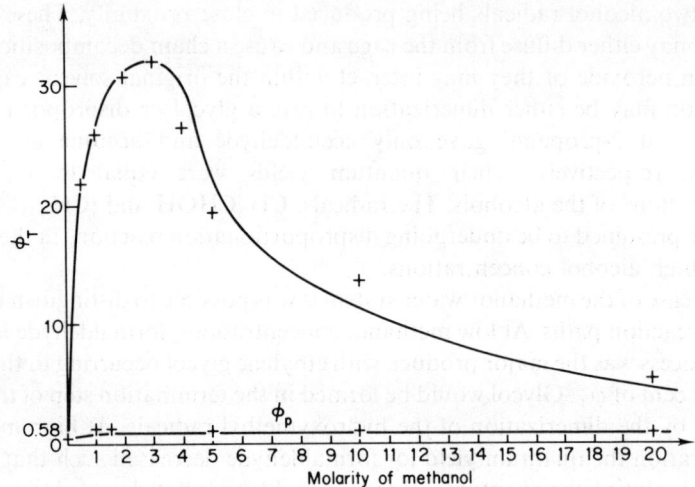

Figure 26 Quantum yield for Peroxodisulphate decomposition in aqueous methanol.

It is suggested that the initial chain reaction is progressively replaced by a process interpreted as a solvent cage effect. The presence of more than one alcohol molecule in the solvent cage could lead to two alcohol radicals being produced in close proximity. These may either diffuse or interact within the solvent cage. At high alcohol concentrations it is suggested that the latter case predominates.

Aqueous peroxide photochemical systems have an obvious potential in photoemulsion formulations.

McGinniss and Kah[32] have carried out studies of photoemulsion polymerization reactions with the sulphate radical anion.

For an emulsion possessing negatively charged monomer–polymer droplets, that is an anionic emulsifier, the negatively charged particle–water interface repels the sulphate ion radicals and initiation occurs mainly with the dissolved soluble monomer fraction. The latter is the homogeneous phase of the aqueous solution. The consequence is oligomeric radical formation, probably by the mechanism below:[33–34]

$$S_2O_8^{2-} \xrightarrow{h\nu} 2SO_4^-$$

$$SO_4^- + M \rightarrow SO_4^-M\cdot$$

$$SO_4^-M\cdot + M \rightarrow SO_4^-MM\cdot$$

$$SO_4^-MM\cdot + nM \rightarrow SO_4^-M(M)_nM\cdot \text{ etc.}$$

$$SO_4^-M_x\cdot + SO_4^-M_y\cdot \rightarrow SO_4^-M_{(x+y)}SO_4^-$$

(polymer or oligomer)

The polymer–water interface is the site where the oligomer radicals grow and absorb. The radical is orientated towards the monomer-swollen polymer phase and the sulphate end-groups solvated by the water phase.[35]

The rate of quenching by various components used in emulsion polymerization reactions was found to depend mainly upon the ionization potential of the dissolved material.

McGinniss and Kah[36] extended this work to investigate the photochemistry of the peroxymonosulphate ion (HSO_5^-) with respect to emulsion polymerization by flash photolysis.

The aqueous photochemistry is thought to proceed via the following set of equations:

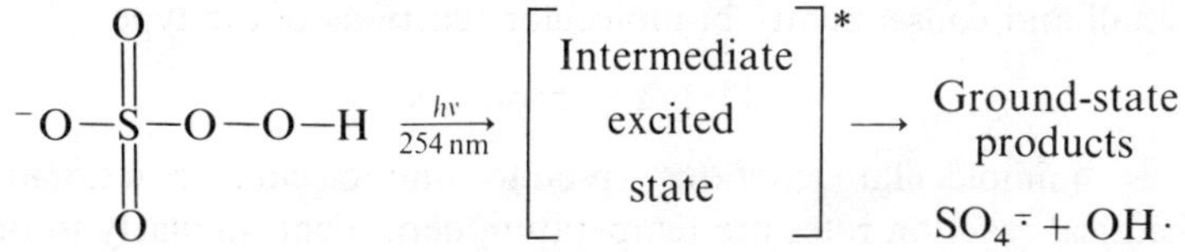

$$^-O-\underset{\underset{O}{\|}}{\overset{\overset{O}{\|}}{S}}-O-O-H \xrightarrow{h\nu, 254\,nm} \begin{bmatrix} \text{Intermediate} \\ \text{excited} \\ \text{state} \end{bmatrix}^* \rightarrow \begin{array}{c} \text{Ground-state} \\ \text{products} \\ SO_4^{-\cdot} + OH\cdot \end{array}$$

$$^-SO_5H + OH\cdot \rightarrow SO_5^{-\cdot} + H_2O$$
$$SO_4^{-\cdot} + H_2O \rightarrow HSO_4^- + OH\cdot$$
$$2OH\cdot \rightarrow H_2O + \tfrac{1}{2}O_2$$
$$SO_4^{-\cdot} + SO_4^{-\cdot} \rightarrow \text{short-lived (30–40}\mu s\text{)}$$
$$\text{bimolecular decay process}$$

Introduction of monomer M primarily adds the following competition reactions to the above:

$$SO_4^{-\cdot} + M \rightarrow \text{(quenching)}$$
$$SO_4^{-\cdot} + M \rightarrow {}^-SO_4-M\cdot \quad \text{(initiation)}$$
$$OH\cdot + M \rightarrow \text{(initiation)}$$
$$SO_4^{-\cdot} + M \rightarrow \text{(oxidation reactions)}$$

1.3 KINETICS AND THERMODYNAMICS OF PHOTOCHEMICAL REACTIONS

1.3.1 Kinetics

The Arrhenius model[37] is frequently used in classical kinetic theory. From this model the rate of chemical reaction (i.e. products P formed from reactants M and N) is:[7]

$$\text{Rate of products formation} = \frac{d(P)}{dt} = K(M)(N)$$
$$= A\,e^{-(E_a/RT)}(M)(N)$$

when K is the product of a pre-exponential factor A and the activation energy factor which includes E_a, the activation energy that has to be overcome for the occurrence of the reaction. RT is the thermal agitation energy at temperature T^0. The exponential factor represents the probability of one encounter complex M...N possessing enough energy to overcome the energy barrier E_a and result in reaction. A, the pre-exponential factor, represents the frequency of potentially reactive encounters.

Bi-molecular collisions do not necessarily lead to reaction including when $E_a = 0$. If a specific geometrical position of the molecules with respect to each other is necessary for reaction to occur, A can be regarded as composed of a frequency factor f, and a probability factor p. For $p = 1$ and $E_a = 0$, then a diffusion-controlled reaction occurs where every encounter leads to reaction. For the majority of photochemical reactions the concentrations of excited states M* is very small and consequently bi-molecular reactions of the type

$$M^* + Y \rightarrow \text{products}$$

behave as unimolecular reactions (pseudo unimolecular reactions). Most photochemical reaction rates are temperature dependent, similarly to ordinary ground-state processes.

Free-radical polymerization occurs when a free radical attacks an unsaturated bond of the type $\diagdown\!\!\!\diagup\!\!\!\!\!\text{C}=\text{C}\!\!\!\diagdown\!\!\!\diagup$ of a monomer, causing the monomer units to add successively to form long polymer chains.

There are several basic steps which include:[38-39] (1) initiation reaction; (2) propagation and/or chain transfer to monomer; (3) termination reaction. Two other important reactions are: (4) inhibition and retardation reactions; (5) auto-acceleration reactions.

1.3.1.1 Initiation reaction

Free radicals are produced photolytically generally by homolytic but sometimes heterolytic fission of a covalent bond. They can also be produced thermally (different mechanism) or by redox systems.

For a photochemical reaction the only meaningful rate constant is the one which defines the chemical transformation of the excited state

$$M^* + N \xrightarrow{k} P$$

The rate dP/dt of product formation is, however, a function of the quantum yield ϕ_p of the reaction and the intensity of light I_a absorbed. That is,

$$\frac{d(P)}{dt} = \phi_p I_a$$

where t is the time during which a compound is irradiated to produce product P.

1.3.1.2 Propagation and chain transfer reactions

(a) *Propagation*. Chain growth occurs via reactions of the type:

$$R_1{\cdot} + M \rightarrow R_2{\cdot}$$
$$R_2{\cdot} + M \rightarrow R_3{\cdot}$$
$$R_n{\cdot} + M \rightarrow R_{(n+1)}{\cdot}$$

The rate of propagation is K_p MR·

where R· is the total radical concentration

K_p is the propagation rate constant

M represents a monomer such as $XCH=CH_2$, and X is a substituent

n is an integer.

Propagation has two important configuration aspects:

(i) It occurs in a 'head-to-tail' manner (head refers to substituted end of monomer). This is because the activation energy is lowest for this method of addition. The transition-state complex is resonance-stabilized and has low steric hindrance. The radicals themselves are resonance-stabilized by the substituents.

(ii) Polymer chains contain asymmetric carbon atoms giving rise to d and l conformations. Hence the polymer can be of the following types:

Isotactic   dddddddddddddd... or llllllllllll...
Syndiotactic   dldldldldldldldldldldl...
Atactic   dddlldldldddldldlldlllldd...   i.e. random configuration

Free-radical polymerization produces essentially atactic polymers with a tendency towards syndiotacticity, especially at low temperatures.

(b) *Chain transfer to monomer.* An alternative to (a) involving hydrogen abstraction and formation of 'dead' polymer.

$$R_n· + M \rightarrow polymer + M·$$

The rate of transfer is $K_{fm}MR·$

K_{fm} is the transfer rate constant.

Chain transfer can occur to any molecule in the system and not just monomer, e.g. to solvent, polymer, initiator, or a specially added transfer agent as well as monomer.

Transfer to polymer does not affect the average chain length but it does produce a branched polymer.

A transfer agent is very effective in small concentrations.

The use of transfer agents is the principal method of chain-length (molecular weight) control.

1.3.1.3 Termination reactions

There are two possibilities:

(a) Combination $R_n· + R_m· \rightarrow P_{(n+m)}$
(b) Disproportionation $R_n· + R_m· \rightarrow P_n + P_m$ (P is polymer)

Both can occur depending on the monomer's physical and chemical properties and the temperature of polymerization.

Probably monomers of the type

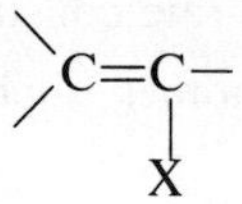

$$\begin{array}{c}\diagdown\\ \diagup\end{array}\!\!C\!=\!\!\underset{\underset{X}{|}}{C}\!-$$

tend to terminate by combination, whereas those of the type

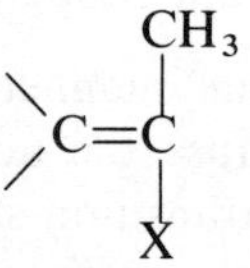

$$\begin{array}{c}\diagdown\\ \diagup\end{array}\!\!C\!=\!\!\underset{\underset{X}{|}}{\overset{\overset{CH_3}{|}}{C}}$$

tend to terminate mainly by disproportionation and the hydrogen abstraction is from the CH_3 and not from the chain CH_2 unit.

Bimolecular termination is of the reaction form

$$R_n\cdot + R_m\cdot \rightarrow \text{polymer}$$

The rate of termination is: $K_t(R\cdot)^2$
K_t is the termination rate constant.

There are three assumptions made for the above type of kinetic scheme. These are that:

(a) the transfer is negligible compared to propagation rate;
(b) the rate constants are independent of radical chain length;
(c) a stationary state in radical concentration exists.

1.3.1.4 Inhibition and retardation reactions

Inhibitors stop the polymerization while retarders reduce the rate of polymerization. The distinction between the two effects is merely one of degree and a molecule may inhibit certain monomers yet only retard others. The effect is produced by small (trace) amounts of additive. Inhibitors do not usually affect the chain length and once used up, polymerization proceeds as normal. Retarders may or may not affect the chain length.

Inhibitor examples are: hydroquinone, *para*-benzoquinone, *p*-methoxy phenol, oxygen, nitro compounds.

1.3.1.5 Auto-acceleration

Many monomers show a sudden and rapid increase in the rate of chain length during the polymerization. For some thermally induced systems it occurs at about 25 per cent conversion for methacrylates and vinyl acetate, and about 2 per cent for acrylates. The rate increase is accompanied by an increase in polymer chain length, an increase in the viscosity of the medium, and a rise in temperature.

The effect is delayed to higher conversions by the use of a solvent, higher temperatures, and higher rates of initiation. It is brought on at lower conversions by the introduction of 'prepolymer'.

The effect is due to poor polymer–solvent (i.e. monomer) interaction which produces severe coiling of the chain radicals and eventually occlusion of the radical ends, thus preventing termination but allowing propagation.

In certain cases (e.g. acrylonitrile, vinyl chloride) the monomer is a non-solvent for the polymer. Precipitation of the polymer occurs immediately polymerization starts and these reactions have auto-acceleration from the start. Occlusion is severe and 'living' polymer can be formed.

There are four chief types of polymerization:

1. Bulk (mass) – monomer and initiator.
2. Solution – monomer and initiator and solvent.
3. Suspension – monomer and initiator and suspension medium (water).

The above are kinetically similar.

4. Emulsion – monomer and water and initiator and emulsifier. This is kinetically different from the first three types as the monomer occurs in three phases: (i) solution in water; (ii) as emulsion; (iii) as suspension. Polymerization occurs principally in the emulsified monomer but initiation occurs in the aqueous phase.

There are three chief types of copolymers (X and Y):

1. Block – X...XY...YX...X
2. Graft – X...Y...X
3. Mixed – XYXXYXYXXYYXYXXYY...

For 3 the units can be alternating, random, or in short sequences to varying degrees.

Copolymerization of two monomers (X and Y) can occur in four propagation steps:

1. ∼∼∼X· + X → ∼∼∼XX·
2. ∼∼∼X· + Y → ∼∼∼XY·
3. ∼∼∼Y· + Y → ∼∼∼YY·
4. ∼∼∼Y· + X → ∼∼∼YX·

In the process of copolymerization the monomer composition generally changes owing to different rates of incorporation into polymer and consequently the polymer composition changes during the reaction.

An exception is the azeotropic composition when rates of incorporation for the two monomers are the same.

1.3.2 Thermodynamics

The thermodynamics of photochemical systems has been described in a review by Suppan.[7] The main points of this are described below.

For a reaction $M + N \rightleftharpoons P$, thermally induced, the equilibrium constant

$$K = \frac{(P)}{(M)(N)}$$

is related to the free-energy change ΔF (or energy available ΔG) from M + N to P,

$$-\Delta G = RT\ln K$$

ΔG therefore determines the maximum yield of products which can be obtained at a given equilibrium.

It is unreliable to adopt this outlook for photochemical reactions for the following reasons:

1. (M) + (N) and P are nearly always never at equilibrium as the energy change is so large that $P \rightarrow M + N$ is practically negligible.
2. There is doubt about the nature of the thermodynamic properties of light. For example, when trying to compare the ground state and excited-state processes,

$$M + N \rightarrow P$$
$$M^* + N \rightarrow P$$

it is a speculative matter as to whether the excess energy of M* compared with M should be regarded as free energy ΔG or enthalpy ΔH, but it is probably insignificant as equilibrium is never obtained and the fate of M is decided by kinetic considerations rather than thermodynamic.

The feasibility of a photochemical reaction occurring is often based on elementary considerations of the bond energies gained and lost in the reaction, the balance assumed to be made up by the excitation energy. To obtain a meaningful result the resonance stabilization energy of the resulting free radicals should be accounted for.

Examples of the above are shown below:

The dissociation of the C—H bond in any molecule can be regarded as unlikely to occur directly on excitation with light of wavelength above 300 nm as this photon corresponds to the CH bond energy of 22.7 J mol^{-1}, but this does not account for the resonance stabilization energy of the resulting free radicals.

Common photopolymerization initiators are aromatic ketones of universal formula:

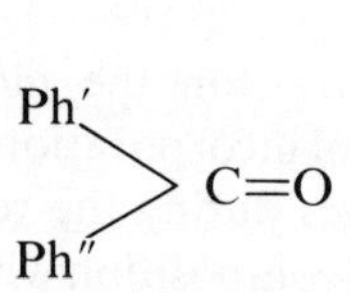

$$\begin{array}{c} Ph' \\ \diagdown \\ C=O \\ \diagup \\ Ph'' \end{array}$$

If this reacts by hydrogen abstraction in a medium RH, then

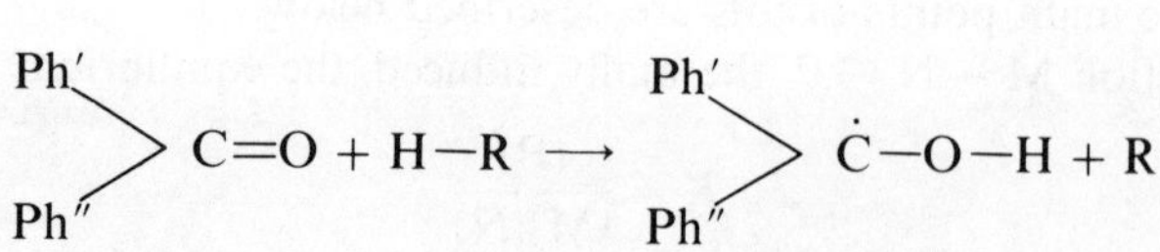

$$\begin{array}{c} Ph' \\ \diagdown \\ C=O + H-R \longrightarrow \\ \diagup \\ Ph'' \end{array} \quad \begin{array}{c} Ph' \\ \diagdown \\ \dot{C}-O-H + R\cdot \\ \diagup \\ Ph'' \end{array}$$

The balance of energy is therefore

$$E = E(\pi_{CO}) + E(\sigma_{RH}) - E(\sigma_{OH}) - (E_{res})$$

As a ground state reaction this is endothermic (positive heat taken in) to approximately

$$14.4(\pi_{CO}) + 22.7(\sigma_{RH}) - 25.2(\sigma_{OH}) - 4.78(E_{res}) = 17.16 \text{ kJ mol}^{-1}$$

The reaction may be possible in the excited state as the CO π bond may be weaker owing to an anti-bonding electron in the π^*_{CO} orbital.

Two factors therefore, determine the energetic possibility of the reaction:

1. The excited state from which reaction can occur has to be at least 7.16 kJ mol^{-1} over the ground state (equivalent to 950 nm).
2. The excited state must have an electronic configuration of the form that the anti-bonding electron is strongly localized on the CO bond.

Condition (1) is not insurmountable as the energy is very small for an electronic excitation energy and indicates the paramount importance of the electron distribution in the excited state.

The above only gives an energy or enthalpy change. To obtain the free energy ΔG, the entropy change ΔS must be included; the latter is, for the above reaction, very small.

1.4 REFERENCES

1. Cundall, R. B., *J.O.C.C.A.*, **59**, 95–101 (1976).
2. Morrison, R. B., *Concise Physics*, Edward Arnold Ltd., pp. 535–536 (1962).
3. Richards, W. G., and Scott, P. R., *Structure and Spectra of Atoms*, John Wiley & Sons, pp. 7–13 (1976).
4. Asimov, I., *Guide to Science. I. The Physical Sciences*, pp. 403–6 (1972).
5. Coulson, C. A., *Valence* (2nd edn), Oxford University Press.
6. Hulme, B. E., *Paint Manufacture*, p. 9, (March 1975).
7. Suppan, P., *Principles of Photochemistry*, The Chemical Society, London, (1972).
8. Press, R. E., *The Chemical Electron*, Longmans, Green & Co. Ltd., p. 15 (1969).
9. Barrett, J., *Introduction to Atomic and Molecular Structure*, John Wiley & Sons, p. 97 (1970).
10. Banwell, C. N., *Fundamentals of Molecular Spectroscopy*, McGraw-Hill, (1966).
11. Cundall, R. A., and Gilbert, A., *Photochemistry*, Nelson, p. 10, (1970).
12. Cox, A., and Kemp, T. J., *Introductory Photochemistry*, McGraw-Hill, London, (1971).
13. Ledwith, A., *J.O.C.C.A.*, **59**, pp. 157–165, (1976).
14. Lee, J. D., *Concise Inorganic Chemistry*, Van Nostrand, pp. 24–29, (1964).
15. Calvert, J. G., and Pitts, J. N., *Photochemistry*, John Wiley, New York, (1966).
16. Kan, R. O., *Organic Photochemistry*, McGraw-Hill, New York, (1966).
17. Turro, N. J., *Molecular Photochemistry*, W. A. Benjamin Inc., New York, (1965).
18. Neckers, D. C., *Mechanistic Organic Photochemistry*, Reinhold Publishing Company, New York, (1967).
19. Style, D. W. G., *Photochemistry*, Methuen & Co. Ltd., (1930).
20. Walker, S., and Straw, H., *Spectroscopy*, Vol. 2, Chapman & Hall, London, (1962).
21. Moore, W. J., *Physical Chemistry*, Longmans, Green & Co., (1956).
22. Thomas, L. B., and Gwenn, W. D., *J. Am. Chem. Soc.* **70**, 2, 643, (1948).
23. Noyes, R. M., *J. Am. Chem. Soc.* **77**, 2,042, (1955).

24. Franck, J., and Rabinowitch, E., *Trans. Faraday Soc.*, **30**, 120, (1934).
25. Noyes, R. M., *Z. Electrochem.* **64**, 153, (1960).
26. Noyes, R. M., in Porter, *Progress in Reaction Kinetics*, Vol. 1, Section 5, p. 153.
27. Levy, M., and Szwarc, M., *J. Am. Chem. Soc.* **76**, 5, 981, (1954).
28. Barrett, J., Mansell, A. L., and Ratcliffe, R. J. M., *Chemical Communications*, No. 1194, p. 48, (1968).
29. Baxendale, J. H., and Wilson, J. A., *Trans. Faraday Soc.* **53**, 344, (1957).
30. Barrett, J., and Baxendale, J. H., *Trans. Faraday Soc.* **54**, 37, (1960).
31. Roffey, C. G., Ph.D. Thesis, London University, (1970).
32. McGinniss, V. D., and Kah, A. F., *J. Coatings Technol.*, **49**, No. 634, Nov., 61–76, (1977).
33. Ono, H., and Saeki, H., *Brit. Polymer J.*, **7**, 21, (1975).
34. Van den Hul, H. J., and Vanderhoff, J. W., *Brit. Polymer J.*, **2**, 121, (1970).
35. Vanderhoff, J. W., Symposium 'Advances in the Characterization of Metal & Polymer Surfaces', Centennial ACS Meeting, New York, April 4–9, (1976).
36. McGinnis, V. D., and Kah, A. F., *J. Coatings Technol.*, **51**, No. 654, July, pp. 81–86, (1979).
37. Laidler, K. J., *Reaction Kinetics*, Vol. 1 *Homogeneous Gas Reactions*, Pergamon Press Ltd., pp. 45–48, (1963).
38. Dyson, R. W., 'Principles of Polymer Science', Course Lecture Summary, Borough Polytechnic, (1973).
39. Bawn, C. E. H., *Chemistry of High Polymers*, Butterworths Scientific Publications Ltd., (1948).

2
Ultraviolet lamp systems

2.1 THE ELECTROMAGNETIC SPECTRUM

Lamps are used to provide light energy, which is one form of radiation contained in the electromagnetic spectrum.[1] This spectrum is shown in Figure 27.

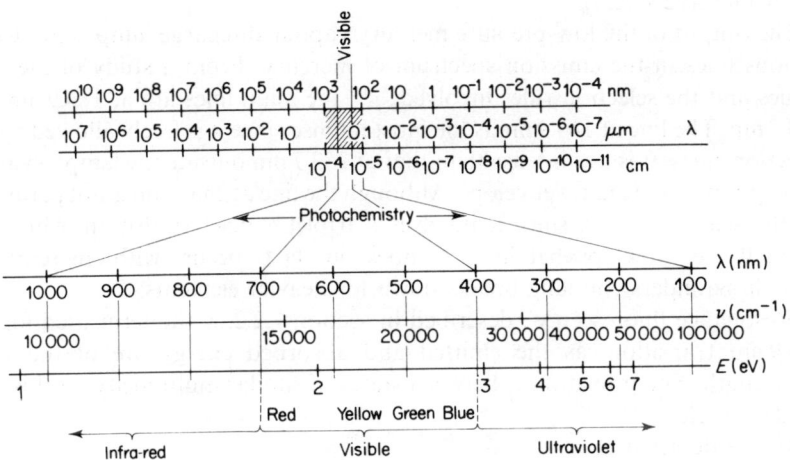

Figure 27 The electromagnetic spectrum.[67] (Reproduced by permission of the Royal Society of Chemistry.)

2.2 LAMPS

An ultraviolet (UV) curing unit can be comprised of the four following components: lamps; reflector units and shielding; a control unit; a cooling system (fan, ducting, air or water, etc.)

A number of radiation lamp systems have been examined and tested for ultraviolet curing of printing inks and coatings. The main types include:

1. Mercury vapour arcs (quartz envelope). These fall into three main categories:
    (a) low pressure
    (b) medium pressure
    (c) high pressure.
2. Plasma arcs

3. Laser beams
4. Electrodeless lamps, such as microwave and radio frequency powered mercury lamps.

Plasma arcs[2-4] and laser beams[5-7] are mainly of research interest and are not as yet a commercial proposition for technological and/or economic reasons as are the electrodeless lamps.[8]

2.2.1 Mercury Arc Lamps

Ultraviolet curing has centred chiefly around the mercury arc lamp.[9] The bulk of commercial lamp units currently in use are of the high-intensity, medium-pressure type, and these show distinct advantages over the low-pressure kind.

At near room temperature, low-pressure mercury vapour lamps operate at about 10^{-6} bar (1 bar = 0.987 atmospheres = 750 mm Hg) and 30 watts per linear inch (12 W/cm).

The output of the low-pressure mercury vapour discharge lamp is limited to various lines in the emission spectrum of mercury. From a study of the term values and the selection rules involved,[9-10] the main lines are at 184.9 nm and 253.7 nm. The line at 184.9 nm is the most intense since it is freely allowed by the selection rules. It is less intense than that at 253.7 nm outside the lamp, owing to absorption by the quartz envelope. Although the line at 253.7 nm is not permitted by the selection rules, since it involves a triplet–singlet transition, which are normally of low probability, it does in fact occur with mercury as Russell–Saunders coupling breaks down for heavier elements.

'Resonance fluorescence', described in section 1.2.2, is the term used for the 184.9 nm transition, as the emitted and absorbed energy are of the same wavelength. The transition is between states of similar multiplicity, in this case singlet–singlet.

This is depicted as

$$\text{Hg}(^1P_1) \rightarrow \text{Hg}(^1S_0) + h\nu \text{ (184.9 nm)}$$

'Resonance phosphorescence' is the term used for the 253.7 nm transition as it is between states of different multiplicities.[11] This is of the triplet–singlet type, and may be written as:

$$\text{Hg}(^3P_1) \rightarrow \text{Hg}(^1S_0) + h\nu \text{ (253.7 nm)}$$

No transitions occur between the $6(^3P_2)$ and $6(^3P_0)$ states to the $6(^1S_0)$ since these are forbidden by the selection rules for J. These are called metastable states because the atom cannot go from them to the condition of lowest energy without the aid of a collision or some other perturbing factor. These lamps are essentially free from continuum and therefore are often employed as sources of monochromatic radiation of wavelength 253.7 nm. (The region where the spectrum becomes a continuum arises because the convergence limit depicts the case where an atomic electron has absorbed the minimal energy required from radiation to escape from the nucleus with no velocity. It may, however, absorb energy greater than this and therefore escape with higher velocities. The kinetic energy of an electron moving in free space is not quantized and therefore any energy above the ionization energy can be absorbed. Continuous absorption or emission occurs rather than a line series which is a result of quantization.) They are, as a result, too low in intensity for present application to UV ink curing alone, but have been used in conjunction with the medium-pressure mercury vapour type, the former giving a surface cure and the latter a bulk cure, with appropriate photo-initiator selection.[12] In this manner, in the coatings field, a bulk film cure is obtained with a medium-pressure lamp and correct photo-initiator, leaving the film surface relatively undercured. This would allow additives such as wax, slip agents, etc., time to rise to the surface and also permit flow-out to occur. The final surface cure can then be achieved at low energy expenditure with a low pressure lamp/photo-initiator combination. This gives good levelling and finish (gloss) to the film.

There are two chief types of low-pressure mercury arcs. These are hot cathode and cold cathode.[13] Hot-cathode electrodes need a high temperature for electron emission. These thermionic lamps have the disadvantage that the electrodes are in part consumed during start-up. Cold-cathode electrodes undergo electron emission by way of a high field at the cathode surface. In contrast to the hot-cathode type, a high temperature is not required, but they need a high voltage for start-up and operation. The electrodes are not consumed and they therefore have longer lifetimes (about 10 000 hrs average) compared to hot-cathode lamps (about 2000 hrs average).

Medium-pressure mercury vapour arc lamps function at approximately 1 bar and around 200 watts per linear inch (80 W/cm). Higher-pressure versions, up to 100 bar, can be between 600–900 watts/inch (240–360 W/cm) but their intensity generally fluctuates with time.[9] The lamp then displays a continuum and this renders it inapplicable for the uniform curing of inks.

In contrast to the low-pressure lamps, the medium-pressure type operate at pressures far greater than 10^{-6} bar. This pressure increase creates a reduction in mean free path for electron/atom collisions, therefore increasing the frequency of collisions. This has the consequence of causing several mercury emission wavelengths as well as those at 184.9 nm and 253.7 nm for the low-pressure mercury arc lamp. This is because of an increased population of excited states other than the $6(^1P_1)$ and $6(^3P_1)$ levels, which permits more transitions and hence more emission wavelengths in the UV region. These are principally narrow bands peaking at 265.4 nm, 303 nm, 313 nm, and 365 nm.

Typical transitions for mercury are shown in the Grotrian diagram, Figure 28. Some other emissions occur pronouncedly in the visible region. These are: 578 nm (yellow), 546 nm (green), 436 nm (blue), 405 nm (violet).

The types of energy levels involved for these in the visible spectrum are:[14]

$$\begin{array}{ll} \text{Yellow} & 6(^1D_2) \rightarrow 6(^1P_1) \\ \text{Green} & 7(^3S_1) \rightarrow 6(^3P_2) \\ \text{Blue} & 7(^3S_1) \rightarrow 6(^3P_1) \\ \text{Violet} & 7(^3S_1) \rightarrow 6(^3P_0) \end{array}$$

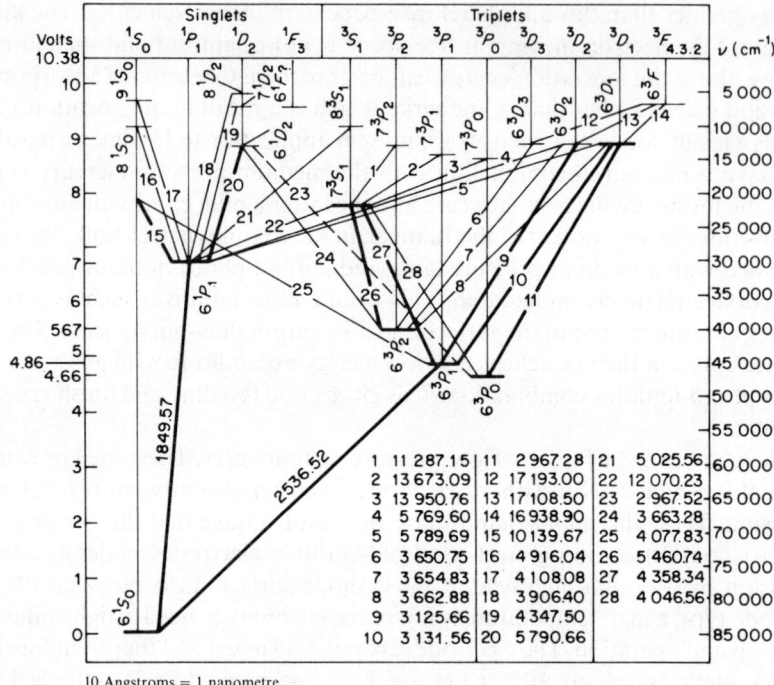

Figure 28   Grotrian diagram showing mercury energy levels.[66] (Reproduced by permission of Thomas Nelson & Sons Ltd.)

Lamps emitting radiation at 184.9 nm produce ozone.[15-16] Ozone gas is the direct result of oxygen absorbing the 184.9 nm radiation and should, therefore, be removed by using an efficient air-exhaust system to the atmosphere where its half-life is short. If the ozone is not removed, this would absorb the 253.7 nm radiation and effectively decrease the efficiency of cure, although presumably in some cases owing to its free-radical nature and oxidative properties it could also promote curing of photopolymerizable systems.

The relatively cool areas surrounding the centre of the arc at these higher pressures causes the 253.7 nm line to become 'reversed'. This means that it is self-absorbed by mercury atoms. Absorption and reversal occur for the single 253.7 nm emission, and instead of emission it appears as a 'non-resonance' line.[9]

The normal guaranteed lamp life at full operating power is of the order of 1000 hours. In practical commercial use, this working period is often greater. Non-ozone producing lamps of the medium-pressure type have been developed; these lamps are 'doped' with a phosphor which filters the radiation responsible for causing ozone to be formed from oxygen.[17]

Three common kinds of chemical bonds in organic molecules are the C—H, C—C, and C—O types. The former has a bond strength energy of about 99 kcal/mole, whereas the other two are of the order 80 kcal/mole. The energy of some of the peak wavelengths of the medium-pressure lamp are shown below in Table 1.

Table 1 Energy of some peak wavelengths of the medium-pressure lamp

Wavelength (nm)	Energy (kcal/mole approx.)
253.7	112
265.0	108
313.0	91
365.0	78

Comparing these figures with the bond energies above, one can therefore expect some bond rupture to be caused in non-photo-initiated irradiated systems (if the molecule absorbs the radiation), and in particular with a high-intensity source. Destructive weathering is often a consequence of this behaviour.[18-19]

The radiated power in watts/inch for the most important lines of the spectrum for a typical high-intensity, medium-pressure mercury vapour arc lamp (undoped) operating at 200 watts/inch (80 W/cm) is shown below in Table 2.[20]

Table 2 Radiated power for mercury arc lamp. (Reproduced by permission of Primarc (Jigs & Lamps) Ltd.)

Spectral line (nm)		Watts/linear inch		Watts/cm
Infra-red				
1367.3		0.9		0.35
1128.7		2.4	A	0.94
1014.0		8.3		3.27
Visible				
578/9	Yellow	14.0		5.51
546.1	Green	15.4	B	6.06
435.8	Blue	12.2		4.80
404.5	Violet	6.1		2.40
Ultraviolet				
365/6.3		12.1		4.76
334.1		1.4		0.55
313.0		5.5		2.17
302.5		3.2		1.26
296.7		2.0	C	0.79
289.4		0.7		0.28
280.4		1.4		0.55
275.2		0.4		0.16
270.0		0.7		0.28
265.2		1.5		0.59
257.1		2.1		0.83
253.7		1.2	D	0.48
248.2		1.3		0.51
240.0		1.2		0.48
Total		94.0		37.00

The nominal input rating of these lamps is 200 watts per linear inch (80 W/cm). In practice, as can be seen from Table 2, only 94 watts (47% of nominal value) per linear inch (37 W/cm) is emitted as output energy. The remaining energy is lost as heat in the ballast, etc., used to obtain a stable arc intensity. Only 35 watts ($17\frac{1}{2}$% of nominal value) per linear inch (14 W/cm) is delivered as ultraviolet energy.

The groups A and B in Table 2 are respectively, infra-red energy and visible light. These are not directly used in the UV curing process, although visible curing reactions can be achieved by dye sensitizing.[21] In the C band, photo-initiators/sensitizers are used to convert light energy into useful chemical energy. For the D area, the energy is generally directly absorbed by the vehicle in UV curing inks, as well as the photo-initiators/sensitizers.

Photo-initiators/sensitizers are incorporated in the inks to match the chief radiation peaks of the lamp source in bands C and D. These must be selected carefully to obtain the maximum conversion of light energy to chemical energy as only light which is absorbed can result in a photochemical effect in accordance with the Grotthus–Draper law of photochemistry.[22–23]

2.2.2 Metal Halide Lamps

The mercury spectrum of the medium-pressure mercury discharge lamp can be modified by introducing other metals into the arc. The use of a particular metal for arc spectrum modification depends on whether it will radiate over the spectral region required and will have adequate vapour pressure at the coolest part of the lamp wall to enter the arc. The latter requirement is difficult to meet but often can be met if the halides of metals, in particular iodides, are introduced into the mercury lamp.[24–27] Metal halides have much higher vapour pressures than the corresponding metal.

Metal halide lamps exist for plate-making, silkscreen (photoemulsion) and photo-resist purposes, with outputs in the 400–420 nm region of the spectrum but are of little value for the rapid curing of lithographic printing inks.

Magnesium iodide has been found suitable for a magnesium iodide mercury lamp, and claims from some manufacturers are that the radiation at about 385 nm is enhanced by 500 per cent and by about 55 per cent at 280 nm, giving some 30 per cent overall gain in cure speed with some systems. Non-ozone producing lamps of this type are under development. Ozone-producing radiation, principally below 210 nm, can be absorbed using doped fused silica and the UV attenuation up to about 275 nm, will vary depending on the quartz type, thickness, operating temperatures, etc. The amount of energy radiated below 230 nm is small, about 1.5 per cent.

Ozone itself starts to absorb radiation at about 245 nm with increasing attenuation until the vacuum UV is reached.

2.3 REFLECTOR UNITS AND SHIELDING

Reflector designs have been described by Gamble.[28] Three chief types of reflector are now commercially available. These are: flat (non-focussing) reflectors; elliptical reflectors; parabolic reflectors.

2.3.1 Non-focusing Reflectors

These are shown schematically in Figure 29 where the substrate is depicted moving perpendicularly into the plane of the page. Several lamps are arranged longitudinally and are mounted equidistant from one another at a predetermined optimum height above the substrate.

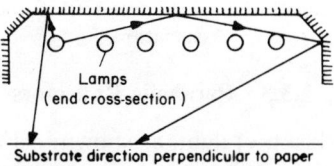

Figure 29   Non-focusing reflector.[28] (Reproduced by permission of the Oil & Colour Chemists' Association.)

The power to each lamp can be individually controlled and the unit can therefore run with varying degrees of power across the substrate according to the print design. This could afford an energy saving in some cases. The radiation is not focused and is simply reflected from a polished aluminium surface.

2.3.2   Elliptical Reflectors

As an ellipse has two foci, the UV lamp is positioned at the first focal point of the reflector so designed, and the substrate so arranged as to pass through the second, as depicted in Figure 30.

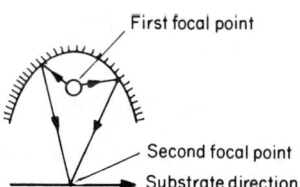

Figure 30   Elliptical reflector.[28] (Reproduced by permission of the Oil & Colour Chemists' Association.)

The printed material has to be protected from the heat generated from the infra-red portion of the radiated power of the lamp. This is achieved by using a shutter mechanism operated by solenoids automatically closing at the instant the press stops (for whatever reason). The lamps can also simultaneously be switched to reduced power.

Shutters are generally of the type where a section of the elliptical design moves, as shown in Figure 31.

The focusing elliptical reflector allows a high concentration of actinic radiation to fall upon the substrate with a simultaneous generation of free radicals.

Figure 31 Shutters.[28] (Reproduced by permission of the Oil & Colour Chemists' Association.)

2.3.3 Parabolic Reflectors

This provides a parallel beam of light, as shown in Figure 32.

Generally, the input power to the tube is automatically halved when the shutters are closed, which permits the tube to be maintained at its working temperature of about 800 °C (the electrodes are at about 300 °C) so that it can be switched up to full power when the press is restarted. Switching off the tube supply during a stoppage would cause a production loss, as time is needed for the tube to cool before it can be restarted.

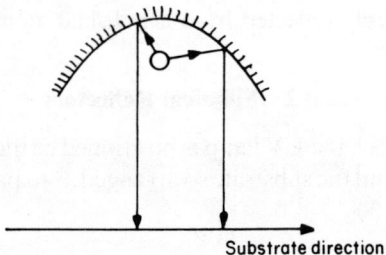

Substrate direction

Figure 32 Parabolic reflector.[28] (Reproduced by permission of the Oil & Colour Chemists' Association.)

2.3.4 Shielding

Apart from electrical and fire hazards, there are two other possible dangers. These are ozone gas and UV radiation.

Ozone, which leads to 254 nm radiation attenuation, is also poisonous. It may readily be eliminated, as mentioned in the cooling section.

Ultraviolet radiation, especially of short wavelength, is known to cause dermal irritation and in some cases of prolonged exposure, skin cancer. 'Arc-eye' is well known in the welding trade and is very painful. Serious damage can result if the retina has exposure to UV radiation, and UV goggles should always be worn whenever any unshielded UV light is emitted. Interlock switches should be fitted to all access doors so that the radiation is adequately screened and cannot either be deliberately or accidentally viewed.

There are several factors affecting shielding design,[29] including:

1. The inverse square law, where radiation incident per unit area decreases with the square of the distance from the lamp.

2. Air absorbs short-wavelength radiation, therefore the further the radiation has to travel the weaker it becomes.
3. Poor reflection occurs from most surfaces.

Reflection of UV radiation is a property of the surface concerned. Some examples are shown below:

Aluminium	about 85%
Stainless steel	about 60%
Brass	about 30%
A matt black surface	less than 5%

Adequate shielding permits the maximum light-path distance of travel by distortion and therefore increases the number of reflections.

Stray radiation leakages can result from a shielded UV curing unit from two prime sources. These are the entrance and exit openings for the substrate when fed into a tunnel design.

The problem and the solution are shown in Figure 33.[30]

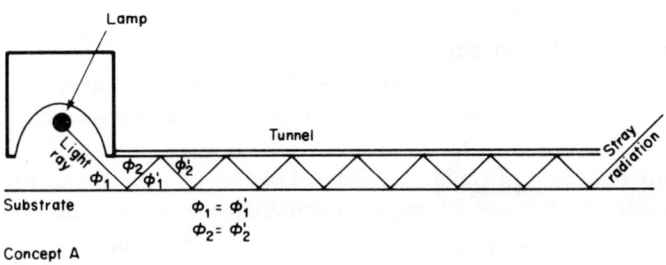

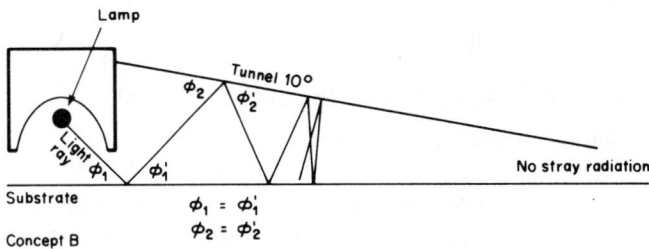

Figure 33 Elimination of stray radiation. (Reproduced by permission of Technology Marketing Corporation.)

Stray radiation is observed in the tunnel using a horizontal reflecting surface parallel to the substrate. This is eliminated by using a shallow slope of the order 10° which forces the UV light to alter direction.

2.4 LAMPS AND CONTROL UNITS

Emission is started from medium-pressure lamps by applying a high voltage across the two electrodes causing ionization of the starting gas. This produces current flow and development into an electric arc.

The commercial mercury arc tube is a sealed cylindrical tube about 2.5 cm in diameter, made of fused quartz and filled with mercury vapour. Lengths are variable to suit press widths.[31] The working voltage depends on the tube length, and is generally of the order between 240 V and 2000 V a.c. Most commercial tubes operate at mains frequency and under normal operating conditions consume about 80 watts per linear centimetre (200 watts per linear inch).

Mercury vapour tubes do not obey Ohm's law.[32] They have a voltage/current characteristic which is negative and non-linear. As the arc intensity increases, so the electrical resistance goes down and, without input power control, the ever-increasing current would inevitably destroy the tube. For this reason a constant-wattage ballast unit comprising a predominantly positive series impedance is used to supply each tube.[33]

The ballast unit can have a capacitor switching facility incorporated in it to allow the UV output to be reduced of the order 50 per cent in 10 per cent steps. This has three chief consequences:

1. It extends useful life of the tube.
2. It reduces the power consumption and hence the heat generated.
3. It matches the curing to press speed if the press is to be run at a reduced speed.

The emitted intensity from an arc may be sensitive to current density flowing through the arc. In the case of low-pressure mercury arcs producing the 253.7 nm line, the intensity is at low current values linearly proportional to the current. At high current values it is, however, non-linear in proportion to current. The intensity is independent of heating the arc envelope. In contrast, heating the electrodes creates a change in the current density. Cooling the electrodes to below ambient temperature will cause the intensity to decrease.

For medium pressure mercury arcs, the following relations have been observed:[9,34-35]

$$\text{Intensity at constant current} = \text{pressure} \times (\text{voltage})^2$$

$$\text{Intensity at constant pressure} = (\text{current})^2$$

Lamp lifetime is affected by the frequency of stopping and starting from cold, the operating temperature, the output intensity, and the number of idling periods. On ageing, the fall off in useful UV output is gradual and of the general order 15 per cent down after 1000 hours. The operator would decide by print inspection whether or not the tube should be replaced. Efficiency can be lost if replacement tubes are not handled with cotton gloves, as perspiration from finger marks deposited on the surface of the tube can become etched into the quartz.

2.5 THE COOLING SYSTEM

There are five major items which require cooling in a lamp drying system: the body of the lamp; the lamp terminals; the lamp reflector; the shutters; the substrates being printed on and any exposed press machinery.

The lamps can be either air- or water-cooled. In the case of the latter, some form of air heat exchanger is still necessary since water cooling of the reflector and shutter does not avoid the requirement of air cooling for the lamp. This will normally take the form of a fan or blower in the apparatus. Air cooling also serves the purpose of promoting safety in the sense of removing the hazardous gas ozone. Ducting to the air outside the factory allows easy removal of ozone, as this gas has a very short half-life and rapidly degrades again to oxygen. Ozone is not a lingering atmospheric pollutant.

Rubber should be excluded from the drying system as ozone viciously attacks it, owing to its powerful oxidative potential. Corrosion problems have, however, been found to be only slight, generally a slight surface defect on metal components. Normal surface finishes give adequate protection.

Cooling requirements can often oppose each other and a balance must be obtained. Overcooling of the lamp reduces its performance and causes its output to be unstable. If the shutter and reflector are undercooled, mechanical problems may result.

The temperature of the outside wall of the tube is important.[36] It should be kept at about 800 °C as if it falls below about 600 °C or beyond the mercury-vapour condensation level, the subsequent pressure drop will result in a change in current density of the arc. It is difficult to restart medium- and high-pressure lamps while hot[37] after switching off, because different voltage requirements are needed at the higher temperature and pressure compared to a cold lamp (because of the lower vapour pressure of mercury).

2.6 INSTALLATION OF UV DRYING SYSTEMS

The number of lamps used in a system is determined by many factors such as printing speed, number of colours, and substrate type. As the printing speed increases, the cure rate will decrease for a given number of lamps. To obtain a desired cure rate at a faster speed will require more lamps if the formulation or lamp-substrate distance is not altered. The number of colours is important, as multi-colour layers are more difficult to cure, if they are not cured separately. The type of substrate also causes variable cure rates for a given number of lamps, depending on whether it is non-porous, (e.g. tin plate and plastic) or porous (e.g. paper). Reflectance is also an important parameter, especially for the non-porous substrates, as it increases the path length of the light radiation and gives a better cure. For this reason tin plate is a good substrate, and also a white substrate is preferred to a black one.

The installation design depends upon the type of printing involved. Web and sheet-fed printing are the two major forms of lithographic printing. In the case of the former, web printing, where the paper is fed via rolls, plenty of space is available and installation of lamps, shielding, etc., is relatively simple. The web can easily be diverted by rollers to a conveniently placed UV unit. As the web can be fed through a very fine gap, the shield can be very compact and very effective.

Sheet-fed presses, on the other hand, represent the most difficult type of press to convert. If the lamps are to go into the press they are in one type of installation inserted between the gripper bar chains before the stacker to avoid shadow from these bars. The delivery may be extended by about a metre, which tends to be complicated and expensive in some presses. An alternative form of installation is to buy a completely independent UV dryer with belt and stacker. This latter choice has the advantage of portability from press to press but several disadvantages as it occupies a large area compared with other installations and involves new capital investment. This portable form also needs a power point both for ventilation and power to the drying unit at each press position. The separate unit is not likely to be an alternative whenever it is possible to fit the installation on the press.

The two main positions for lamp units for non-portable curing systems are at the end of a press, where multi-colour printing can present problems with the thick films produced by 'wet on wet' ink printing, or by an interdeck arrangement where the lamp units are between each colour unit on the press requiring UV curing.[38] This allows setting of each colour before the next is put down. It is important here for this 'wet on dry' ink process, as when set, the ink is then 'dry' in the sense of its surface being tack-free and thus receptive to being overprinted with the next wet, tacky ink. This eliminates the need for tack grading of inks as in a four-colour set drying by thermal means, as for 'wet on wet' ink printing each successive colour should be of lower tack than the former to allow trapping of the following ink. Varnishes may then be applied (conventional or UV) and the print delivered to stack.

Typical types of installation[39] are shown in Figure 34.

2.7 IRRADIATION INTENSITY

The intensity of spectral lines is dependent upon factors such as the population of the initial level, statistical weight effects, and the intrinsic probability of the transition.[40]

The initial population is dependent upon the Boltzmann factor, that is, the excitation energy and temperature of the initial state as defined by the Boltzmann distribution law equation:

$$n_i = n_0 \, e^{-E_i/kT}$$

where n_i is the number of atoms in a specific (non-degenerate) atomic energy level labelled i.

n_0 is the number of atoms in the lowest level (ground state).

E_i is the excitation energy to the level i

k is the Boltzmann constant

T is the temperature in degrees Kelvin.

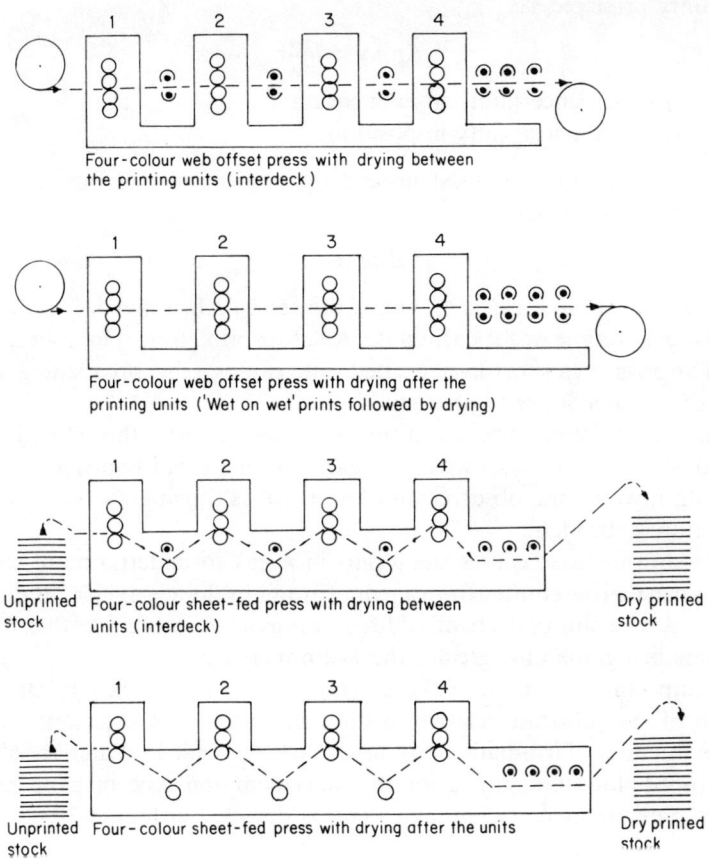

Figure 34   Lamp installation on the press.

Statistical weight effects are introduced into this equation by the symbol g_i which is the statistical weight of the level i. The equation then becomes:

$$n_i = n_0 g_i \, e^{-E_i/kT}$$

The transition's intrinsic probability I is depicted by:

$$I = \int \psi_1 R \psi_2 \, d\tau$$

where ψ_1 and ψ_2 are the wave functions of the initial and final states. R is the operator for the most frequently observed electric dipole transitions.

The Beer–Lambert absorption coefficient, E (see section 1.2.1) is the practical measurement of the transition intensity and may be regarded as the probability of an atom/molecule collision with a photon.

Spectral line widths are not infinitely sharp owing to three factors:[40] natural linewidth; Doppler effect; pressure effect.

Each line has a natural linewidth as a consequence of the Heisenberg uncertainty principle, i.e.

$$\Delta p \, \Delta x \approx h/4\pi$$

where Δp is the uncertainty in momentum
Δx is the uncertainty in position

Energy and time may be used instead of momentum and position and the equation may be written as

$$\Delta E \, \Delta t \approx h/4\pi$$

The natural width of a spectral line, expressed by ΔE, is inversely proportional to the atom's lifetime in a specific state. An absorption from a long-lived ground state of an atom to a short-lived excited state causes a line broadening owing to the small Δt value for the upper state.

Light waves being emitted from an atom exhibit the Doppler effect, analogously to sound waves which appear to be of higher frequency if the source is moving towards the observer and lower if it is moving away. This effect is temperature dependent.

A pressure increase causes the atoms in a gas to undergo more collisions. Photons can then be emitted from a pair of atoms in close proximity at the time of emission. A widening of the band width occurs as a consequence of this, the closer the atoms being, then the greater the width increase.

It is important to optimize the moving web or sheet when printed, as the position of the substrate relative to the light emission source determines the incident intensity of irradiation. Studies have been made by Vanderhoff[41] along three co-ordinate axes x, y, z for the most common type of lamp used, the medium-pressure mercury arc lamp. This is depicted in Figure 35.

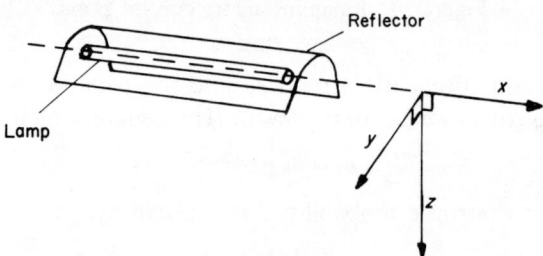

Figure 35   Three co-ordinate axes for intensity studies on a medium-pressure mercury arc lamp.[42] (Reproduced by permission of the Federation of Societies for Coatings Technology.)

Two studies have been made,[42,43] one with a semi-elliptical reflector and the other with an elliptical type, both yielding similar observations.

For the semi-elliptical type depicted in Figure 36, for the vertical z axis, the diagram indicates that the relative intensity decreased, then increased to a

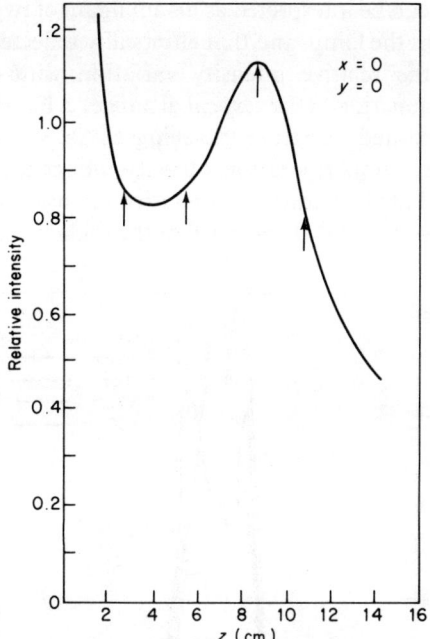

Figure 36 Relative UV irradiation intensity variation with distance along vertical z axis for a lamp with a semi-elliptical reflector.[42] (Reproduced by permission of the Federation of Societies for Coatings Technology.)

maximum at the focal distance, and then again decreased. Figure 37 is similar for an elliptical lamp reflector.

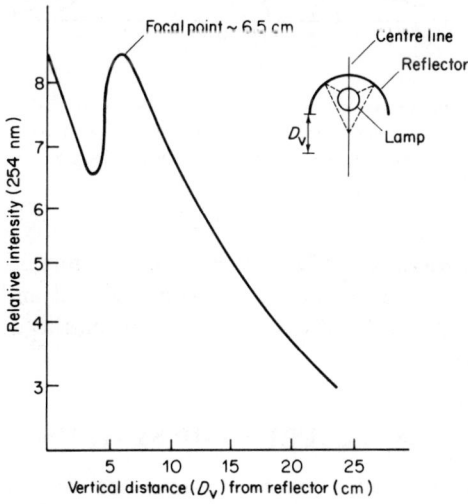

Figure 37 Relative intensity variation of UV irradiation with vertical distance from elliptical reflector (distance along vertical axis z in Figure 36).[42] (Reproduced by permission of the Federation of Societies for Coatings Technology.)

This type of curve can be interpreted as an amalgam of two types of radiation: emission directly from the lamp, and that elliptically reflected from the reflector.

Figure 38 gives the relative intensity variation with distance along the horizontal y axis as a function of the vertical distances z. Figure 38 shows (as areas are similar) that a printed substrate travelling at the same velocity would be subjected to the same overall irradiation intensity, independently of the value of z. Figure 39 indicates that the variation of relative intensity along the x axis was about constant independently of the x value throughout the arc length.

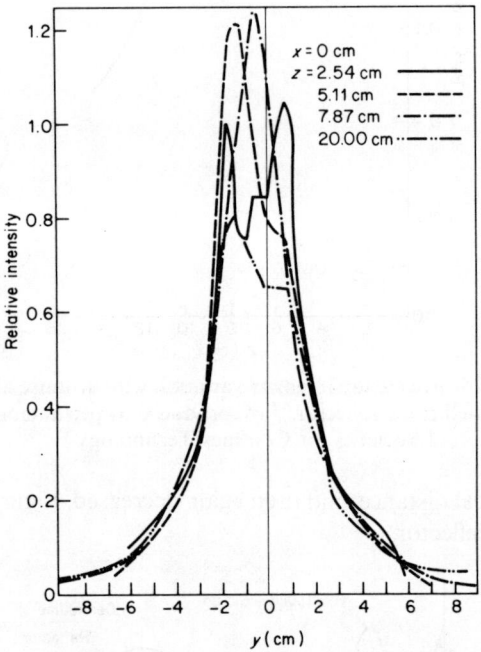

Figure 38   Relative intensity of UV irradiation variation with distance along horizontal y axis as a function of vertical z distance.[42] (Reproduced by permission of the Federation of Societies for Coatings Technology.)

Figure 40 indicates that the variation of relative intensity and lateral distance from the centre line was also similar compared with Figure 38. That is, two peaks at distances less than the focal distance, single peaked and high at the focal distance and single-peaked and low and broad when the distance was greater than the focal distance.

2.8   FUTURE LAMP SYSTEMS

As described for the medium-pressure arc lamp, the efficiency of UV radiation ($17\frac{1}{2}\%$) is low. More powerful lamps, of the order of 300–400 W, are under development. The increase in nominal input rating is however, at the expense of the lamp life owing to electrode wear. Nevertheless, more powerful lamps will

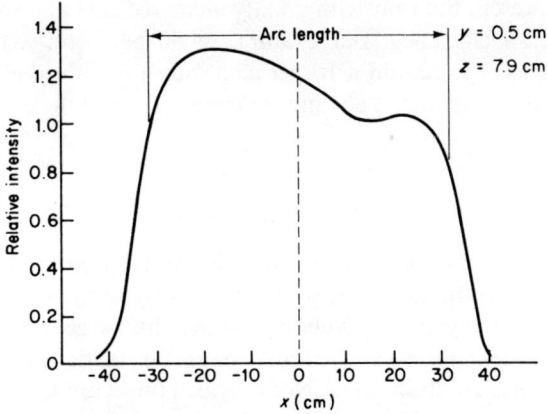

Figure 39 Relative intensity variation of UV irradiation along x axis of lamp.[42] (Reproduced by permission of the Federation of Societies for Coatings Technology.)

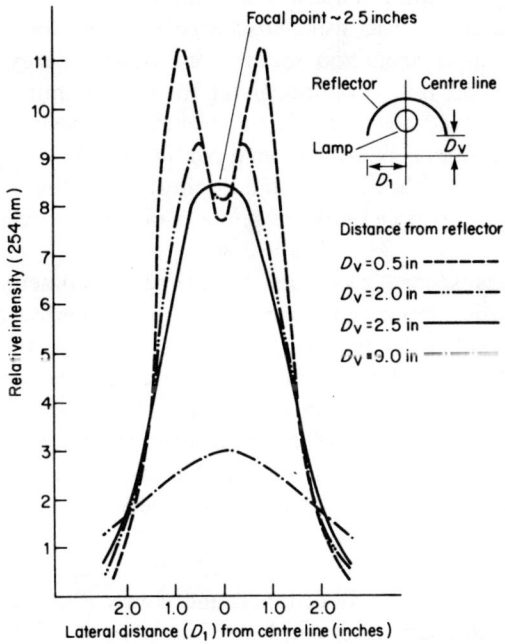

Figure 40 Relative intensity variation of UV with distance from centre line (distance along horizontal axis in Figure 38).[42] (Reproduced by permission of the Federation of Societies for Coatings Technology.)

appear and heat recycling may be applied to conserve energy. The present UV medium-pressure lamps have remained unchanged for about 30 years. The further application of doped lamps will be a prime development area, as the addition of metallic elements or metal halides can increase the output energy by several magnitudes.[34,44–46] Some marginal increases in power consumption can be obtained at present but only with greatly increased cost and sometimes at the expense of overall efficiency. For example, a 50 per cent increase in power consumption is used to obtain a 10 per cent increase in cure efficiency.

Perhaps the most significant advance in lamp systems will come from impulse drying.

2.8.1 Impulse Drying

The pulsed Xenon lamp, developed originally by Dr Edgerton[47] of MIT, has received much attention as a light source in photography.

Impulse systems are generally Xenon gas-filled discharge tubes (but they may be filled with another inert gas, mercury, or metal halide salts) which can be operated on a 'pulse' or 'flash' basis. Not only can they cure UV surface coatings but they may also cause the accelerated drying and curing of a wide range of conventional vehicles such as alkyds, polyesters, polyurethans, etc., which compose most oleoresinous inks.[48]

In the case of UV curing materials, an impulse system provides a high flux density. As an alternative to using discharge tubes in continuous operation, special circuitry is used throughout the a.c. cycle to store energy. This is permitted to escape over a short time duration, giving the term 'impulse' or 'flash' drying. The 'pulse' or 'flash' width is restricted in order that the pulse peak is augmented to far higher levels than is possible from a continuous lamp. This creates a dead time between pulses, although suitable duplication of the sources and phase shifting of the power supplies ensures uniform radiation coverage over printed sheets.

The system has special reflectors possessing a geometry such that two intersecting radiation beams are produced.[49] These embrace the wavebands across the UV/visible/shortwave IR areas of the spectrum. The longer-wavelength IR which could be detrimental in the sense of causing the substrate to overheat, is not so efficiently condensed because it is radiated from the tube walls (see Figure 41). Powerful cooling fans convect this radiation away. At high peak flux, this mixed radiation band is responsible for curing photopolymerizable systems, oxidation drying systems, and a wide variety of solvent- and water-based thermoplastic and thermoset systems.

The flux density remains constant, with the exception of the effects of atmospheric absorption, at any distance, and consequently irregularly shaped objects as well as flat substrates can be irradiated and cured.

Flash lamps have the advantage as a light source of possessing a high peak intensity, short pulse duration, and variable radiant output.[50–52]

A Xenon tube may be operated in the continuous mode or with pulsed high peak intensity. With the pulsed-mode operation and high current density, a greater output of energy efficient in the 350–500 nm region results compared to that in the same region of the 350–1100 nm spectral range, when continuous.[53–55]

A typical flash duration to half peak would be of the order 20–40 μs. This would release some 10^{20} quanta per flash in the spectral range 200–400 nm.[14]

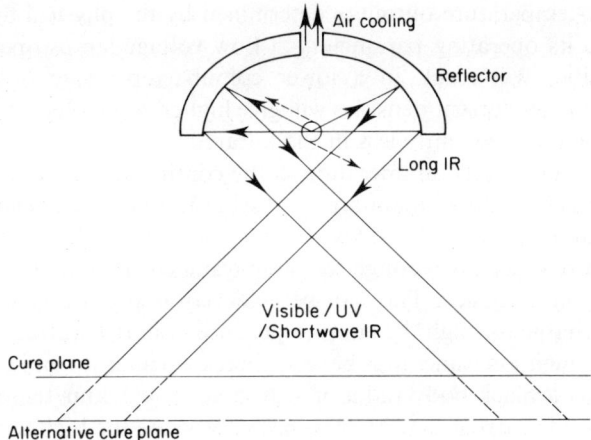

Figure 41   Impulse curing unit.[49] (Reproduced by permission of Coates Brothers and Company Limited.)

Air cooling can be used for these lamps. There is no danger of overcooling or light-intensity fluctuation, in contrast to mercury vapour lamps which are temperature sensitive.

2.8.1.1 Spectral Output

The spectral output of a pulsed Xenon lamp[14] depends upon: lamp dimensions; envelope material; pressure of Xenon gas; and type of driving circuit. Michalaski[56] has published a review on pulsed Xenon lamps, as described below.

The lamp dimensions are important as its impedance depends upon the shape, diameter, and length of tubing, as well as the electrode material and structure. Small-diameter, long tubing or arc length with high gas pressure provides a higher-impedance lamp. In contrast, a larger diameter, shorter arc length and lower gas pressure gives a lower lamp impedance. Linear lamps can be up to one metre in length and have an output in the range 100–10 000 W.

The basic lamp construction material is from quartz tubing or borosilicate glass with one electrode at each end. For d.c. operation, the lamp has a positive anode electrode and a negative cathode electrode. For a.c. operation it has identical electrodes.

The envelope is filled with Xenon gas at a little below or a little above one bar.

Instantaneous ignition occurs with these lamps, in contrast to the mercury arc lamps, because Xenon is a gas at ordinary temperatures (compared to liquid mercury which needs first to be vaporized).

These lamps provide high luminous efficiency output with a constant colour temperature closely matching daylight, independent of cooling, gas pressure, or voltage variations, which are important properties in colour-separation work with these lamps for the graphic arts.

Of the inert gases, xenon gives the greatest conversion efficiency of electrical energy into light.

The colour-temperature output is determined by the physical dimensions of the lamp and its operating parameters.[56] Low-voltage lamps operated at low current densities will result in a lower colour-temperature output. Higher voltages and higher current densities will give higher, more blue and ultraviolet-rich output and consequently less in the infrared.

A line spectrum results on operation in the continuous mode with a low d.c. potential across the lamp to maintain the discharge. Only a small amount of UV is produced and the total electricity-to-light conversion efficiency is poor. An increase in current density through the lamp results in a higher UV content and total conversion efficiency. This can be achieved as previously mentioned by capacitor discharge through the lamp to produce a short duration, high current pulse. Even though some lines may be seen, there is an intense continuum. The arc created acts as a black-body radiator with a specific colour temperature that increases with the current density. A penalty for using the high current density needed to maximize efficiency is a reduction in lamp life. As well as determining the lamp life, the driving circuit also determines the lamp efficiency.[57–59]

The addition of mercury to a Xenon lamp results in an increase in the UV output in the region of mercury spectral lines. This kind of Xenon/mercury lamp is a practical source for UV curing with the same disadvantage as the mercury lamp, i.e., it needs a warm-up period for the UV output to reach its peak even though the Xenon spectral output exists at the moment of ignition.

A comparison of the spectral outputs of a typical Xenon lamp at various capacitance, voltage, and inductance combinations and a Primarc 80 W/cm medium-pressure mercury lamp arc shown in Figure 42.[60] The output energy of a pulsed Xenon lamp is given by

$$E = \tfrac{1}{2}CV^2$$

where E is the energy in watt-seconds (Joules),

C is the capacitance (in farads), and V is the voltage to which the capacitor is charged.

Pulse duration[61] is dependent upon both inductance (L) and capacitance (C) and is found to be proportional to $(LC)^{1/2}$.

These two relationships indicate that a varied range of pulse energies and pulse widths are feasible.

A large voltage applied to the lamp terminals causes the gas in the lamp to break down and become ionized, giving a glow discharge of light at very low efficiency.

For high efficiency and high peak intensity flash, a capacitor is charged to a voltage below the self-breakdown potential of the lamp. Ionization is then obtained by a high voltage and high-frequency pulse applied externally or in series with the lamp. The capacitor is then discharged through the lamp, resulting in a high peak intensity, short duration flash of light.

As high-efficiency operation needs a pulsing circuit, the capacitor is charged through a current-limiting impedance in the form of a choke. The stored energy is then discharged into the lamp through a switch which can be a saturable

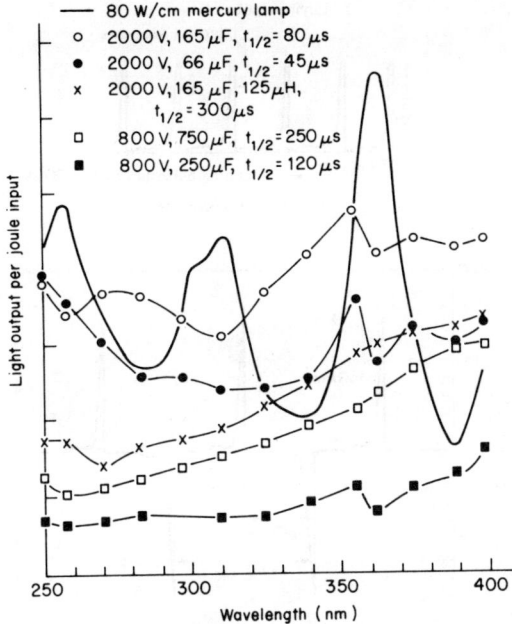

Figure 42 Spectral outputs of Xenon lamp (8 nm bore Spectrosil envelope 150 mm arc Wingent lamp) and mercury lamp (primarc 80 W/cm medium pressure).[60] (Reproduced by permission of the Oil & Colour Chemists' Association.)

magnetic inductor or a solid-state switching device of the thyristor type. On operation in synchronization with the frequency of the power line, the pulsing occurs on each half cycle and the resultant flash rate is twice the power line frequency. For d.c. operation, the rate of flashes per second can be varied and monitored with a timer or alternative pulsing device.

Pulsed Xenon power supplies generally use a starting circuit which generates high voltage and high-frequency starting pulse for the initial lamp ionization. When the discharge is initiated, the starting pulse is removed and the lamp stays ionized between the pulses as there is a sustaining current flowing through the lamp. Other systems do use individual starting pulses for each discharge and then the lamp is deionized between each pulse and there is no current flow.

Instant starting at full intensity eliminates the necessity of mechanical shutters and consequently there is no loss of production time.

A disadvantage of pulsed Xenon lamp operation is the effect of reciprocity failure which can occur if the flash duration is too short, but this can be corrected by carefully selecting the values of the capacitance and operating voltage. Also, at its best, the Xenon lamp efficiency is comparable to that of a mercury lamp.

Some parameters of pulsed Xenon lamps are shown in Figure 43.

2.8.1.2 Kinetics of Impulse Drying

A review concerning the kinetics of 'flash' or 'pulse' drying as applied to surface coatings containing photo-initiators has been made by Phillips.[60] Some of the

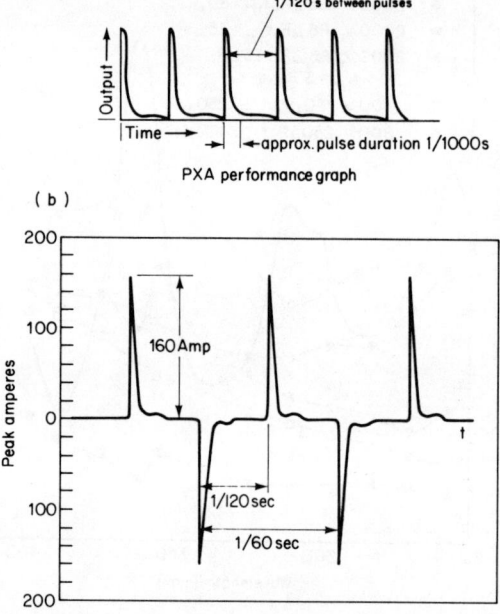

Figure 43 Some parameters in pulsed Xenon lamps.[56] (Reproduced by permission of the Society of Manufacturing Engineers.)

main points are summarized below:

The rate of a free radical polymerization reaction[61] (thermally induced) can be governed by the equation:

$$\frac{-d(M)}{dt} = K_p \left\{ \frac{K_d f(S)}{K_t} \right\}^{1/2} (M)$$

where M = monomer concentration

S = initiator concentration

K_p = rate constant for propagation

K_d = rate constant for decomposition of initiator molecules into radicals

f = fraction of radicals escaping the solvent/medium cage

K_t = rate constant for termination

A modification of the above for photopolymerization[62] is:

$$\frac{-d(M)}{dt} = K_p \left\{ \frac{I_0 \phi l E(S)}{K_t} \right\}^{1/2} (M)$$

where I_0 = incident light intensity

ϕ = quantum yield

l = path length of light

E = extinction coefficient of the photo-initiator

Flash kinetics can be understood from this expression, as two important points emerge from this equation:

1. The polymerization rate is proportional to the square root of the incident intensity.
2. The amount of polymerization resulting from a high-intensity flash for a short time is lower than that from a low-intensity flash for a longer time.

The steady-state theory, which requires that primary radical production is exactly matched at all times by the rate of loss of radicals by termination processes, is assumed for the equation. This is justified for low intensity and long polymerization times but not likely to be so for irradiations during fractions of a second simultaneously with large photon flux variations.

A temperature rise will occur because the reaction is adiabatic and exothermic, and absorption of the spectrum by formulation components occurs.

The viscosity increase resulting from the liquid-to-solid state phase change influences the rates of some of the processes occurring, such as cage reactions described in Chapter 1.

The relative rates of propagation and termination reactions are an example of the 'Tromsdorff effect'.[63-64] An auto-acceleration takes place when vinyl monomers are polymerized either in the absence of solvent or in a solvent in which the polymer is not soluble, i.e., when there is a rapid increase in viscosity on polymerization. The viscosity increase causes an increase in $K_p/K_t^{1/2}$ which increases the polymerization rate.

The main steps in photopolymerization are: primary radical formation; initiation; propagation; and termination.

(i) Primary radical formation

The production rate (r_p) of free radicals from the photo-initiator is $r_p = I_0 \phi lE(S)$.
Homolytic reactions should give two primary radicals,

$$\text{i.e., photo-initiator} \xrightarrow{h\nu} R_a\cdot + R_b\cdot$$

with one quantum absorbed, and ϕ should therefore be 2. In reality, it is generally less, owing to the presence of a solvent 'cage' preventing radicals escaping, leading to recombination within the cage and hence a decreased quantum yield. The crosslink density and primary radical size will determine the magnitude of the 'cage' effect. Primary radical separation is diffusion controlled and temperature dependent. I_0 is high in flash systems and leads to a high r_p, but it will be short-lived.

(ii) Initiation

This is primary radical addition to monomers:

$$R\cdot + M \xrightarrow{K_i} RM\cdot$$
$$r_i = K_i(R\cdot)(M)$$

(iii) Propagation

$$M + RM_n\cdot \xrightarrow{K_p} RM\cdot_{(n+1)}$$
$$r_{pr} = K_p(\Sigma RM_n\cdot)(M)$$

where r_{pr} is the production rate for products.

The peak rates of initiation and propagation will be high and a rise of temperature occurs as they are adiabatic and exothermic reactions. Propagation has an activation energy as it is not diffusion controlled. Steric effects are significant for a polyfunctional monomer system that has undergone a large percentage conversion to polymer.

(iv) Termination processes

These are radical–radical interactions limiting the amount of polymerization.

There are four main termination processes:

(a) Primary radical recombination outside the cage;
(b) Primary radical reaction with growing chain radicals;
(c) Inter-reaction of growing chain radicals;
(d) Oxygen reaction with primary or growing chain radicals.

(a) *Primary radical recombination outside the cage.* The reaction rate is proportional to the square of the primary radical concentration. This process explains the formation of benzil which is yellow when two benzoyl radicals combine outside the 'cage', and leads to the observed phenomenon of yellowing of some clear varnishes on cure.

As radical combination reactions are diffusion controlled and have no activation energy requisites, they are prevalent in low-viscosity systems. A high rate of radical recombination results if these are irradiated with a short intense pulse.

(b) *Primary radical reaction with growing chain radicals.* These are diffusion controlled, especially by the diffusion of primary radicals. The rate will be proportional to the product of the concentrations of the two radical types.

(c) *Inter-reaction of growing chain radicals.* These are diffusion controlled but with limited mobility because of 'steric' interference between polymer chains.

Flash curing is dependent on the above termination processes, as for every increase in radical concentration, the termination rate will be augmented by the square of that increase. All of these processes are most rapid in thin, low-viscosity systems as they are all diffusion controlled.

(d) *Oxygen reaction with primary or growing chain radicals.* As oxygen is a ground-state triplet free radical, the rate at which it terminates a reaction is dependent upon its diffusion rate to the surface and the film interior.

In many systems such as thin films, oxygen inhibition probably causes poor surface cure. A high photon flux can often surmount this.

The effects of termination processes may be reduced by irradiation with several 'pulses' or 'flashes', with a sufficient time interval between them to allow partial polymerization to occur. A single flash has been found in general to yield poor cure of a surface coating.[60]

Impulse drying has received much attention for the surface coatings industry.[65]

2.9 REFERENCES

1. Dyer, R. J., *Applications of Absorption Spectroscopy of Organic Compounds*, Prentice Hall, Inc., N.J., p. 3, (1965).
2. U.S. Patent 3,364,387 (1968).
3. Anderson, J. E., Eschenbach, R. C., and Trone, H. H., *Applied Optics*, **4**, 1435, (1965).
4. Eschenbach, R. C., and Trone, H. H., *Plasma Arc Source for Plasma and Radiation Analysis*, AIAA Plasma dynamics Conference, Monterey, pp. 66–186, March (1966).
5. Wilson, R. M., *Journal of Paint Technology*, **47**, (609), 43, (1975).
6. Leone, S. R., and Moore, C. B., *Laser Sources in Chemical and Biological Applications of Lasers* (Ed. C. R. Moore), Academic Press, New York, (1974).
7. Ross, M., *Laser Applications*, Vols. 1 & 2, Academic Press, New York, (1971, 1974).
8. Weissberger, A., *Techniques of Organic Chemistry, Catalytic, Photochemical and Electrolytic Reactions*, (2nd edn) Vol 11, Interscience Publishers, New York, (1956).
9. Calvert, J. G., and Pitts, J. N., *Photochemistry*, J. Wiley, New York, (1966).
10. Herzberg, G., *Atomic Spectra and Atomic Structure*, Dover Publications, Ch. IV, (1944).
11. Noyes, W. A., Jnr., Hammond, G. S., and Pitts, J. N., Jnr., *Advances in Photochemistry*, Vol. 1, New York, Interscience Publishers, (1963).
12. Holman, R. J., and Rubin, H., *J.O.C.C.A.*, **61**, 189–194, (1978).
13. *General Electric Lamp Bulletin, Applications Engineering Bulletin* LD-1, January 1956.
14. McGinnis, V. D., *U.V. Curing: Science and Technology*, (Ed. S. P. Pappas), Technology Marketing Corporation Publication, Ch. 4, 'Light Sources', (1978).
15. *Sylvania Engineering Bulletin*, 'Germicidal and Short Wave U.V. Radiation', 0–342.
16. Hulme, B. E., Tioxide of Canada Ltd., Technical Service Dept. D.8667 G.C., *The Curing of Coatings with UV Radiation*, Part 11, August (1974).
17. Lienhard, O. E., *SME Technical Paper*, FC 75-335, (1975).
18. McNeill, I. C., 'Photodegradation of Polymers', *J.O.C.C.A.*, **59**, 231–236, (1976).
19. Rabek, J. F., and Ranby, B., *Photodegradation, Photo-oxidation and Photostabilization of Polymers*, New York: Wiley, (1975).
20. Technical Data, Courtesy Primarc Jigs and Lamps Ltd.
21. Oster, G., *Nature* **173**, 300, (1954).
22. Srinivasan, R., and Roberts, T. D., *Organic Photochemical Synthesis*, Vol. 1, J. Wiley, Interscience, New York, (1971).
23. Kosar, J., *Light Sensitive Systems*, J. Wiley, (1965).
24. Tyrrell, A., *Basics of Reprography*, Focal Press, (1972).
25. Knight, R. E., *J.O.C.C.A.*, **61** (4), 119–122, April (1978).
26. Beeson, E. J., and K. F. Furmidge, *Professional Printer* **20** (1), (1976).
27. Henderson, S. T., and Marsden, A. M., *Lamps and Lighting*, (2nd edn) pp. 270–277.
28. Gamble, A. A., *J.O.C.C.A.*, **59**, 240–244, (1976).
29. Jezuit, L. J., *Radiation Curing*, **1** (3), August, 19–23, (1974).
30. Kirk, N., *Radiation Curing*, **1**, (3), August, 24–29, (1974).
31. Schooley, W., *SME Technical Paper*, FC 75-334, (1975).
32. Koller, L. R., *U.V. Radiation*, John Wiley, New York, (1965).
33. Copeland, P., and Sparing, W. H., *J. Appl. Phys.*, **16**, 302, (1945).
34. Fitzgerald, J. M., *Analytical Photochemistry and Photochemical Analysis*, Marcel Dekker Inc., New York, (1971).

35. Heidt, L. J., and Boyles, H. B., *J. Am. Chem. Soc.* **73**, 5728, (1951).
36. Coppinger, C., *Radiation Curing*, **3**(2), 4, (1976).
37. Hankins, W. C., *J.O.C.C.A.*, **60**, 300–306, (1977).
38. Shulman, J. J., *Paperboard Packaging*, **34**, Sept. (1974).
39. Knight, R. E., *J.O.C.C.A.*, **59**, 237–239, (1976).
40. Richards, W. G., and Scott, P. R., *Structure and Spectra of Atoms*, J. Wiley, (1976).
41. Vanderhoff, J. W., 'Status of UV Light Cured Printing Inks', *J. Rad. Curing*, **1**, (4), Oct. (1974).
42. Rubin, H., *J. Paint Tech.* **46**, (No. 588), 74–81, (Jan. 1974).
43. Bassemir, R. W., and Bean, A. J., 'Parameters of UV Printing Inks', *Taga Proceedings*, Techn. Assn. of the Graphic Arts (Rochester, New York), p. 133, (1974).
44. *Radiation Curing* **2** (1), 8, (1975).
45. ICI Petrochemical Division, *Technical Bulletin*, 91-9828, (Nov. 1974).
46. GTE Sylvania, *Technical Bulletin*.
47. Edgerton, H. E., *Electronic Flash Strobe*, McGraw Hill, New York, (1970).
48. Dotzel, W., *Polymers, Paint and Colour Journal*, p. 990, Dec. 3, (1975).
49. Coates Brothers, *Accelerated Drying* – a technical review.
50. L. R. Panico, *SME Technical Paper* FC 75-326, (1975).
51. Ketley, A. D., *SME Technical Paper* FC 75-331, (1975).
52. Bordzol, L., *SME Technical Paper* FC 75-338, (1975).
53. Gonz, J. H., and Newell, P. Bruce, *J. Opt. Soc. Am.* **56**, 87, (1966).
54. Newell, P. B., and O'Brien, J. D., IEEE, *J. Quantum Electronics*, (correspondence), 291, (1968).
55. Oliver, J. R., and Barnes, F. S., IEEE, *J. Quantum Electronics*, QE5, 232 (1969).
56. Michalaski, M., *SME Technical Paper*, FC 337, (1975).
57. Barnes, F. S., *Journal of the SMPTE*, **73**, 569, (1964).
58. Gonz, J. H., and Park, S. W., *Microwaves*, **7**, 34, (1965).
59. Emmett, J. L., and Schawlow, A. L., *Applied Physics Letters*, 204, (1963).
60. Phillips, R., *J.O.C.C.A.*, **61** (7), 233–240, (1978).
61. Oster, G., and Yang, N., *Chemical Reviews* **68**, 2, (1968).
62. Mc.Guiniss, V. D., and Dusek, D. M., *J. Paint Tech.*, **46**, 23, (1974).
63. Norrish, R., and Smith, R., *Nature* **150**, 336, (1942).
64. Tromsdorff, E., Kohle, H., and Lagally, P., *Makromol Chem.* **1**, 169, (1948).
65. Hildebrand Information, No. T-31-511, (March 12, 1973).
66. Cundall, R. A., and Gilbert, A., *Photochemistry*, Nelson, (1970).
67. Suppan, P., *Principles of Photochemistry*, The Chemical Society, London, (1972).

Note. Parts of the articles in question are reprinted from *Journal of Radiation Curing*®, Vol. 1, 3, 1974, published by Technology Marketing Corporation, 17 Park Street, Norwalk, CT 06851, USA. Copyright © 1974 Technology Marketing Corporation.

3
Photo-initiators and photo-sensitizers

3.1 DEFINITIONS

A photopolymerizable coating essentially consists of a polymerizable vehicle (resin binder) and a light-sensitive compound able to convert absorbed light energy into another useful form capable of causing the binder to polymerize into a hard solid mass. This compound may be called a photo-initiator/sensitizer.

A distinction is necessary between the terms photo-initiator and photo-sensitizer, which are often commonly incorrectly regarded as interchangeable. In some cases, however, depending on the environment, compounds may act as either.

A photo-initiator is an additive present to facilitate the initiation reaction in a photopolymerization. This additive may absorb actinic radiation and break to form the primary reactive species, usually free radicals, though not necessarily so.

The photo-initiator is consumed in the reaction. On the other hand, an additive may function as a sensitizer in the strictest sense, that is, absorb then transfer energy to another molecule that forms a primary reactive species. In this type of energy-transfer mechanism, the sensitizer is not consumed or structurally altered.

A 'sensitizer' may be defined as a compound having a positive influence on the photochemical reaction rate as described above, and in this sense, the photo-sensitizer may therefore be regarded as a photo-catalyst.

The additive may also be excited to a state with unpaired electrons (i.e. triplet state) which may accept or donate an electron from (or to) another molecule which, now having an unpaired electron, acts as the reactive species in subsequent steps. This mechanism is known as electron transfer and the additive is reduced or oxidized.

An alternative mode of action for the additive is the formation of a complex that is readily excited. This may then undergo fission to form reactive species, transfer its energy, or undergo electron transfer.

There are also compounds which may be termed photo-optical sensitizers. These give a red shift to polymers, that is, they extend their optical sensitivity to radiation of a longer wavelength.

Efficient photo-initiation depends upon, given an appropriate light source, several factors, including:

1. Suitable absorption coefficients and wavelength sensitivities for the initiator molecule.

2. Important initiation quantum yields in the range 0.1–1.0.
3. The initiator molecule or any of its photo-fragments should not function as chain transfer agents or terminating agents.

3.2 GENERAL MECHANISMS FOR PHOTO-INITIATOR/SENSITIZER ACTION

Classification of photo-initiator/sensitizer reactions according to the mechanism involved is difficult in many cases, owing to inadequate quantitative evidence. It is often common to find reactions grouped according to the chemical constitution and nature of the photo-initiator/sensitizer.

The number of photo-initiators/sensitizers used in photopolymerization reactions are now legion. There is, however, a certain number which have become commercially viable and where known, their probable modes of action are recorded.

The most important mechanistic interpretations of photo-initiator/sensitizer action are summarized in the broad categories below.

The chief reactions are likely to be:

1. Fragmentation reactions
2. Hydrogen abstraction reactions
3. Ionic initiation reactions
4. Photo-crosslinking reactions
5. Triplet energy transfer reactions.

3.2.1 Fragmentation

Initiating free radicals can be produced by fission of a covalent bond in a molecule either via a singlet or triplet excited state. Fragmentation is essentially a simple cleavage of the molecule as is depicted below:

$$X-Y \xrightarrow{h\nu} (X-Y)^* \rightarrow X\cdot + Y\cdot$$

The cleavage can be either: (a) homolytic ($X = Y$), or (b) heterolytic ($X \neq Y$). In the former case the scission results in two symmetrically identical free radicals. For the latter, two asymmetric free radicals occur. The formation of free radicals is directly from an excited state.

These reactions are popularly used for the curing of unsaturated polyesters, acrylated epoxy, or isocyanate-modified acrylic systems.

3.2.2 Hydrogen Abstraction

Free radicals are formed after an excited state molecule has abstracted a hydrogen atom from a monomer or UV reactive resin. Hydrogen is readily abstracted owing to the low bond dissociation energy of the C—H bond (~ 85 kcal/mole)

$$X \xrightarrow{h\nu} X^* \xrightarrow{RH} XH + R\cdot$$

Aromatic ketones react in this way,[1-2] the photopolymerization taking the following general pathway.

Initiation:

$$Ph_2C=O \xrightarrow{h\nu} {}^1(Ph_2C=O)^* \underset{n\to\pi^*}{} \to {}^3(Ph_2C=O)^* \underset{n\to\pi^*}{} \xrightarrow{R_1H} Ph_2\dot{C}-OH + R_1\cdot$$

$$R_1\cdot + M \longrightarrow P_1\cdot \quad \text{where } P_1\cdot = R_1(M)\cdot$$

Propagation:

$$P_n\cdot + M \longrightarrow P_{(n+1)} \quad \text{where } P_{(n+1)} = P_n(M)\cdot$$

Termination:

$$P_m\cdot + P_n\cdot \longrightarrow polymer$$

$$Ph_2\dot{C}-OH + Ph_2\dot{C}-OH \longrightarrow Ph_2\underset{\underset{OH}{|}}{C}-\underset{\underset{OH}{|}}{C}-Ph_2$$

$$Ph_2\dot{C}-OH + P_n\cdot \longrightarrow polymer$$

3.2.3 Ionic Initiation

Ionic initiators can be generated via the photon from reactions involving electron or charge transfer. These are electron donor/acceptor complex formation, and exciplex formation. 'Exciplex' is the term usually used to define a complex that is stable only in the excited state. 'Electron donor/acceptor' complexes (DA's) is the term used frequently for systems which also show ground-state interactions.

3.2.3.1 Electron Donor/Acceptor Complexes

Ionic initiators are formed in solution by the dissociation of a ground state complex into radical anions and cations by the absorption of light.[3-4]

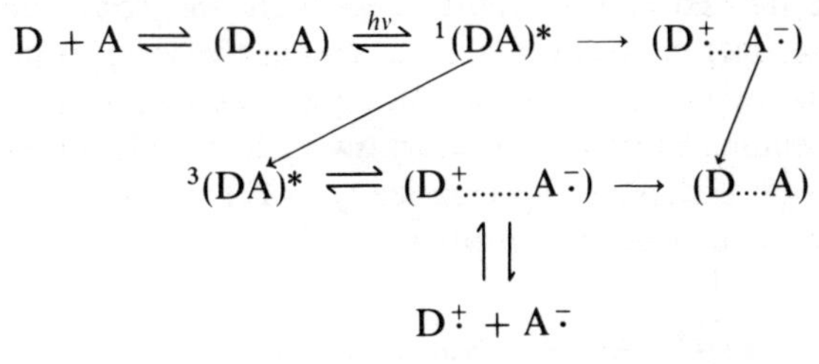

$$D + A \rightleftharpoons (D....A) \xrightarrow{h\nu} {}^1(DA)^* \longrightarrow (D^{+\cdot}...A^{-\cdot})$$

$${}^3(DA)^* \rightleftharpoons (D^{+\cdot}........A^{-\cdot}) \longrightarrow (D....A)$$

$$D^{+\cdot} + A^{-\cdot}$$

The excited states produced from a ground-state interaction (non-light induced) leads to formation of the ionic species in this kind of reaction mechanism. α-methylstyrene/tetracyanobenzene react typically in this way. Both singlet and triplet DA excited complexes, here, allow radical-cation formation, which subsequently causes cationic polymerization of the α-methylstyrene.

Tetrahydrofuran and maleic anhydride form such complexes, and the absorption of actinic radiation by them may induce both cationic and free-radical polymerization in different monomer systems.[5]

Also, oxygen and styrene create a complex of this type which results in photopolymerization of the latter.[6]

An important consequence of this type of reaction is the formation of Lewis acids upon photolysis. Complex diazo salts are used for this method of photopolymerization, the Lewis acid leading to further production of ionic species by reaction with UV resins and causing subsequent photopolymerization.[7]

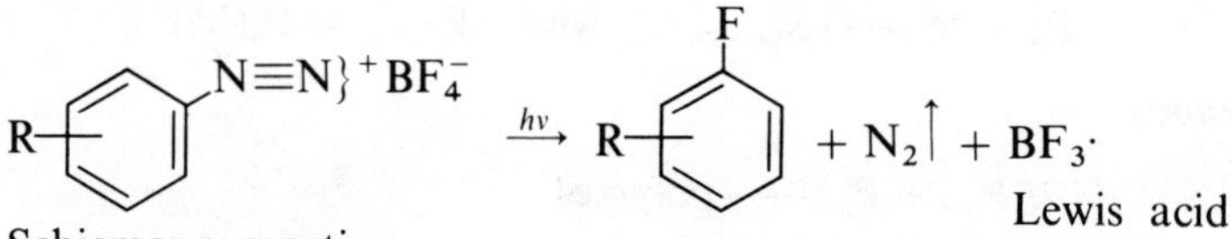

$$R\text{-}C_6H_4\text{-}\{N\equiv N\}^+ BF_4^- \xrightarrow{h\nu} R\text{-}C_6H_4\text{-}F + N_2\uparrow + BF_3.$$

Schiemann reaction
Lewis acid

$$R\text{-}C_6H_4\text{-}\{N\equiv N\}^+ PF_6^- \xrightarrow{h\nu} R\text{-}C_6H_4\text{-}F + N_2\uparrow + PF_5.$$

Lewis acid

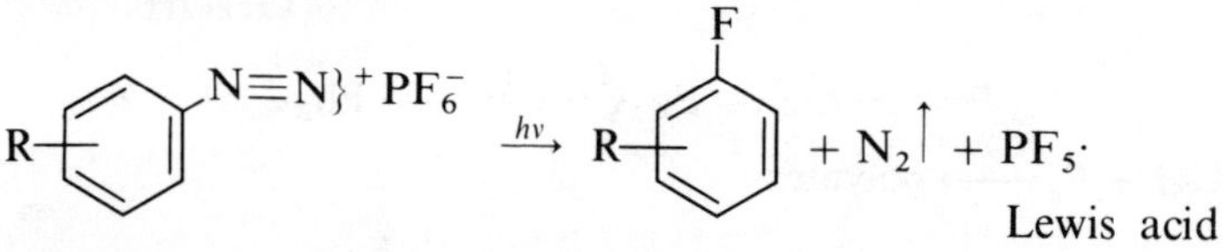

Oxygen inhibition does not occur in these kind of reactions.

The liberation of Lewis acids may cause corrosion problems and pot stability is in general poor. Stabilizers can be Lewis bases such as acetonitrile; alternatives have included ureas, sulphoxides and amides.

Thick films may be problematic owing to the liberation of nitrogen gas and can cause bubbling in the final films.

3.2.3.2 Exciplex Formation

In the case of appropriate electron donor/acceptor systems, interaction between an electronically excited molecule and a ground-state molecule of another type may form an excited-state complex, termed an 'exciplex'.[8] These complexes are often stabilized by charge transfer, and can in solution emit actinic radiation at wavelengths on the red side of fluorescence from the non-complexed excited molecule. The exciplex emission is structureless as the complex does not possess binding energy in the ground state and consequently dissociates within the period of a vibration, leading to broad emission owing to the Heisenberg uncertainty principle. Emission is depicted schematically in Figure 44.

An elementary kinetic scheme proposed is:[9]

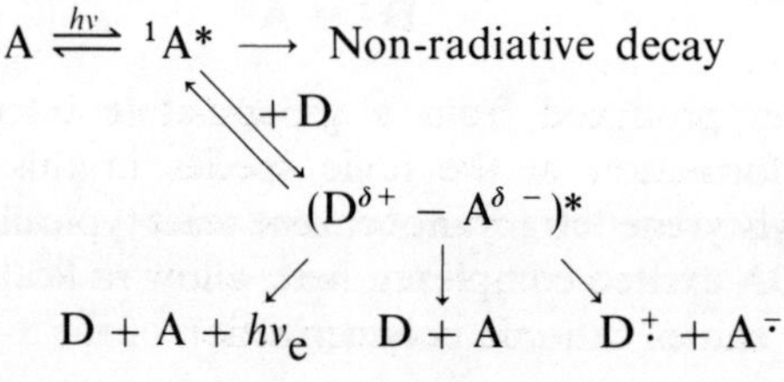

$$A \underset{}{\overset{h\nu}{\rightleftharpoons}} {}^1A^* \longrightarrow \text{Non-radiative decay}$$

$$\searrow +D$$

$$(D^{\delta+} - A^{\delta-})^*$$

$$\swarrow \quad \downarrow \quad \searrow$$

$$D + A + h\nu_e \quad D + A \quad D^{\ddot{+}} + A^{\ddot{-}}$$

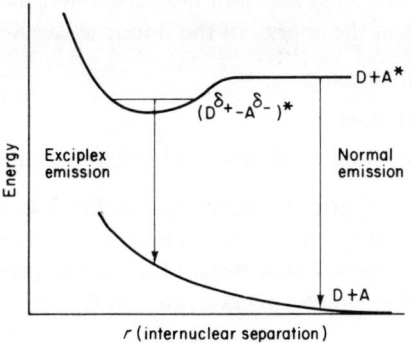

Figure 44 Exciplex formation between electron donor D and acceptors A.[9] (Reproduced by permission of the Oil & Colour Chemists' Association.)

Photo-excited aromatic carbonyl compounds such as benzil, fluorenone, 2-chlorothioxanthone with tertiary amines give exciplex interaction.[10-11] This may be depicted[12-13] for the active carbonyl constituent:

$$>C=O \xrightarrow[n \to \pi^*]{h\nu} >C=O^*$$

$$>C=O^* + RCH_2NR'_2 \longrightarrow (>\dot{C}-\bar{O}, R\dot{C}H_2\overset{+}{N}R'_2)$$
$$\text{Exciplex}$$
$$\downarrow$$
$$>\dot{C}OH + R\dot{C}HNR'_2$$

Similarly N-vinyl carbazole (VCZ) and thiobenzophenone ($Ph_2C=S$) exhibit exciplex formation:[15]

$$Ph_2C=S + VCZ \xrightarrow[\lambda = 589\,nm]{h\nu} (Ph_2C=S)^{\bar{}} + VCZ^{\dot{+}}$$
$$\downarrow nVCZ$$
$$\text{VCZ cationic polymer}$$

3.2.4 Photo-crosslinking Reactions (see Chapter 6)

3.2.5 Triplet Energy Transfer Reactions

Triplet–triplet energy transfer reactions primarily involve triplet energy transfer from excited molecules to monomer or other reaction components to create triplet excited monomer.[14]

i.e.

$$P^*_{(T_1)} + M_{(S_0)} \to P_{(S_0)} + M^*_{(T_1)}$$

This occurs when the energy acceptor molecules (M), have a triplet state of an energy similar to that of the energy of the donor molecule (photosensitizer, P).

* represents an excited state

S_0 is a singlet ground state

T_1 is the lowest triplet state of the molecule in question

Triplet states are important intermediates for the reaction routes that consequently cause fragmentation and electron transfer reactions. Electron transfer reactions lead to a single free radical being produced in contrast to dissociative processes in which two radicals result.

3.3 AROMATIC CARBONYL COMPOUNDS

Photo-initiation of ultraviolet-sensitive systems is predominantly carried out by aromatic carbonyl compounds.[15] These molecules possess electrons in orbitals linked with an oxygen atom that are not participating in the molecule bonding system. Carbonyl compounds have two electrons in each of the oxygen atom's non-bonding orbitals. Ledwith[1] has described this in some detail, which is outlined below.

The outcome of the absorption of actinic radiation may be an n → σ^* or n → π^* transition by promoting one electron into a σ^* or π^* orbital. The types of electronic transition involved and the corresponding differences between them are important for understanding the ultimate chemistry.

As an approximation, the bonding and anti-bonding molecular orbitals energy content ascends in the order below:[1]

$$\sigma < \pi < n < \pi^* < \sigma^*$$

This indicates that the most common electronic transitions are likely to be n → π^* and $\pi \to \pi^*$ in nature. Generally, the lowest energy transitions are the n → π^* rather than the $\pi \to \pi^*$ type. The positions may be reversed depending on the degree and kind of substitution present. As the lowest energy transition is generally n → π^* in nature, it is not to be confused with the lowest energy triplet state, which should not necessarily be so defined.

Significant differences exist in the electronic structures and chemical reactivities of (n, π^*) and (π, π^*) excited states. Electron density is removed from the oxygen atom when an electron is promoted from an n orbital to a π^* anti-bonding orbital with the consequence that (n, π^*) excited states (in particular triplet) show similar reactivity to that of alkoxyl radicals which are known for hydrogen abstraction reactions.

In contrast, an increase in electron density at the oxygen atom occurs when a bonding electron is promoted to a π^* anti-bonding orbital. As a result, the polar nature of the carbonyl group increases. Solvent polarities affect the energies of transitions as a consequence of the different electron distributions of (n, π^*) and (π, π^*) excited states. For $\pi \to \pi^*$ transitions, augmenting solvent polarity results in an energy decrease and hence a red shift occurs, that is, to longer wavelengths.

This happens predominantly by increased solvation of polar solvents, reducing the energy of the more polar excited state. Alternatively, for n → π* transitions, a blue shift occurs, that is, to shorter wavelengths owing to higher energies, by increasing solvent polarity, in particular by hydrogen bonding solvents or more generally by dipolar interactions which cause a decrease in energy of the n orbital. The solvation environment can be critical in determining whether the lowest (n, π*) or (π, π*) excited states are formed, if a relatively small energy separation occurs between the highest energy n and π orbitals. Xanthone and acetophenone have this small energy difference. There is a higher probability of intersystem crossing, that is conversion of singlet (S_1) state → triplet (T_1) state, with (n, π*) states relative to (π, π*) states. There are two main factors determining this. Firstly, selection rules partly forbid the (n → π*) excitation and consequently the reverse transition will be partially forbidden. This results in a greater lifetime tendency of S_1(n, π*) states relative to S_1(π, π*) states, and therefore greater probability of conversion to the triplet state for the former. The second determining parameter is the often quite small energy separations between S_1 and T_1 states of the (n, π*) type, which are of the order 1500–5000 cm^{-1}. For (π, π*) states they are of the magnitude 10 000–15 000 cm^{-1}.[3]

Carbonyl (n, π*) triplet states react easily by hydrogen abstraction reactions with high quantum yields.

In the cases of benzophenone and quinone, which have the lowest energy n → π* transitions, intersystem crossing is basically quantitative, hydrogen abstraction occurring readily with alcohols and ethers.

There are two kinds of aromatic carbonyl compounds which form lowest energy excited states not having (n, π*) character. A commonly used photoinitiator such as Michler's ketone, which is a *para*-amino substituted benzophenone, forms excited 'charge transfer states' which are neither (n, π*) nor (π, π*) in nature.

The excited state structure can be written[1] as:

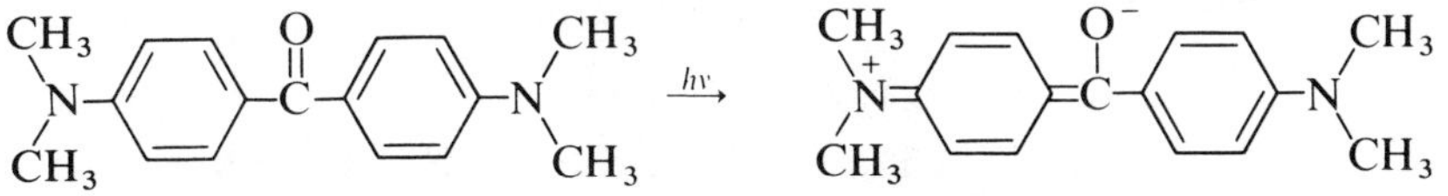

In compounds such as xanthone and fluorenone where the degree of conjugation is more extensive than, for instance, the Ph—C=O group, the lowest lying triplet excited states are often (π, π*) in nature and these compounds choose to react by mechanisms other than hydrogen abstraction with alcohols and ethers, but will tend to do so by the presence of electron donors such as amines and sulphur compounds which act as synergistic agents.

3.3.1 The Norrish Reaction[16]

Excited organic carbonyl compounds undergo three chief types of photochemical reaction, named after Norrish.

3.3.1.1 Norrish Type I Reaction

Homolytic cleavage occurs in the primary process, severing the bond between the carbonyl group and adjacent α-carbon atom:[17]

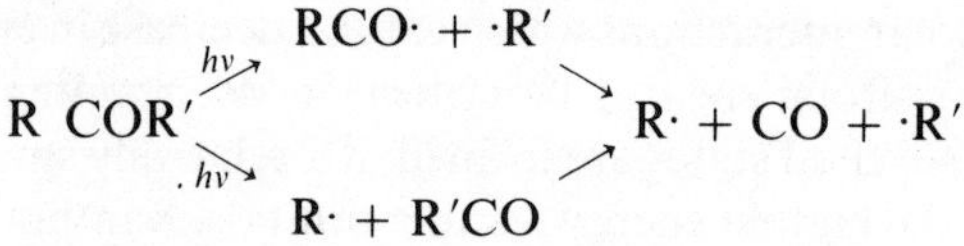

$$R\ COR' \xrightarrow{h\nu} \begin{array}{l} RCO\cdot + \cdot R' \\ \\ R\cdot + R'CO \end{array} \longrightarrow R\cdot + CO + \cdot R'$$

3.3.1.2 Norrish Type II Reaction

This is an intramolecular non-radical process leading to the creation of a six-membered cyclic intermediate. This subsequently decomposes by hydrogen abstraction from the α-carbon to yield an olefin and an alcohol or an aldehyde:

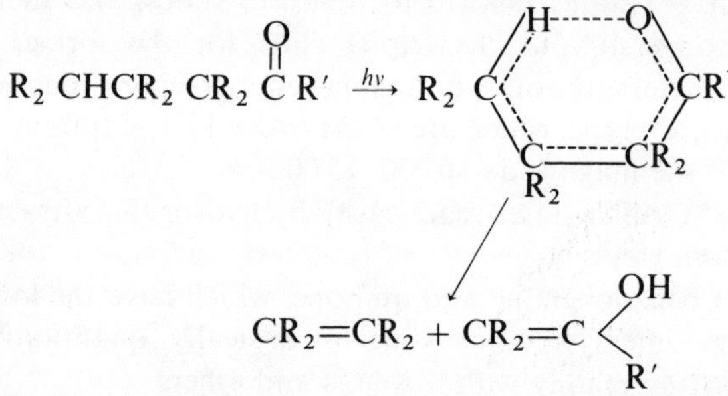

$$R_2\ CHCR_2\ CR_2\ \overset{O}{\overset{\|}{C}}\ R' \xrightarrow{h\nu} R_2\ C \underset{\underset{R_2}{C-----CR_2}}{\overset{H-----O}{\diagup \diagdown}} CR'$$

$$\swarrow$$

$$CR_2{=}CR_2 + CR_2{=}C \diagup \overset{OH}{\underset{R'}{}}$$

An intramolecular process within a longer chain segment may occur with polymers.

3.3.1.3 Norrish Type III Reaction

This is another intra-molecular non-radical process which involves a β-hydrogen atom transfer. This leads to the formation of an aldehyde and an olefin in a carbon–carbon bond scission next to the carbonyl:

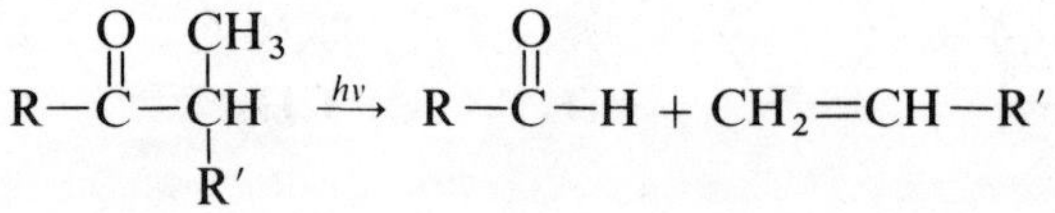

$$R-\overset{O}{\overset{\|}{C}}-\overset{CH_3}{\underset{R'}{\overset{|}{C}H}} \xrightarrow{h\nu} R-\overset{O}{\overset{\|}{C}}-H + CH_2{=}CH{-}R'$$

3.4 CLASSES OF PHOTO-INITIATORS/SENSITIZERS

Some of the more common classes of commercial photo-initiators/sensitizers in use for photopolymerization are listed in the following sections.

3.4.1 Photo-ionic Polymerizing Compounds

Photo-ionic polymerization occurs in two forms, *cationic* and *anionic*. For this discussion, the main area of interest is photo-induced cationic polymerization. An excellent review of cationic polymerization has been published by J. V. Crivello in *UV Curing: Science and Technology*.[18]

Photopolymer preparation occurs generally via three major routes. These are: light-induced polycondensations;[19-20] light-induced radical polymerizations;[21-22] light-induced ionic reactions.

Of these three, photo-induced ionic reactions have some advantages over the rest, in particular over free radical–initiated polymerizations. Some of these are:[23]

1. They can be used to cure saturated monomers of the epoxide, cyclic ether, sulphide, acetal, lactone, as well as vinyl-type monomers. The latter is the type to which photo-initiated free radical polymerization is alone limited. The advantage of curing saturated monomers such as epoxy resins over the unsaturated types is that the former have only a small volatility, good flow owing to low rheological characteristics, no significant colour, negligible toxicity, and superb physical and chemical properties.
2. Cationic photopolymerization is insensitive to aerobic conditions and inert blanketing is not needed.[24]
3. On removal of the actinic radiation, these systems continue to polymerize thermally.

Photocationic systems have several disadvantages such as:

1. Pot stability was poor with some early types and these systems were mainly two-pack in nature, although current systems have good stability.
2. Lewis and Brønsted acids are produced on exposure to actinic sources. This causes handling problems owing to their corrosive nature.

Two main kinds of action have been observed for these systems. These are *photocationic olefinic polymerization,* and *photocationic ring-opening polymerization.*

1. *Photocationic olefinic polymerization.* Only monomers possessing vinyl-type unsaturation are photo-curable in this manner. Systems of this type are:

    N-Vinyl carbazole[25-28] – photopolymerized by gold(III) halides and silver perchlorate.
    Vinyl ethers and styrene[29] – photopolymerized by dicyclopentadienyl titanium dichloride.
    Isobutylene[30-31] – photopolymerized by catalysts of the Friedel–Craft type such as VCl_4, $TiCl_4$, $TiBr_4$, TiI_4.

    Charge transfer complexes such as α-methyl styrene/tetracyanobenzene formed between the electron-donating monomer and electron acceptor molecule are an alternative type of system.[32]

    Further examples of photoionic initiators are discussed in section 3.4.11.
2. *Photocationic ring-opening polymerizations.* Many strained-ring systems may be opened by photocationic curing compounds. Such rings include acetals, cyclic ethers, epoxides, and β-lactones.[33-34]

    Apart from Friedel Craft and simple metal halide compounds previously mentioned, photo-initiators for cationic polymerization are generally of four chief types: aryldiazonium compounds; diaryliodonium compounds; triaryl sulphonium compounds; and triaryl selenonium compounds.

3.4.1.1 Aryldiazonium Compounds

On photolysis these produce an aryl halide via the Schiemann reaction.[35–37] The Lewis acid formed tends to be specific in action, notably so towards epoxy resins. This is depicted below:[38–39]

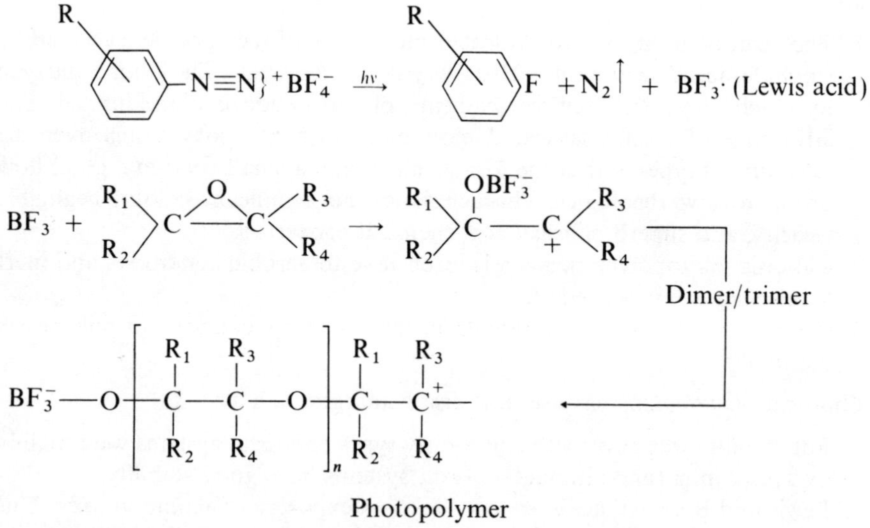

$$\text{R-C}_6\text{H}_4\text{-N}\equiv\text{N}\}^+ \text{BF}_4^- \xrightarrow{h\nu} \text{R-C}_6\text{H}_4\text{-F} + \text{N}_2\uparrow + \text{BF}_3 \text{ (Lewis acid)}$$

$$\text{BF}_3 + \underset{R_2}{\overset{R_1}{>}}\text{C}\underset{}{\overset{O}{-}}\text{C}\underset{R_4}{\overset{R_3}{<}} \longrightarrow \underset{R_2}{\overset{R_1}{>}}\overset{\overset{OBF_3^-}{|}}{\text{C}}-\overset{+}{\text{C}}\underset{R_4}{\overset{R_3}{<}} \longrightarrow \text{Dimer/trimer}$$

$$\text{BF}_3^- - \text{O}\left[\begin{array}{cc} R_1 & R_3 \\ | & | \\ \text{C}-\text{C} \\ | & | \\ R_2 & R_4 \end{array}-\text{O}\right]_n \begin{array}{cc} R_1 & R_3 \\ | & | \\ \text{C}-\overset{+}{\text{C}} \\ | & | \\ R_2 & R_4 \end{array}$$

Photopolymer

The anions present may be BF_4^-, PF_6^-, AsF_6^-, SbF_6^-, FeCl_4^-, SbCl_6^-, etc.[40–41] On photolysis these would give the following Lewis acids: BF_3, PF_5, AsF_5, SbF_5, FeCl_3, SbCl_5. Dianions such as BCl_5^{2-} and SnCl_6^{2-} may also be used.

The photoefficiency of aryl diazonium compounds is dependent upon both the cationic and ionic structures present.[40] For example, PF_6^- is often more efficient than BF_4^-.

Para substitution of a dimethylamino group tends to give a slow curing time. Other slower anions for epoxy cure are: difluorophosphate, phosphotungstate, phosphomolybdate, tungstogermanate, silicotungstate, and molybdosilicate anions. Generally, for these a post-bake is needed to obtain adhesion properties onto metals.[42]

Electron-withdrawing substituents on the benzene ring of the azocompound have been found to be advantageous.

3.4.1.2 Diaryliodonium Compounds

These salts[18,43–47] have the general structure

$$\text{Ph}_2\text{I}\}^+ \text{MeX}_n^-$$

Where MeX_n^- is a complex metal halide and Ph may be a substituted aromatic ring. This substitution affects the photoresponse wavelength. The photolytic mechanism is thought to have two possible paths, one predominating over the other.[43–44]

The predominant reaction is

$$Ph_2I^+MeX_n^- \xrightarrow{h\nu} (Ph_2I^+MeX_n^-)^*$$

$$\downarrow$$

$$Ph\text{-}I^+ + Ph\cdot + MeX_n^-$$
$$\downarrow RH$$
$$Ph-I^+-H + R\cdot$$
$$\downarrow$$
$$Ph\text{-}I + H^+$$

The lesser reaction is to about 5 per cent and proceeds as:

$$(Ph_2I^+MeX_n^-)^* + RH \xrightarrow{h\nu} (Ph - RH)^+ + PhI + MeX_n^-$$
$$\downarrow$$
$$Ph - R + H^+$$

The overall photolysis reaction is therefore:

$$Ph_2I^+MeX_n^- + RH \xrightarrow{h\nu} PhI + Ph\cdot + R\cdot + HMeX_n$$

In this case, strong Brønsted acids for cationic curing such as HBF_4, $HAsF_6$, HPF_6, $HSbF_6$, may be photochemically formed.[48,49] Diaryliodonium salt photolysis is unaffected by oxygen. The cationic photopolymerization probably proceeds as:[50]

$$Ph_2I^+ MeX_n^- \xrightarrow[RH]{h\nu} PhI + Ph\cdot + R\cdot + HMeX_n$$

$$M + HMeX_n \longrightarrow HM^+ McX_n^-$$
$$\downarrow +M$$
$$HMM^+ + MeX_n^-$$
$$\downarrow mM$$
$$HM(M)_{m+1}^+ MeX_n^-$$

The rate of photodecomposition in a series of diaryliodonium salts having the same cation structure is independent of the anionic part of the molecule but the rate of polymerization is dependent on the nature of the acid produced.

The reactivity order is often found to be

$$SbF_6^- > AsF_6^- > PF_6^- > BF_4^-$$

This is thought to be due to the spacing of the ionic pair. A negatively charged ion is the more loosely bound and the more active the propagating cationic species is in the polymerization. In the above, SbF_6^- is the largest anion and consequently the most loosely bound, whilst BF_4^- is the smallest and most tightly bound anion.[51]

Diaryliodonium salts are transparent to visible light and are photo-initiated only in the UV region of the electromagnetic spectrum. They may, however, be photosensitized in the visible range by certain dyes such as benzoflavin, acridine orange and yellow, phosphine R and Setoflavin T, used singly or in conjunction with each other.[52–53] Visible light has the bonus of being cheaper and not so hazardous as UV radiation.

Some diaryl halonium salts such as diarylchloronium and diarylbromonium compounds are reported as being possibly photosensitive[54–55] (especially in their complex metal halide salts).

The UV spectra of some substituted diaryliodonium salts are shown in Figure 45.

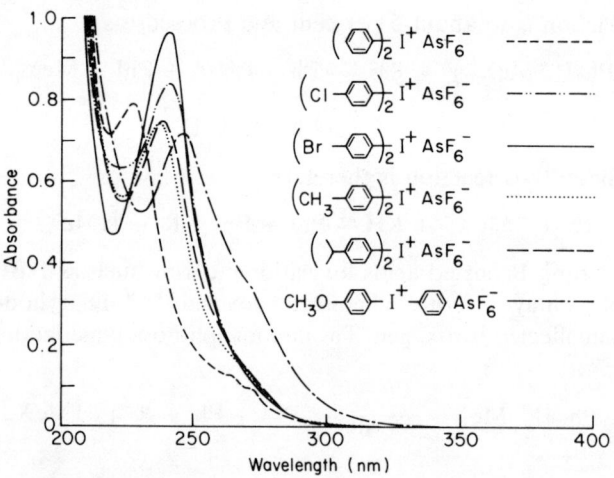

Figure 45   UV spectra of substituted diaryliodonium salts.[44] (Reproduced by permission of the General Electric Company and the American Chemical Society.)

3.4.1.3   Triarylsulphonium Compounds

These salts[55–56] have the structure

$$Ph_3S\}^+ MeX_n^-$$

where X is a halogen and Ph may be substituted, Me is a metal, and n is an integer. They have a photochemical mechanism similar to that proposed for diaryliodonium salts,[55] but have greater thermal stability. Again, they possess the properties of not undergoing air inhibition, neither are they temperature sensitive nor affected by other radical inhibitors. Also, the photo-reactivity is not quenched by triplet-state quenchers and is not accelerated by radical photo-initiators, which are properties in common with the diaryliodonium compounds. In contrast to the latter, however, only one particular dye is known to confer significant photosensitization of these compounds in the visible region and this is perylene.[18,57]

This class of compounds almost approaches an ideal for photo-initiators.

The complex salt 2,5-dimethoxy,4-tolyl-thio benzene diazo tetrafluoroborate and the diethoxy compound, which are sulphur-containing cationic curing compounds without a metal atom present, are also effective curing agents:

e.g. $CH_3-\langle\bigcirc\rangle-S-\langle\bigcirc\rangle-N\equiv N\}^+ BF_4^-$
with OC_2H_5 and H_5C_2O substituents

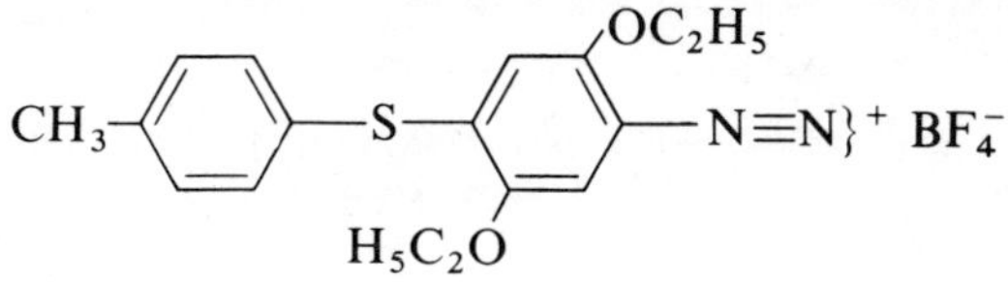

The diethoxy compound

3.4.1.4 Triarylselenonium Compounds

These have a similar structure to those in previous section[58,18]

$$Ph_3Se\}^+ MX_n^-$$

and are analogous in photomechanism to them.

They cure slightly slower than the triarylsulphonium compounds and are particularly good for curing oxirane compounds.

3.4.2 Benzoin/acetophenone and Derivatives

Acyloin compounds of prominent photochemical interest in the surface-coatings industry centre around benzoin and its derivatives.

A general formula for this group can be written as

$$Ph-\underset{\underset{R''}{|}}{\overset{O}{\overset{\|}{C}}}-\underset{}{\overset{R'}{\underset{|}{C}}}-R'''$$

The chief compounds of commercial importance can be classified as:

Benzoin $R' = H$
                                $R'' = OH$
                                $R''' = Ph$

Benzoin alkyl ethers $R' = H$
                                $R'' =$ any alkoxy groups
                                   such as OCH_3, OC_2H_5, etc.
                                $R''' = Ph$

Benzil ketals $R' = R'' =$ any alkoxy radical
                                $R''' = Ph$

Acetophenone derivatives

(i) *Dialkoxyacetophenones* $R' = R'' =$ any alkoxy radical.
                                $R''' = H$

(ii) *Di- and tri-chloroacetophenones* $R' = R'' = Cl$
                                $R''' = Cl$ or H

Conversely, all the above could be regarded as acetophenone derivatives.

3.4.2.1 Benzoin

Dissociation probably occurs via dissociation of the excited triplet state to form free radicals (Norrish type I cleavage).[1,17,59]

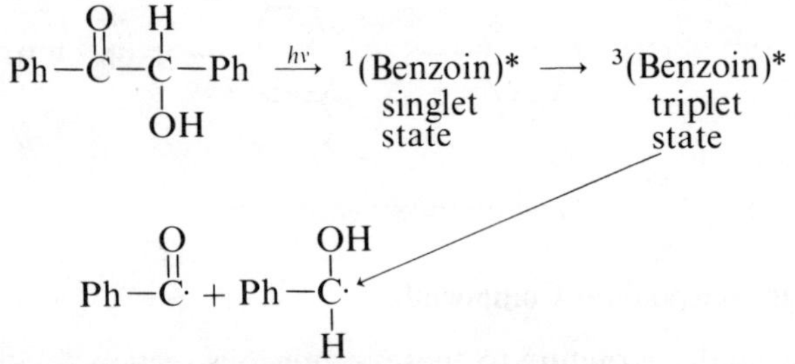

$$\text{Ph}-\overset{\text{O}}{\underset{\text{OH}}{\text{C}}}-\overset{\text{H}}{\underset{}{\text{C}}}-\text{Ph} \xrightarrow{h\nu} {}^1(\text{Benzoin})^* \longrightarrow {}^3(\text{Benzoin})^*$$

singlet state   triplet state

$$\text{Ph}-\overset{\text{O}}{\text{C}}\cdot + \text{Ph}-\overset{\text{OH}}{\underset{\text{H}}{\text{C}}}\cdot$$

Evidence for this mechanism is given by the following:[60–62]

1. The radical pair created on irradiation of benzoin has been found to be identical to that formed on irradiation of benzaldehyde.
2. The use of diamagnetic scavengers below as spin traps for radicals produced on photolysis of benzoin and benzoin methyl ether. In both cases, PhĊO and PhĊHOR (R = H, Me) radicals were trapped and characterized from the electron spin resonance (ESR) spectra of the stable nitroxide radicals formed.[1]

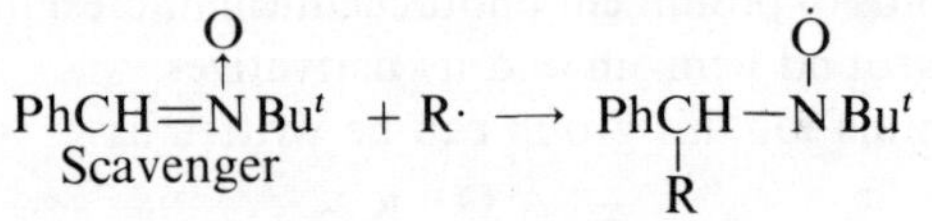

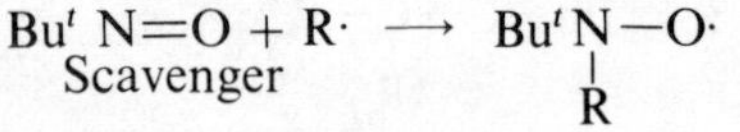

$$\underset{\text{Scavenger}}{\text{PhCH}=\overset{\uparrow\text{O}}{\text{N}}\text{Bu}^t} + \text{R}\cdot \longrightarrow \text{PhCH}-\overset{\overset{\cdot}{\text{O}}}{\underset{\text{R}}{\text{N}}}\text{Bu}^t$$

$$\underset{\text{Scavenger}}{\text{Bu}^t\text{N}=\text{O}} + \text{R}\cdot \longrightarrow \text{Bu}^t\underset{\text{R}}{\text{N}}-\text{O}\cdot$$

3. Similar results occurred for reactions performed in benzene and in methanol, those in the latter indicating that fragmentation of photo-excited benzoin and benzoin methyl ether occurs relatively rapidly compared to hydrogen abstraction from methanol.
4. The products formed on photolysis of benzoin alkyl ethers (methyl, ethyl, isopropyl) in benzene, and these too were indicative of the intermediary of benzoyl and alkoxybenzyl radicals.

The clear disadvantages of benzoin would seem to be the short-term dark-storage stability of UV photo-initiated systems.

The reactivity of benzoin derivatives is strongly dependent on the nature of the substitution of the α-carbon atom next to the carbonyl group. Those substituents with the ability of stabilizing an adjacent positive charge are the most effective in accelerating α-cleavage.[63]

3.4.2.2 Benzoin Alkyl Ethers

To obtain better performance, modification of the benzoin molecule has thus included α-alkylation i.e.

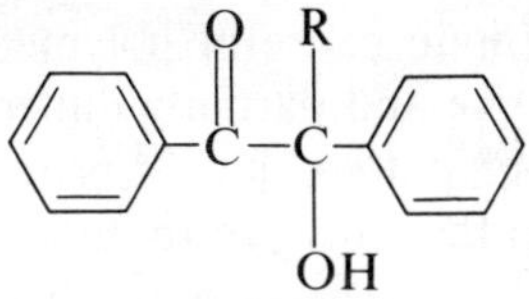

$$\text{Ph}-\underset{\underset{}{\overset{\overset{O}{\|}}{C}}}{}-\underset{\underset{OH}{|}}{\overset{\overset{R}{|}}{C}}-\text{Ph}$$

These compounds are the benzoin alkyl ethers. α-Methylolation has also been tried,

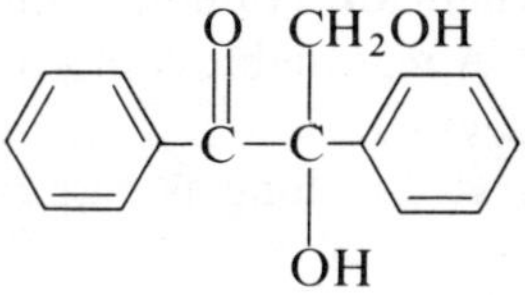

$$\text{Ph}-\underset{\underset{}{\overset{\overset{O}{\|}}{C}}}{}-\underset{\underset{OH}{|}}{\overset{\overset{CH_2OH}{|}}{C}}-\text{Ph}$$

and also the benzil ketals,

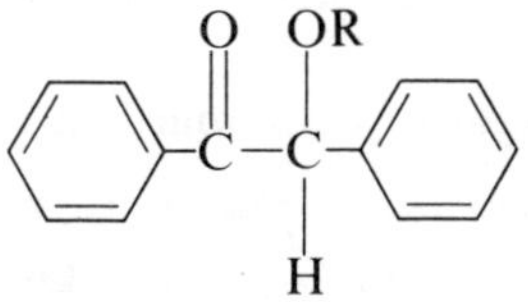

$$\text{Ph}-\underset{\underset{}{\overset{\overset{O}{\|}}{C}}}{}-\underset{\underset{H}{|}}{\overset{\overset{OR}{|}}{C}}-\text{Ph}$$

As opposed to benzoin, which undergoes fragmentation from the excited triplet state, these were thought to fragment from singlet-excited states.[62] This may be the cause of benzoin exhibiting higher quantum efficiencies for fragmentation in monomers having high triplet energies than it would in low triplet energy monomers, where triplet–triplet energy transfer may be in direct competition with fragmentation of the excited benzoin. Quantum yields for photoinitiation by benzoin derivatives are often in the region 0.2–0.3. Later work, however, indicates a reactive triplet state.[64]

Norrish type I cleavage of benzoin, benzoin alkyl ethers and α-alkylated benzoins will form benzoyl radicals,[65] which can be expected to be reactive entities and efficient photo-initiators.

Photo-initiation[66] is thought to proceed as follows:

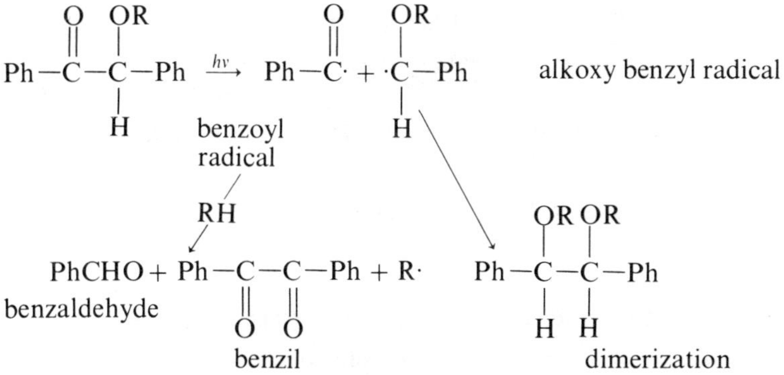

$$\text{Ph}-\underset{\underset{H}{|}}{\overset{\overset{O}{\|}}{C}}-\underset{}{\overset{\overset{OR}{|}}{C}}-\text{Ph} \xrightarrow{h\nu} \text{Ph}-\overset{\overset{O}{\|}}{C}\cdot + \cdot\underset{\underset{H}{|}}{\overset{\overset{OR}{|}}{C}}-\text{Ph} \quad \text{alkoxy benzyl radical}$$

benzoyl radical

RH ↓

PhCHO + Ph−C−C−Ph + R·
benzaldehyde   ‖ ‖
                O O
              benzil

Ph−C−C−Ph (OR OR / H H)
dimerization

The two radicals formed: (i) benzoyl radical, and (ii) alkoxy benzyl radical, are of different reactivity.

Benzoyl radicals are the main cause of polymer chain initiation. α-alkoxy benzyl radicals are less reactive and partially dimerize.

Benzoin ethers lead to poor pot stability,[63] probably owing to the activated hydrogen in the α position of the ether group. Ethers with this structure readily react with oxygen, forming hydroperoxides.[67] This intermediate may result in thermal instability, particularly in the presence of transition metals which are often present in filler materials. The substitution pattern of the benzoin ethers can also affect the pot life, short-chain alkyl or non-branched ethers, such as benzoin methylether, being the worst. A reasonable compromise between reactivity and shelf life is benzoin isopropyl ether.

Stabilizers can be incorporated as long as cure rate is not affected.

3.4.2.3 Benzil Ketals

Perhaps the most important benzil ketal commercially used at the present is dimethoxy-2-phenyl acetophenone.[1]

Its mode of action is thought to be primarily a Norrish type I cleavage:[68]

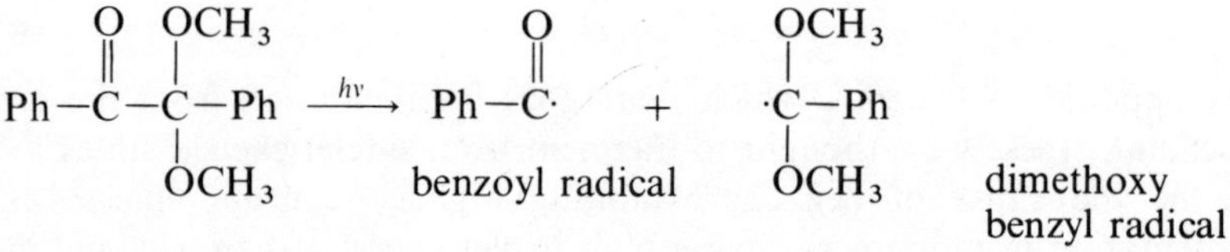

$$\underset{\underset{OCH_3}{|}}{\overset{\overset{O}{\|}}{Ph-C-}}\overset{OCH_3}{\underset{|}{C}}-Ph \xrightarrow{h\nu} \overset{O}{\underset{}{Ph-\overset{\|}{C}\cdot}} + \underset{\underset{OCH_3}{|}}{\overset{OCH_3}{\overset{|}{\cdot C-Ph}}}$$

benzoyl radical   dimethoxy benzyl radical

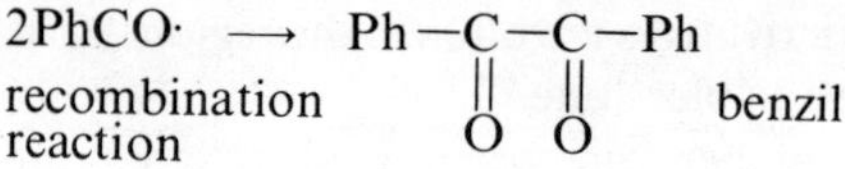

$$2\,PhCO\cdot \longrightarrow Ph-\underset{\underset{O}{\|}}{C}-\underset{\underset{O}{\|}}{C}-Ph$$

recombination reaction   benzil

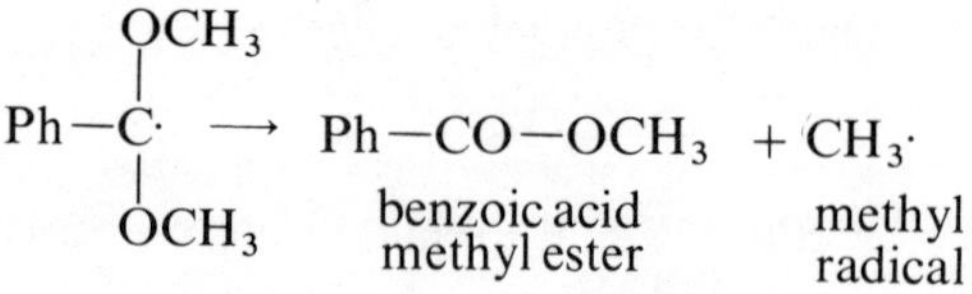

$$Ph-\underset{\underset{OCH_3}{|}}{\overset{\overset{OCH_3}{|}}{C\cdot}} \longrightarrow Ph-CO-OCH_3 + CH_3\cdot$$

benzoic acid methyl ester   methyl radical

This ester formation is strongly temperature dependent.

$$PhCO\cdot + RH \longrightarrow PhCHO + R\cdot$$

benzaldehyde

$$PhCO\cdot + \cdot CH_3 \longrightarrow Ph-CO-CH_3$$

acetophenone

Here, acetophenone is a cage collapse product of the benzoyl and methyl radical.

The pot stability of this compound relative to benzoin ethers is reported to be far greater.[69]

A yellow colour sometimes occurs with this compound on irradiation and could be from:[70]

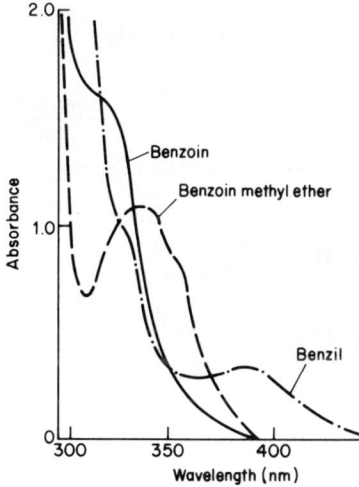

The spectra of benzoin, benzil, and a benzoin alkyl ether are shown in Figure 46.

Figure 46   UV absorption spectra of benzoin, benzoin methyl ether, and benzil in methyl methacrylate. (Reproduced by permission of the Oil of Colour Chemists' Association.)

3.4.2.4 Acetophenone Derivatives

There are two chief classes of acetophenone derivatives for commercial use as photo-initiators. These are dialkoxyacetophenones and chlorinated acetophenone derivatives.

(i) Dialkoxyacetophenones
Dialkoxy acetophenones can be regarded as benzoin ethers. The most popular used is DEAP (diethoxyacetophenone).[68]

DEAP undergoes a Norrish type II cleavage yielding a biradical as the chain-initiating species:[71-73]

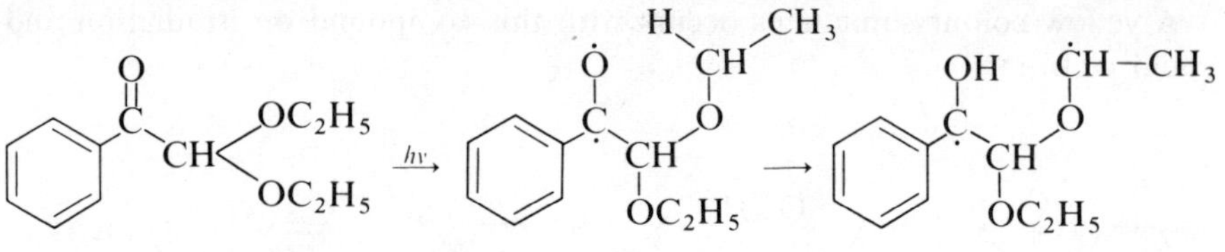

[Scheme showing photochemical reaction of DEAP]

If no reactive double bonds are available, the biradical undergoes internal coupling to form an oxetanol[22] intermediate which disproportionates thermally to acetaldehyde and ω-ethoxy-acetophenone:

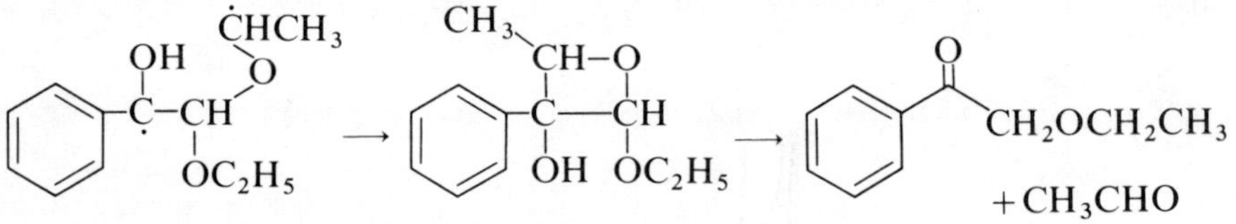

[Scheme showing oxetanol intermediate formation and disproportionation to give ω-ethoxy-acetophenone + CH$_3$CHO]

Norrish type I cleavage also appears to take place:

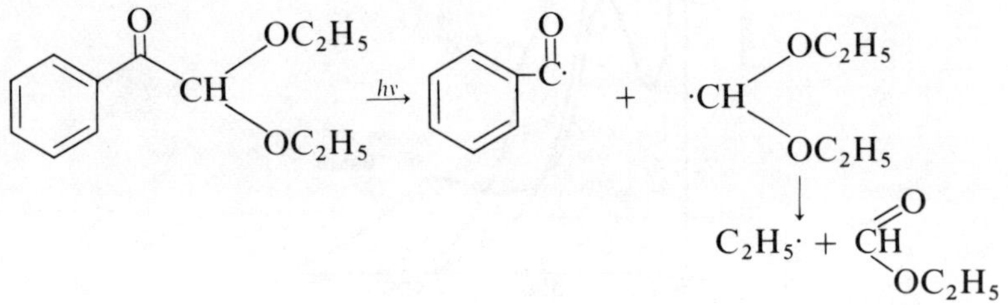

[Scheme showing Norrish type I cleavage producing benzoyl radical and ·CH(OC$_2$H$_5$)$_2$, which fragments to C$_2$H$_5$· + CH(=O)OC$_2$H$_5$]

Norrish type I/Norrish type II is of the order 2:1.

Compounds such as the benzil ketals and dialkoxyacetophenones are substantially faster in cure rate than benzoin alkyl ethers. This could be because of the secondary fragmentation of the alkoxy alkyl radicals which might be a process competing with radical combination reactions (including termination of polymerization) and therefore might give rise to higher overall efficiencies in radical polymerization.[67]

Secondary fragmentation can be written:

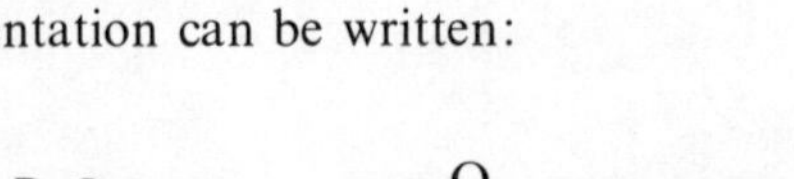

$$R_1O\text{-}C(R)(OR_1)\cdot \longrightarrow R\text{-}C(=O)\text{-}OR_1 + R_1\cdot$$

(ii) Chlorinated acetophenone derivatives

Substituted di- and tri-chloroacetophenones undergo photolysis,[74] such as p-tert-butyl tri-chloro acetophenone (in the general formula below, for this type of compound $R = (CH_3)_3C$). β-cleavage is the prime reaction. X may be either an H or another Cl atom.

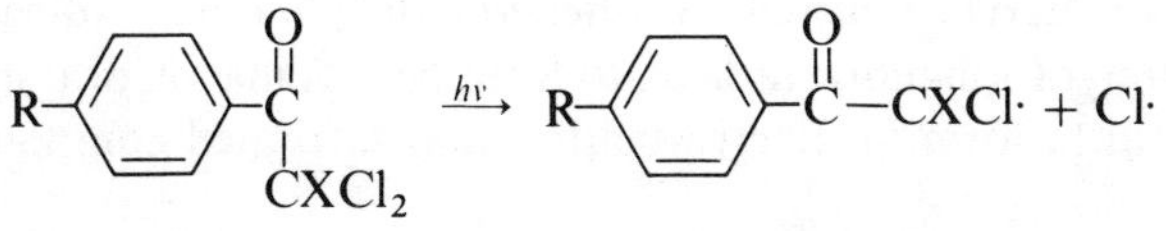

$$R-C_6H_4-CO-CXCl_2 \xrightarrow{h\nu} R-C_6H_4-CO-CXCl\cdot + Cl\cdot$$

$$Cl\cdot + R'H \rightarrow HCl + R'\cdot$$

The liberated chlorine radical is particularly highly reactive and initiates polymerization efficiently, especially ionic curable binders. A disadvantage is that this radical can form hydrochloric acid by hydrogen abstraction from a hydrogen donor and this presents an obvious disadvantage in many areas, such as metal decorating and may also lead to poor pot stability.

A certain amount of Norrish type I reaction also occurs:

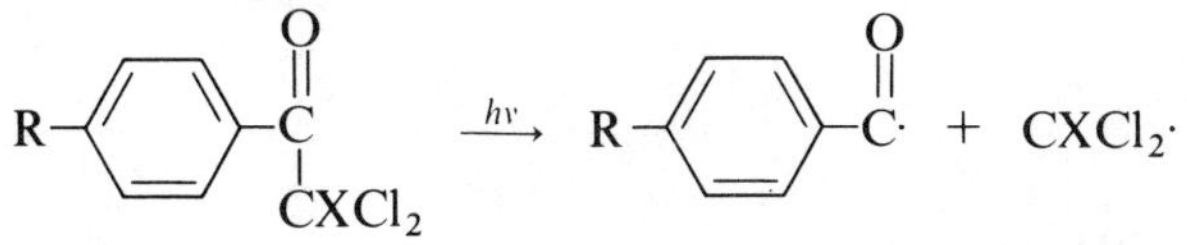

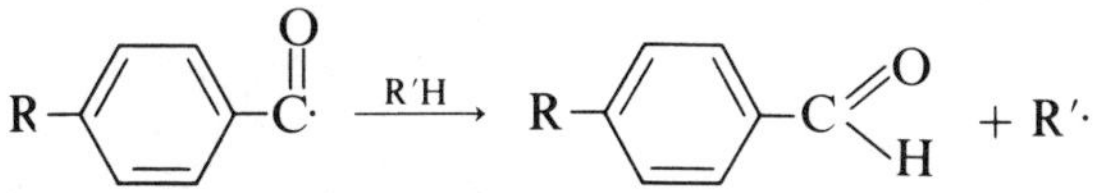

$$R-C_6H_4-CO-CXCl_2 \xrightarrow{h\nu} R-C_6H_4-CO\cdot + CXCl_2\cdot$$

$$R-C_6H_4-CO\cdot \xrightarrow{R'H} R-C_6H_4-CHO + R'\cdot$$

$$CXCl_2\cdot \xrightarrow{R'H} CHXCl_2 + R'\cdot$$
$$CXCl_2\cdot \rightarrow CXCl\cdot + Cl\cdot$$
$$CXCl\cdot \xrightarrow{R'H} HCl + R'CX$$

3.4.3 Aromatic Ketone/Amine Combinations

3.4.3.1 Benzophenone

Benzophenone reacts by hydrogen abstraction and is photo-reduced to benzopinacol in the presence of hydrogen donors.[1,75–77] Benzophenone undergoes a photo-induced excitation predominantly at 340 nm to form a singlet excited state (which is an isomer of the ground state). This singlet state is converted by intersystem crossing in a fraction of a second into the relatively longer-lived excited triplet state. This time period permits the excited electrons adequate time to collide with other atoms. The hydrogen atom is a particularly mobile atom and is readily attracted by this kind of triplet state, and radicals of the surrounding monomers and benzohydrophenone are produced. The monomers polymerize whereas the benzohydrophenone radical reacts with another benzohydrophenone radical, to form benzopinacol, or it may react with another radical to form an electro-neutral and saturated compound.[78–81]

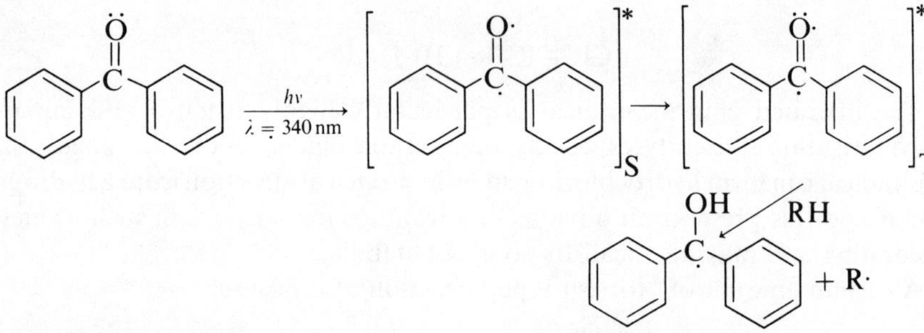

monomer + R· ⟶ polymer

benzohydrophenone radical

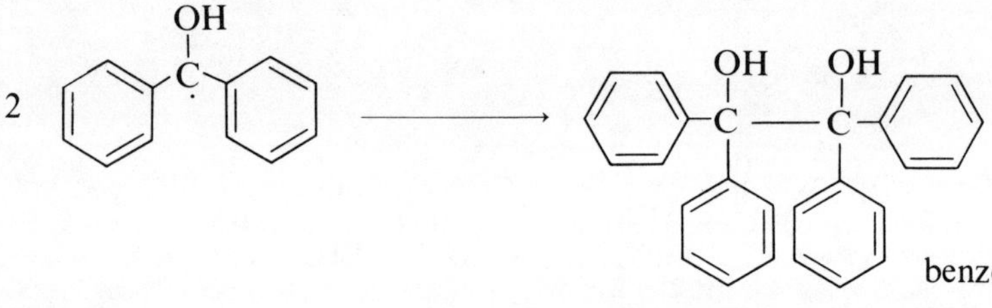

benzopinacol

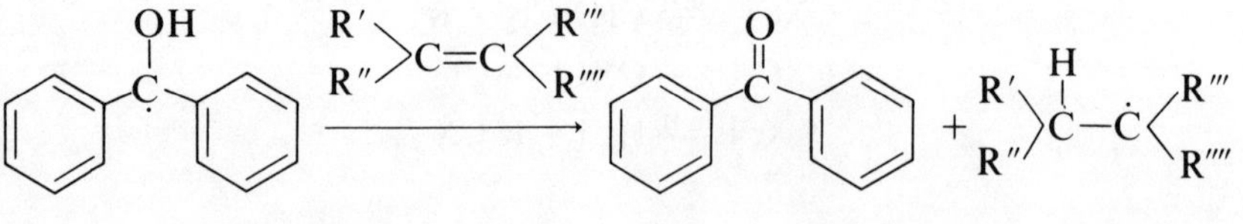

Benzophenone can also react with oxygen on UV exposure. The benzohydrophenone reacts with oxygen, regenerating benzophenone and the hydroperoxide radical is formed. This latter species is well known to produce polymerization or degradation reactions.

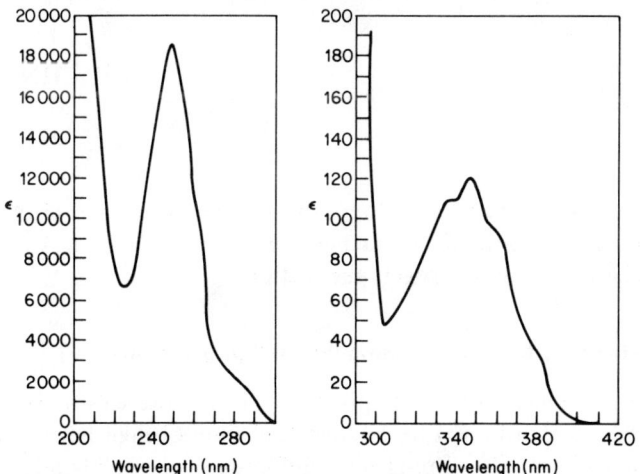

As this reaction uses up the oxygen present in the ink or coating, a small oxygen-inhibitor effect would be expected, when this photo-initiator is used. The absorption spectra for benzophenone are shown in Figure 47.

Figure 47 Absorption spectra for benzophenone in cyclohexane at 25 °C.[37] (Reproduced by permission of Dr. J. G. Calvert)

Near the region of the important mercury line at 254 nm, ε is about 17 600 litre mole^{-1} cm^{-1}.

Benzophenone is often used in combination with Michler's ketone:[82]

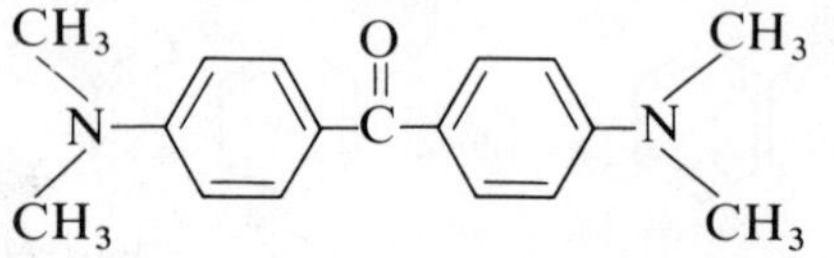

$$(CH_3)_2N-C_6H_4-CO-C_6H_4-N(CH_3)_2$$
Michler's ketone

This latter compound can be considered an independent primary photo-initiator as well as being used as a synergist or secondary photo-initiator.

Amines such as small amounts of triethylamine have been found to enhance the photopolymerization rate in benzophenone/acrylate systems, probably via the formation of an exciplex, which gives rise to free radicals:[13,83]

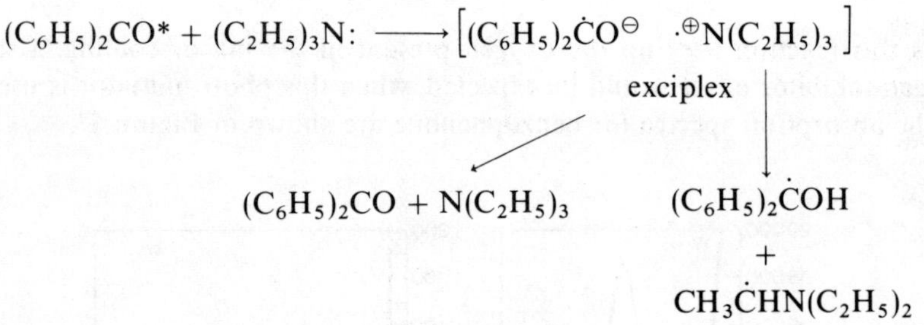

$$(C_6H_5)_2CO^* + (C_2H_5)_3N: \longrightarrow [(C_6H_5)_2\dot{C}O^{\ominus} \quad \cdot^{\oplus}N(C_2H_5)_3]$$
$$\text{exciplex}$$

$$(C_6H_5)_2CO + N(C_2H_5)_3 \qquad (C_6H_5)_2\dot{C}OH$$
$$+$$
$$CH_3\dot{C}HN(C_2H_5)_2$$

In this instance, the excited state of the acceptor is created prior to complex formation, characteristic of the exciplex formation.

Benzil also reacts by hydrogen abstraction.

3.4.3.2 Michler's Ketone (4,4'-dimethyl-amino-benzophenone)

Michler's ketone has markedly improved light response sensitivity for most UV systems when used in admixture with benzophenone, benzil, benzoin alkyl ethers and/or amines.[64,13,82–85] Pre-ground mixtures of the solids added to an ink appear to be more efficient than if they are added individually and milled straight into an ink.

Tertiary amino groups are potent electron-donor sites for photo-excited aryl ketones, and consequently Michler's ketone could act both as an electron donor and excited chromophore. It is possible that an intermolecular exciplex is formed.

There is experimental evidence for the photo-induced self-condensation of Michler's ketone and a possible reaction mechanism[1,86–87] is:

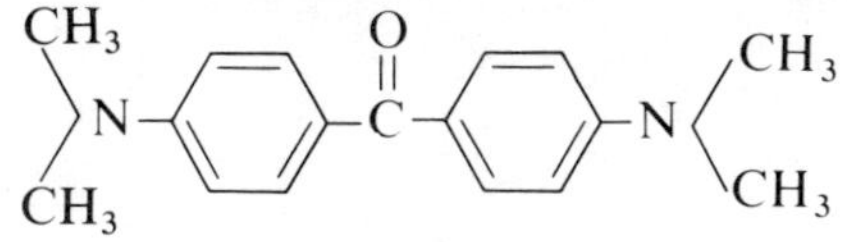

$$(CH_3)_2N-C_6H_4-CO-C_6H_4-N(CH_3)_2 \quad \text{Michler's ketone} = (MK)_{S_0}$$

$$(MK)_{S_0} \xrightarrow{h\nu} (MK)_{S_1} \longrightarrow (MK)_{T_1}$$

$$(MK)_{T_1} + (MK)_{S_0} \longrightarrow \text{exciplex} \longrightarrow P + Q$$

i.e.

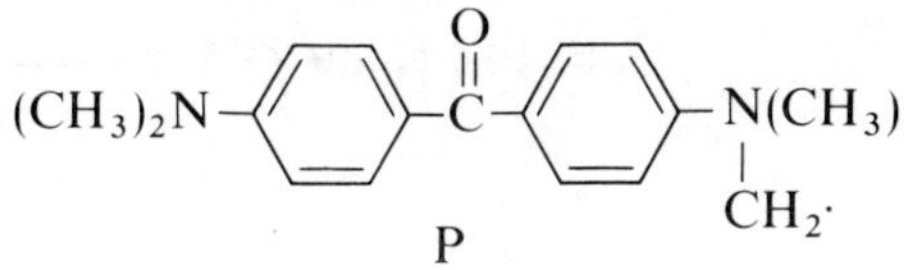

$$P: (CH_3)_2N-C_6H_4-CO-C_6H_4-N(CH_3)(CH_2\cdot)$$

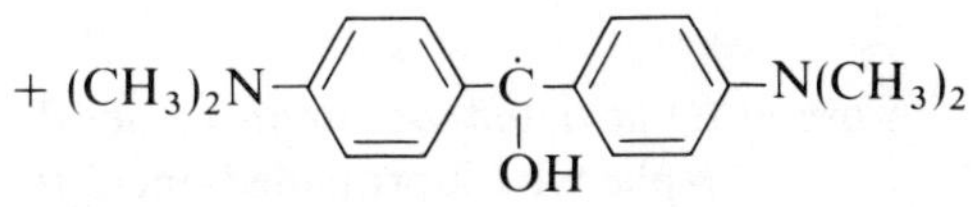

$$+ \; Q: (CH_3)_2N-C_6H_4-\overset{\cdot}{C}(OH)-C_6H_4-N(CH_3)_2$$

$$P + Q \longrightarrow (CH_3)_2N-C_6H_4-C(OH)(CH_2-N(CH_3)-C_6H_4-CO-C_6H_4-N(CH_3)_2)-C_6H_4-N(CH_3)_2$$

monomer ↓ polymer

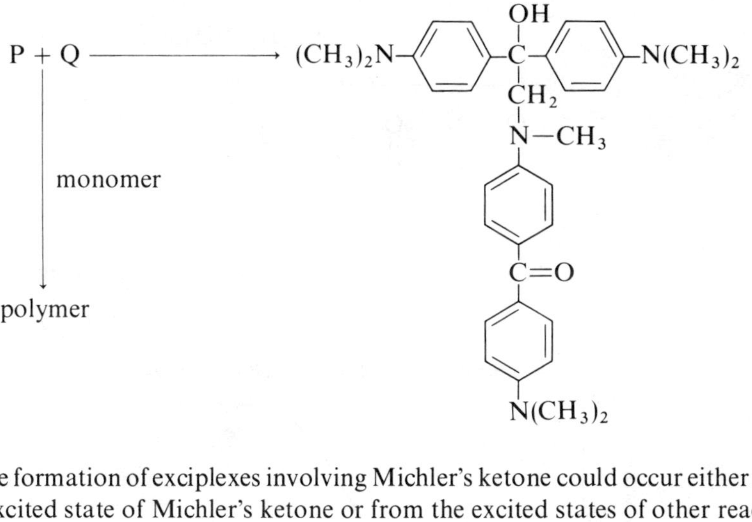

The formation of exciplexes involving Michler's ketone could occur either from the excited state of Michler's ketone or from the excited states of other reaction components such as benzophenone, benzil, etc.[1]

$$(MK)_{T_1} + (Ph_2C=O)_{S_0} \searrow$$
$$(Ph_2C=O)_{T_1} + (MK)_{S_0} \nearrow \text{exciplex} \longrightarrow \text{free radicals}$$

3.4.4 α-Acyloxime esters

A Norrish type I cleavage reaction occurs to form a benzoyl radical and another which undergoes further cleavage.[88]

$$Ph-\underset{O}{\underset{\|}{C}}-\underset{\underset{\underset{O}{\|}}{\underset{O-C-R}{|}}}{\overset{Ph}{\underset{N}{C}}} \xrightarrow{h\nu} PhC\overset{\cdot}{=}O \quad + \quad \underset{\underset{PhCN + CO_2\uparrow + R\cdot}{\underset{O-C-R}{|}}}{\overset{Ph}{\underset{N}{\overset{\cdot}{C}}}\underset{\|}{O}}$$

A useful (O-acylated-α-oximinoketone)[89–90] derivative that has been used is: 1-phenyl-1,2-propanedione-2-(O-ethoxycarbonyl oxime)

$$Ph-\underset{O}{\underset{\|}{C}}-\underset{\underset{O-\underset{\|}{\overset{\|}{C}}-OC_2H_5}{\underset{N}{|}}}{\overset{}{C}}-CH_3 \qquad C_{12}H_{13}NO_4 = 235.2$$

This compound has the spectral characteristics shown in Figure 48.

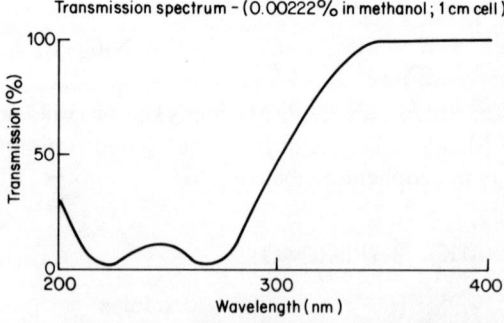

Figure 48 UV transmission spectrum of 1-phenyl-1,2-propanedione-2-(O-ethoxycarbonyl oxime).[90] (Reproduced by permission of Ward Blenkinsop & Co, Ltd.)

Its physical properties are as follows:

Colour	White/off-white crystalline solid
Odour	Nearly odourless
Thermal stability	Stable for at least one year at room temperature. Differential thermal analysis indicates that no decomposition occurs up to 200 °C.
Solvent solubility	(g per 100 ml of solvent/24 °C)

Solvent	g per 100 ml
Toluene	80
Methanol	18
Ethanol	10
Acetone	150
Carbon tetrachloride	67
Hexane	2
Methyl methacrylate	80
1,6-Hexanediol diacrylate	40
Trimethyolpropane triacrylate	14
Hexane (50 °C)	50

Physiological aspects

The oral LD50 (mice) is greater than 1 g per kg body weight and the dermal LD50 (rats) is greater than 0.5 g per kg body weight. The Ames test for mutagenicity (and within the context of this method) indicates that it does not have carcinogenic potential.

3.4.5 Thioxanthone and Derivatives

Many photo-initiators can be regarded as derivatives of benzophenones, in particular thioxanthone and its substituted compounds. The table overleaf[91] shows some of these which can be seen to have a derivation from the formula

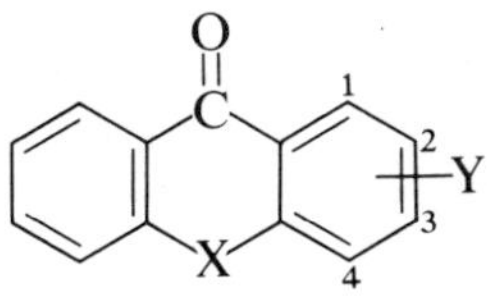

For

Benzophenone, Y = H; X = zero.
Chlorothioxanthones, Y = Cl; X = S.
(Y may occupy any one of the positions, 1, 2, 3, 4, if it represents a chlorine atom replacing one hydrogen atom, or it may itself be a hydrogen atom.)

Table 3 Thioxanthone and derivatives.[90,91] (Reproduced by permission of Ward Blenkinsop & Co. Ltd. and the Oil & Colour Chemists' Association.)

Compound	X	Y
Thioxanthone	—S—	H—
2-Chlorothioxanthone (CTX)	—S—	Cl—
2-Methylthioxanthone	—S—	CH_3—
2-Ethylthioxanthone	—S—	CH_3CH_2—
2-Isopropylthioxanthone	—S—	$(CH_3)_2CH$—
2-tert-Butylthioxanthone	—S—	$(CH_3)_3C$—
2-Phenylthioxanthone (M.Pt.130–131.5 °C)	—S—	C_6H_5—
2-Benzylthioxanthone	—S—	$C_6H_5CH_2$—
2-Cyclohexylthioxanthone (M.Pt.86–88 °C)	—S—	C_6H_{11}—
4-Isopropylthioxanthone (M.Pt.104–106 °C)	—S—	$(CH_3)_2CH$—
2-Acetylthioxanthone	—S—	CH_3CO—
2-Chloroxanthone	—O—	Cl—
Fluorenone	single bond	H—
Dibenzosuberone	—CH_2CH_2—	H—
6,11-Dihydrodibenzo-thiepin-11-one	—CH_2S—	H—
2-Chloro-6,11-dihydrodibenzo-thiepin-11-one	—CH_2S—	Cl—

Substituted thioxanthones have been developed for white pigmented, coatings as two chief problems occur with these relatively thick coatings;

1. The pigment can reflect or absorb incident UV light, diminishing the available light to the photo-initiator.
2. The opacity of thicker pigmented films can result in a poor through-cure.

Titanium dioxide has high reflectivity through the visible range of the spectrum and strong absorption at wavelengths < 340 nm. As the region 350–400 nm represents a transition between the reflection and absorption of light, this 'window' can be used for photo-sensitizers such as the thioxanthone derivatives.

The main properties needed from a photo-initiator from a formulation aspect are:

1. UV curing response.
2. Colour of final cured film.
3. Solubility/ease of incorporation
4. Pot stability of the formulation.
5. Physiological aspects (toxicity, odour, etc.).
6. Cost effectiveness.
7. Patent situation acceptable.

Of particular use are:

1. 2-Chlorothioxanthone (2-CTX)

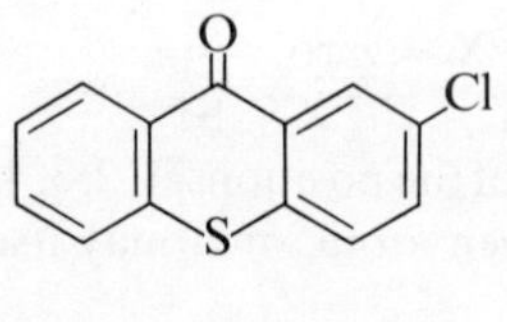

2. 2-Isopropylthioxanthone

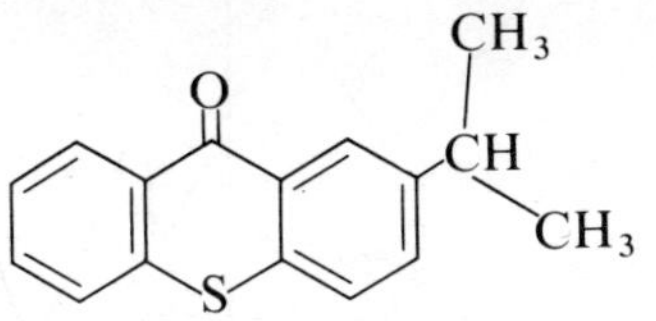

The latter has good solubility (as it is a liquid) in many solvents. 2-Chlorothioxanthone is, however, a solid.

These are used generally in conjunction with one of several accelerators, such as:

1. ethyl-4-dimethylaminobenzoate;
2. ethyl-2-dimethylaminobenzoate;
3. 2-(*n*-butoxy) ethyl 4-dimethyl amino benzoate;
4. 2-(dimethylamino) ethyl benzoate.

2-CTX probably reacts by a hydrogen abstraction method:

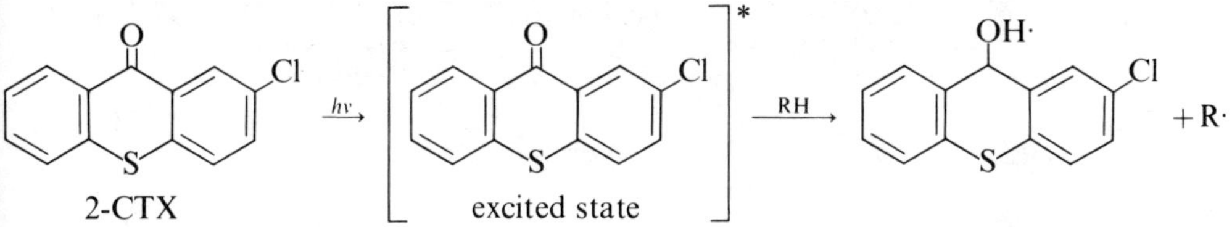

2-CTX has absorption bands at 260 nm and 385 nm.

The spectral characteristics of thioxanthone and some of its important derivatives are shown in Figure 49.

Some synergistic agents (or photoactivators) often used with thioxanthone and its derivatives (or any other convenient photo-initiator responsive to these compounds) are:

1. Ethyl *para*-dimethylaminobenzoate

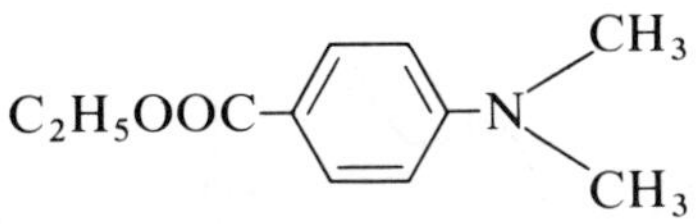

The spectrum of this compound is shown in Figure 50.

2. Ethyl *ortho*-dimethylaminobenzoate

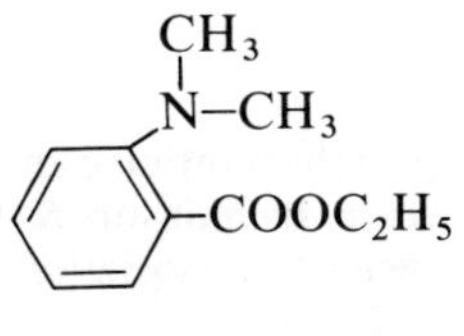

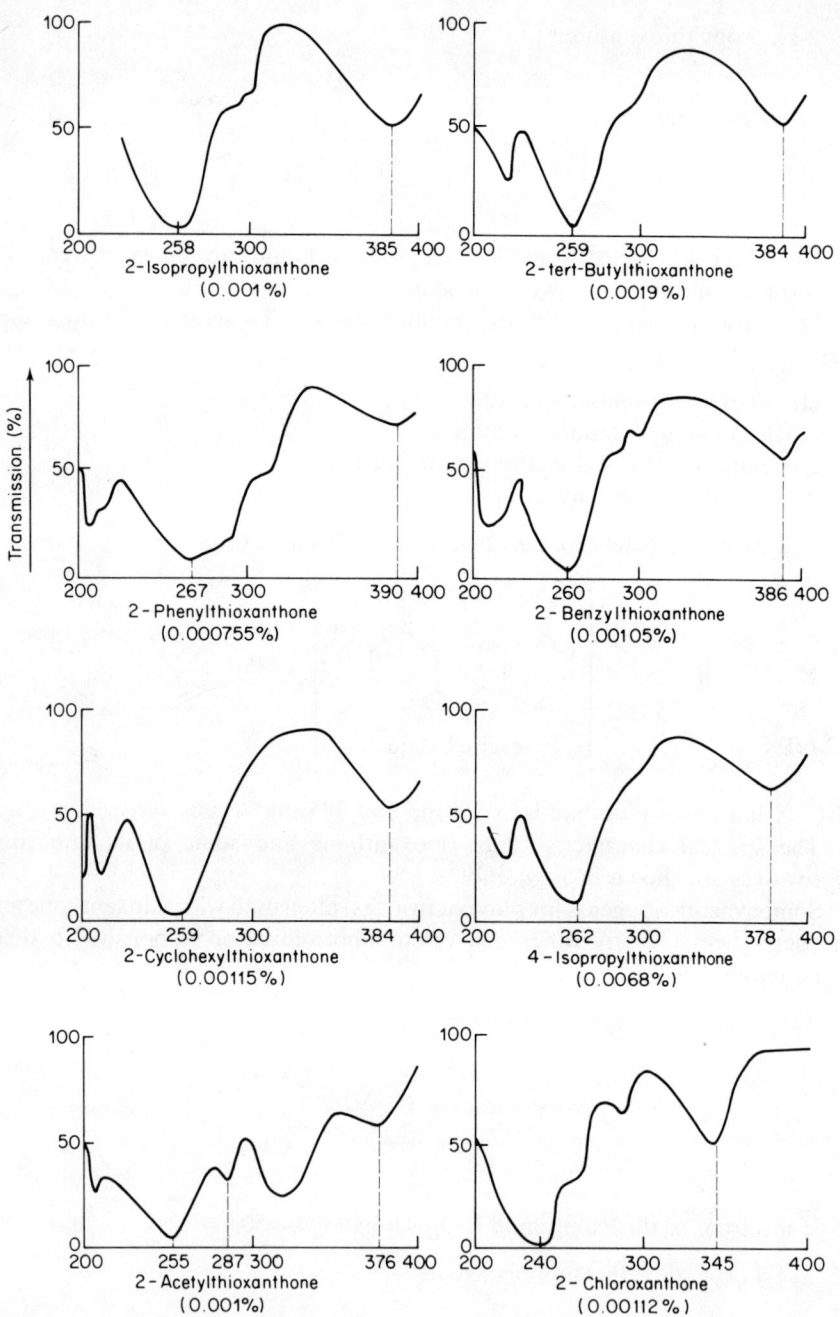

Figure 49 Spectral characteristics of thioxanthone and some of its near derivatives.[90,91] (Reproduced by permission of Ward Blenkinsop & Co, Ltd and the Oil & Colour Chemists' Association.)

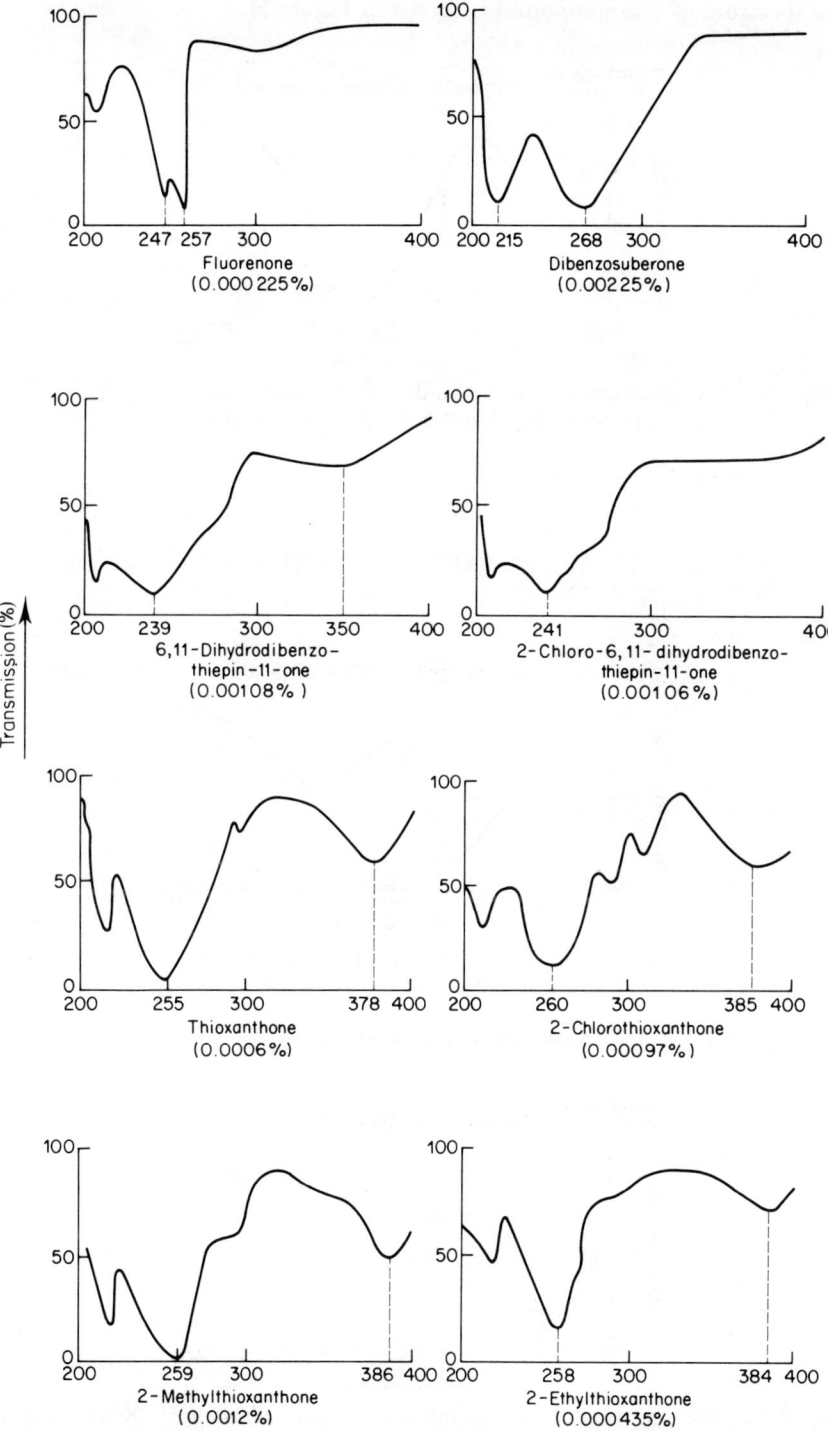

The spectrum of this compound is shown in Figure 51.

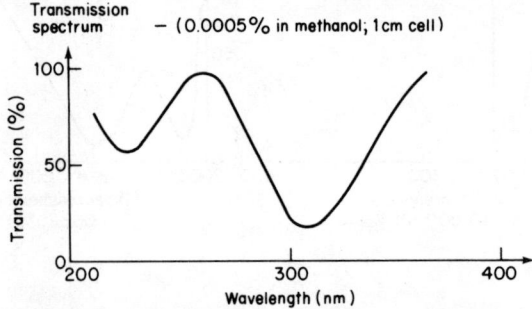

Figure 50 UV spectrum of ethyl *para*-dimethylaminobenzoate.[90] (Reproduced by permission of Ward Blenkinsop & Co, Ltd.)

3. 2-(Dimethylamino) ethyl benzoate

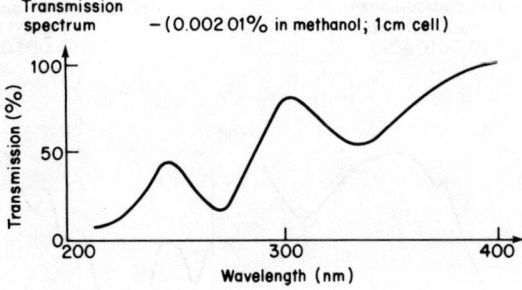

Figure 51 UV spectrum of ethyl *ortho*-dimethylaminobenzoate.[90] (Reproduced by permission of Ward Blenkinsop & Co, Ltd.)

The spectrum of this compound is shown in Figure 52.

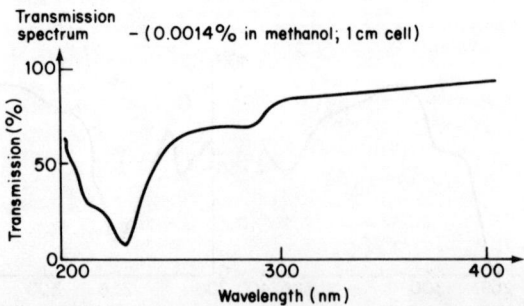

Figure 52 UV spectrum of 2-(dimethylamino) ethyl benzoate.[90] (Reproduced by permission of Ward Blenkinsop & Co, Ltd.)

4. 2-(n-Butoxy) ethyl para-dimethylaminobenzoate

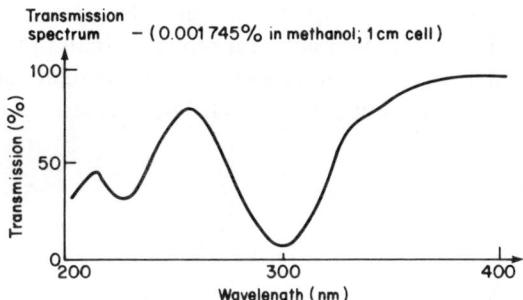

The spectrum of this compound is shown in Figure 53.

Figure 53 UV spectrum of 2-(n-butoxy) ethyl para-dimethylaminobenzoate.[90] (Reproduced by permission of Ward Blenkinsop & Co, Ltd.)

Fluorenone is a useful derivative that may be placed in this class. Fluorenone is particularly active in the presence of amines as synergists, where exciplex formation is thought to occur.[1,92–93]

The photoreduction of aromatic carbonyls by amines is well documented.[1] Many amino compounds efficiently quench singlet and triplet excited states of aromatic ketones and a tendency exists between quenching efficiency and the amines' ionization potential for a specific amine class. Often many tertiary amines are more efficient than the corresponding secondary and primary derivatives, but particular solvation phenomena may reverse this reactivity order.

Amine photoreduction of aromatic ketones occurs at faster rates than the corresponding photo-induced hydrogen abstractions.

The photo-oxidation of amines may be generally depicted as

$$Ph_2C=O \xrightarrow{h\nu} (Ph_2C=O)^* \xrightarrow{RCH_2CH_2NR_2} (Ph_2C=O)^{\cdot -} \ (RCH_2CH_2NR_2)^{\cdot +}$$
$$\downarrow$$
$$Ph_2\dot{C}-OH + RCH_2\dot{C}HNR_2$$

$$Ph_2\dot{C}OH + RCH_2\dot{C}HNR_2 \longrightarrow Ph_2CHOH + RCH=CHNR_2$$

$$RCH_2\dot{C}HNR_2 + Ph_2C=O \longrightarrow RCH=CHNR_2 + Ph_2\dot{C}-OH$$

$$2Ph_2\dot{C}OH \longrightarrow Ph_2\underset{OH}{\overset{}{C}}-\underset{OH}{\overset{}{C}}Ph_2$$

Fluorenone is an aromatic ketone with a low lying (π, π^*) triplet excited state and is notably influenced by the surrounding solvent.[94–95]

It is not directly photoreduced in alcohol, ether, or alkane solvents, but it is in the presence of amines, especially the tertiary variety.[1,96–97]

Triplet-state activity is optimized by decreasing the solvent polarity.[98]

The mechanism depicted above is a simplification as it presupposes direct formation of a radical ion pair subsequent to the collisional interaction of an excited carbonyl compound and a ground-state amine molecule.

The final production of a radical ion pair is a consequence of the formation of several possible intermediates such as the type referred to as an 'exciplex'.

Interaction of a donor and an acceptor molecule both in their ground states form charge transfer complexes and these, on exposure to actinic radiation, have excited states with a significant amount of electron transfer on comparison to the ground state,

i.e.

$$D + A \rightleftarrows D,A \xrightarrow{h\nu} (\dot{D}^+ + \dot{A}^-)^*$$

The charge transfer complex excited state should not be equated with a thermally equilibrated radical ion pair.

Exciplex interactions for the photo-initiation of polymers is reviewed by Ledwith, Bosley, and Purbrick.[99]

A specific advantage of the ketone–amine combination for photo-initiation is its applicability to block and graft copolymerization. Terminal amino groups are likely to be present in most polymers photo-initiated in this manner and it is advantageous to synthesize prepolymers possessing pendent dimethylamino-groupings by copolymerization of 2-dimethylaminoethyl methacrylate. Whether or not the block or graft polymerization process is totally efficient, utilization of prepolymers possessing tertiary amine substituents permits advantages in the design of photoresponsive systems.[1,8] For example:

1. *Block copolymerization.*

FLO = Fluorenone M_1 and M_2 are monomers

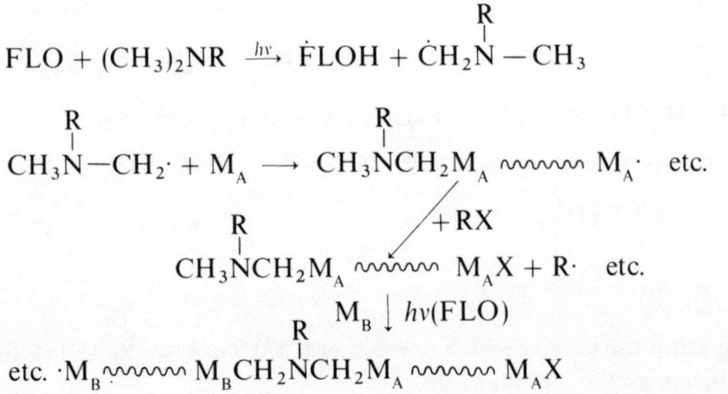

$$FLO + (CH_3)_2NR \xrightarrow{h\nu} \dot{F}LOH + \dot{C}H_2\overset{R}{N}-CH_3$$

$$CH_3\overset{R}{N}-CH_2\cdot + M_A \longrightarrow CH_3\overset{R}{N}CH_2M_A \sim\!\!\sim\!\!\sim M_A\cdot \quad \text{etc.}$$

$$CH_3\overset{R}{N}CH_2M_A \sim\!\!\sim\!\!\sim M_AX + R\cdot \quad \text{etc.}$$

$$+RX$$

$$M_B \downarrow h\nu(FLO)$$

$$\text{etc.} \cdot M_B \sim\!\!\sim\!\!\sim M_BCH_2\overset{R}{N}CH_2M_A \sim\!\!\sim\!\!\sim M_AX$$

2. Graft copolymerization.

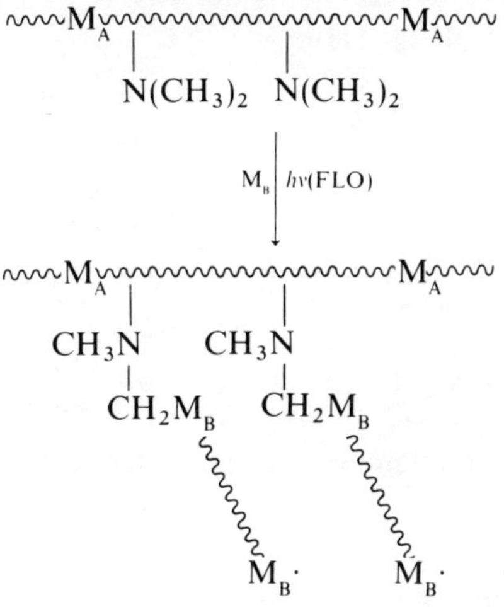

3.4.6 Quinones

The reaction mechanism of these compounds are very similar to benzophenones[100–102] with two notable exceptions:

1. Quinones more readily yield amino species radicals on reduction, especially in an alkaline environment.
2. Quinones show a more predominant and frequently useful involvement of intermediate peroxy compounds in reactions when performed in air.

A typical reaction scheme for anthraquinone in methylmethacrylate and tetrahydrofuran is shown below:[1,103–104]

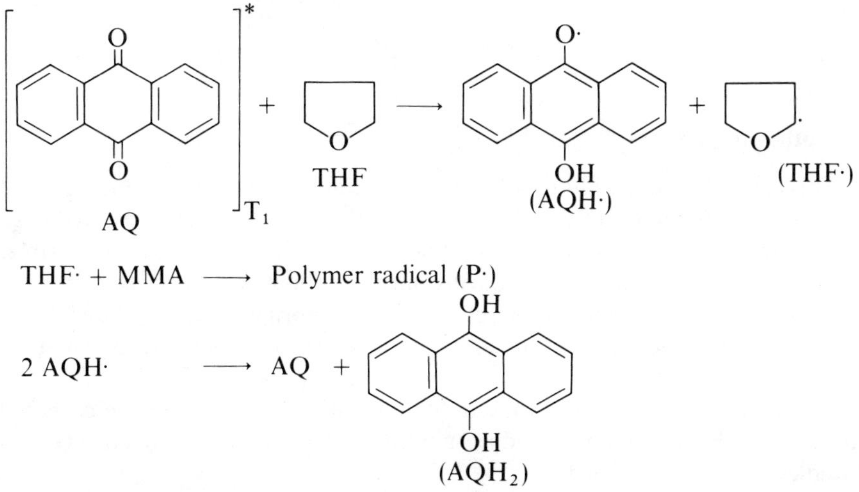

$$[AQ]^*_{T_1} + THF \longrightarrow AQH\cdot + THF\cdot$$

THF· + MMA $\longrightarrow$ Polymer radical (P·)

2 AQH· $\longrightarrow$ AQ + AQH$_2$

$$AQH\cdot + P\cdot \rightarrow \text{Terminated polymer}$$
$$AQH_2 + P\cdot \rightarrow \text{Terminated polymer} + AQH\cdot$$
$$AQH\cdot + THF\cdot \rightarrow$$

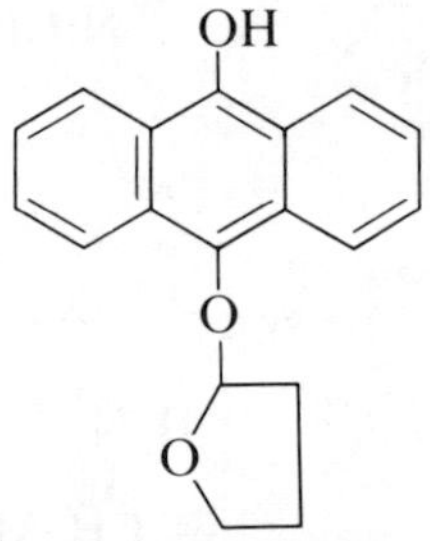

Substituted anthraquinones of particular use for photopolymerization reactions are:

1. 2-ethyl anthraquinone
2. Tertiary butyl anthraquinone
3. Anthraquinone sulphonate

1. 2-ethyl anthraquinone

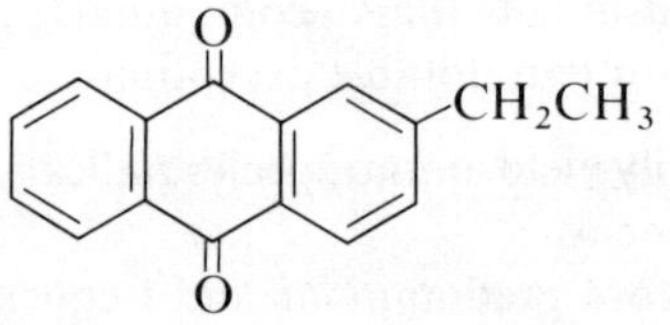

2. Tertiary butyl anthraquinone

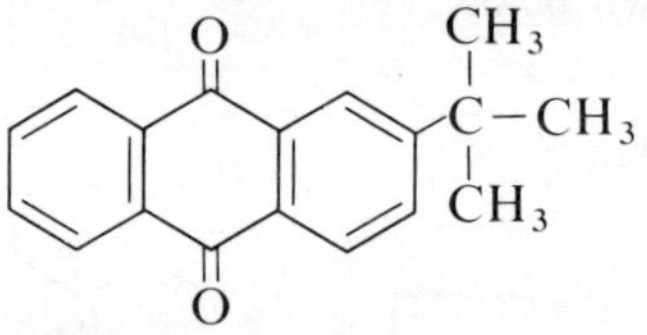

3. Anthraquinone sulphonate.[105]

This can enhance the photopolymerization rate especially if a reducing agent such as chloride ion is present. Anthraquinone sulphonate undergoes excitation to the triplet state by intersystem crossing from a singlet state. The excited triplet interacts with the chloride ion reducing agent, probably by an electron transfer reaction, to give anthraquinol sulphonate and chlorine radicals. The chlorine radicals initiate polymerization and are incorporated into the polymer as terminal groups.[106]

For quinones in general, actinic radiation causes reaction over the carbonyl groups and the ring double bonds. Quinone excited states tend to show typical biradical properties.

p-benzoquinones can also photodimerize and also undergo addition to compounds possessing double or triple bond unsaturation. The carbonyl groups may also undergo addition reaction with excited *p*-quinones.

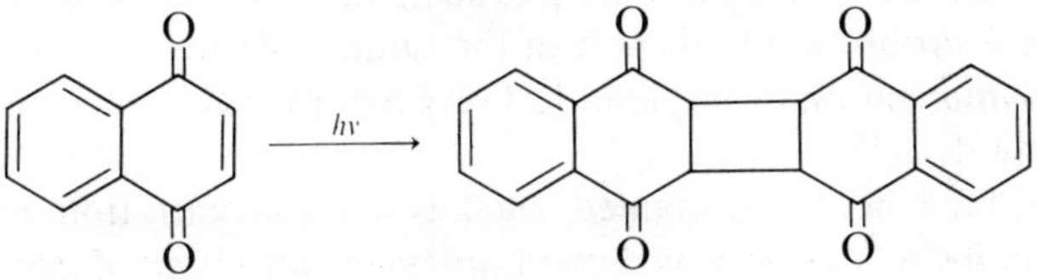

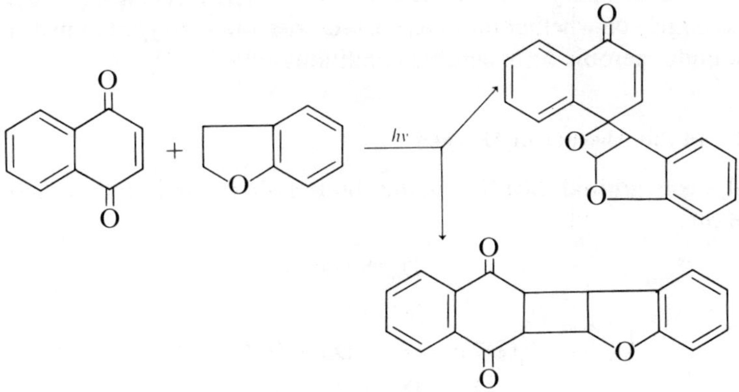

p-benzoquinones are often used as thermal stabilizers in UV curing systems to prevent gelling in the can.

Hydrogen donors can have their hydrogen abstracted from them by *p*-quinones. The photo-oxidation of alcohols (RH) in the presence of *p*-quinones (Q) involves two different, short-lived radical intermediates:

1. Semi-quinone radical ions ($Q \cdot^-$)
2. Monoprotonated neutral semiquinone radicals ($\cdot QH$).

$$Q \xrightarrow{h\nu} {}^1(Q)^* \rightarrow {}^3(Q)^*$$
$${}^3(Q)^* + RH \rightarrow QH\cdot + R\cdot$$
$${}^3(Q)^* + RH \rightarrow Q^- + RH^+$$
$$Q^- + RH \rightarrow Q + RH^-$$
$$QH\cdot + O_2 \rightarrow Q + HO_2\cdot$$
$$R\cdot + O_2 \rightarrow RO_2\cdot$$

With molecular oxygen present, singlet oxygen is probably formed by an energy transfer reaction of the excited triplet state of quinone molecules with molecular oxygen.

$$^3Q + {}^3O_2 \rightarrow Q + {}^1O_2$$

3.4.7 Dye Photo-sensitization

Dye additives can lead to absorption of radiant energy in the whole visible range as well as UV.[107] Dyes absorb the energy, then act by energy transfer or, more commonly, by electron transfer (redox) reactions. Dyes can be used alone or in conjunction with various electron donors (reducing agents). For some dyes, the acceptor and donor groups are both part of the same molecule, e.g., riboflavin, used for photopolymerization of acrylamide. Oxygen is probably a necessary co-additive with some dyes.[108]

Many variables have been investigated, such as dye concentration, reducing agent, and monomers, as well as wavelength, intensity, and time of irradiation.

Dye photochemistry is a complex subject.[109] The mechanism of dye action depends largely on whether the reaction is carried out in oxygen or in its absence, that is, under aerobic or anaerobic conditions.[110]

3.4.7.1 In the Absence of Oxygen

For a dye with ground state (D_0) in the absence of oxygen the following reactions may occur:

$$D_0 \xrightarrow{h\nu} {}^1D^* \to {}^3D^*$$
$${}^3D^* \to D_0$$
$${}^3D^* + {}^3D^* \to D_0 + {}^3D^*$$
$${}^3D^* + D_0 \to D_0 + D_0$$
$${}^3D^* + {}^3D^* \to X + R$$
$${}^3D^* + D_0 \to X + R$$

The semi-reduced state is R, and the singlet and triplet states are designated by the superscripts 1 and 3, respectively.

The semi-oxidized state is X. All excited states are depicted by the asterisk, *.

The photochemistry of dyes is related to their structure and the medium used.

In the presence of electron donors, such as solvents or other organic (RH) compounds, a triplet-excited dye molecule may abstract a proton from the donor and create a semi-reduced radical which can then abstract a second hydrogen atom to give the leuco form of dye. For eosin, a possible mechanism could be:[111–113]

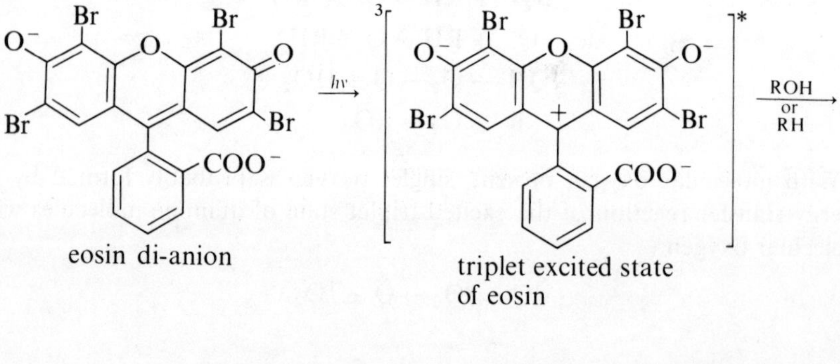

eosin di-anion

triplet excited state of eosin

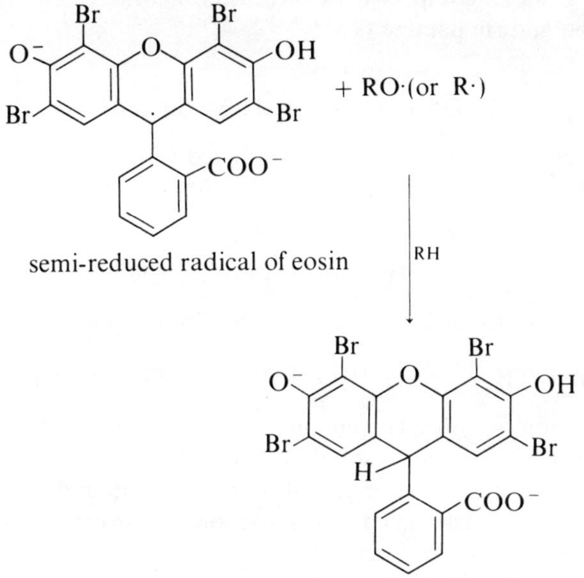

[Structure: semi-reduced radical of eosin] + RO· (or R·)

↓ RH

[Structure: leuco form of eosin]

Some dyes of the leuco ethyl crystal violet type, when in the presence of solvents such as carbon tetrachloride possibly react by an electron transfer mechanism from the excited dye molecules to CCl_4:

$$DH \xrightarrow{h\nu} {}^1(DH)^* \text{ or } {}^3(DH)^*$$

$$CCl_4 \xrightarrow[{}^3(DH)^*]{{}^1(DH)^* \text{ or}} DH^{+} + CCl_4^{-} \longrightarrow D^+ + HCl + \cdot CCl_2 + Cl^-$$

Photobleaching may be seen when dyes in solution are irradiated. The colour disappears after addition of protons (strong acid) to the conjugated system in the dye, which governs the colour. The photobleaching rate increases rapidly with reducing agents such as triethanolamine.

3.4.7.2 With Oxygen Present

The triplet state of dyes may react in the following manner:

1. *Physical quenching reaction.*

$$^3D^* + {}^3O_2 \rightarrow D_0 + {}^1O_2$$

Physical quenching is a process with the absence of chemical reaction. The physical quenching of the triplet state of a dye by molecular oxygen may involve the actual formation of a reactive intermediate (complex) between the excited dye and oxygen molecules. Formation of singlet oxygen occurs in the photo-sensitizing reactions of dyes such as eosin, fluorescein, Rose Bengal, crystal violet, and rhodamine G.[114–115]

2. *Chemical quenching.*

$$^3D^* + {}^3O_2 \rightarrow X + HO_2\cdot \text{ (or } O_2^-)$$

Another reaction seen while a dye is irradiated with light in the presence of oxygen is the fate of semi-reduced and semi-oxidized forms of the dye:

$$R + O_2 \rightarrow D_0 + HO_2\cdot \text{ (or } O_2^-)$$
$$X + HO_2\cdot \text{ (or } O_2^-) \rightarrow D_0 + O_2$$

The reduced form (leuco) of eosin reacts with oxygen and water:

$$(\text{eosin} - H_2) + O_2 + H_2O \rightarrow OH\cdot + H_2O_2 + (\text{eosin} - H\cdot)$$

The formation of hydrogen peroxide is an important reaction seen during irradiation of eosin in aqueous solution. This may be the reason that wet fibres show less resistance to photodegradation processes than do completely dry fibres. The OH· radicals created in this reaction may abstract hydrogen atoms from the polymer.

On the other hand, hydrogen peroxide may also photolyse to form hydroxy radicals.

Dye molecules may also take part in energy transfer reactions as either donor or acceptor.

3.4.7.3 Photoconductive Effect

Solid state photochemistry of dyes and pigments is an important area of study, as the phenomenon of photoconductivity is often observed, which allows an interpretation of a possible mechanism in the area of fading. Roffey has reviewed the fading of coloured paint films.[117] Photo-sensitization may still occur in a dye/medium, even when solid or cured, causing further changes such as a colour change in the dyestuff/pigment or photodegradation of the surrounding medium.

The photoconductive effect in the solid phase, that is the phenomenon of light-stimulated electrical conduction of various pigments used in the ISO Blue wool scale, has been studied by Gates and Patterson.[118] The important idea in these photoconductivity mechanism studies is the concept of the 'exciton'.

An exciton can be regarded as an electron and a positive hole linked together by coulombic attraction and able to migrate throughout the lattice. It is created by the absorption of light by most types of insulators and semiconductors.[119] In a photoconductive process it is thought that the absorption of light raises electrons from the ground state to trapping levels; thermal activation energy represents the extra energy necessary to promote these electrons into the conduction bands.[120–122]

Patterson[118] found that the greater this extra energy, the higher the lightfastness of the dye. The dyes he investigated were found to be *n*-type semiconductors, because the conductivity was greatly reduced by admitting oxygen, an electron scavenger. The charge carriers were therefore considered to be electrons. The primary process was thought to be the formation of an exciton. The trapping of an exciton by a dye molecule which captures both the electron and the positive hole constituting the exciton can be depicted as:

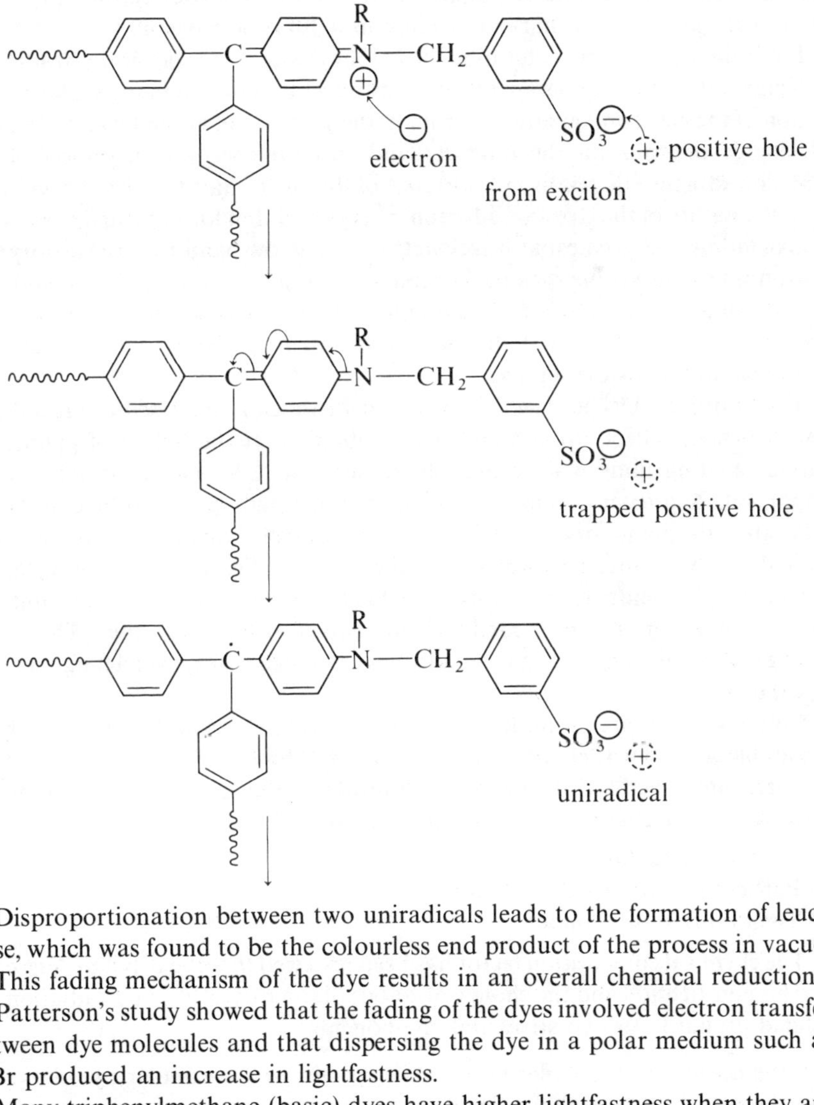

[Structure diagram showing dye molecule with R-N-CH₂ group and SO₃⁻, with electron and positive hole from exciton]

electron   positive hole
       from exciton

[Second structure diagram showing trapped positive hole]

trapped positive hole

[Third structure diagram showing uniradical]

uniradical

Disproportionation between two uniradicals leads to the formation of leuco base, which was found to be the colourless end product of the process in vacuo.

This fading mechanism of the dye results in an overall chemical reduction.

Patterson's study showed that the fading of the dyes involved electron transfer between dye molecules and that dispersing the dye in a polar medium such as KBr produced an increase in lightfastness.

Many triphenylmethane (basic) dyes have higher lightfastness when they are used as pigments after precipitation by phosphotungstic or phosphotungstomolybdic acid than when used unsupported. The anion structure from the complex acid is possibly $(PX_{12}O_{40})^{3-}$ where X = tungsten or molybdenum.[123]

Availability of these trivalent ions in the lattice will make electron transfer between the associated dye cations more difficult and could be the reason for the increase in lightfastness.

For highly insulating environments surrounding dyes and pigments, charge transfer can be reduced by the insulating properties of the medium. This is shown for basic dyes where the lightfastness on polyacrylonitrile fibres is much higher than on wool, and for certain very fugitive dyes such as Rhodamine B when incorporated in melamine–sulphonamide formaldehyde resins to make fluorescent pigments; a marked increase in lightfastness occurs.

Environmental effects determine the colour changes that occur on irradiation of a pigmented surface coating film or dyestuff with radiation in the visible or UV region of the electromagnetic spectrum in the presence of air and water. The rate of change depends on the intensity and spectral energy distribution of the emission source. The relative importance of the visible and UV regions will vary with the nature of the dyestuff. Morton[124] reported that for fugitive dyestuffs the rate of fading was greatest at wavelengths in a narrow band near the absorption maximum (in the visible region). For more lightfast dyes, Blaisdell[107] found that on exposing dyed cellulose acetate film to radiation of wavelengths 303.0, 313.0, 365.0, 405.0, and 435.0 nm, fading was most intense at 303.0 and 313.0 nm and negligible in the visible region.

The Grotthus–Draper law, where only light which is absorbed can result in a photochemical effect, applies and it is probable that the fading of pigmented surface coating films will be notably sensitive to UV. Radiation intensity is important, a greater intensity yielding a faster fading rate. On comparing intensities, the possibility of a higher specimen temperature at the higher intensity should be considered. This would have the opposed effects of accelerating fading from the higher temperature alone, or reducing this acceleration on account of a lower humidity in the air, immediately in contact with the specimen. The actual fading can be the same at a lower intensity as at a higher one, over a longer period of exposure.

Colour changes are promoted by the presence of oxygen and water, it has been debatable as to whether they are essential for fading.

There are several constitutional parameters that determine the photoresponsive colour change of a pigment or dyestuff. These include:

1. Chemical structure.
2. Particle size and its distribution.
3. The nature of the binder.

The chemical constitution factor has been reviewed in some detail by Vesce.[125]

This constitution and the change of colour of some dyestuffs on irradiation are related primarily to two structural components:

(a) the nucleus of the molecule which is specific to a particular dye class;
(b) the substituents in the molecule.

The nucleus tends to determine the most common lightfastness properties of a dye class. Substituents tend to have a secondary effect. The latter sometimes masks the general dye class trends, resulting in a wide variation of lightfastness.

Giles and McKay[126] have demonstrated that the best lightfastness is given by dyes of as uniform and large a particle size as possible. Maikowski[127] confirms this for the pigment field.

For the binders, Vesce[125] and ICI Dyestuffs Ltd,[128] looked at the lightfastness rating of pigments in tints with TiO_2, and different media, such as alkyd, alkyd/melamine, nitrocellulose and acrylic (Vesce) and air-drying medium, stoving medium and an acrylic (ICI). They agreed that fading was least in the stoving medium, and the nitrocellulose was also effective.

A stoving medium is expected to have much lower absorption and diffusion coefficients than, for example, an air-drying alkyd or emulsion medium, and these properties are important in determining rates of fading. Also, air-drying alkyds are most likely to contain hydroperoxy groups that would diminish slowly but probably not disappear completely. The relative performance between a stoving finish, an acrylic, and a nitrocellulose may be a function of conductivity in addition to absorption and diffusion coefficients.

Photochemical effects such as the fading of dyestuffs and polymer photodegradation of polymeric substrates possibly occur mechanistically via a donor/acceptor relationship in which oxygen and moisture may take part as either a donor or acceptor. Excitation states occuring in these reactions are debatable, but the possibility of triplet-state activity is supported from consideration of its longer lifetime than that of the singlet state, and by flash photolysis research in the solution phase.

Oxygen seems to be necessary as a component in the fading of dyes. This in some cases has been shown by identification of oxidation products of dyes on cellulose substrates. Desai and Giles[129] have found that electronegative groups such as nitro and chloro tend to increase the resistance to oxidation of the dye and its lightfastness. Atherton and Peters[130] demonstrated that electropositive groups decrease lightfastness. A substituent's position influences the nature of the electron movement within the molecule and hence the relative importance of the substituent.

An interesting mechanism of fading described by Egerton[131] requires initially the promotion of the dye molecule to an excited singlet state from which potentially triplet states can be generated by intersystem crossing. The excited molecule in either of these two states (the triplet state possesses a longer lifetime) subsequently becomes deactivated. Several paths are available for this deactivation. Two major ones are reaction with water molecules and/or oxygen, and if polymer is present, reaction between the activated dye molecule and polymer. The consequent change in constitution of the dye is accompanied by a change in colour. For lightfast dyes a tendency prevails for the energy possessed by the excited singlet or triplet state to be dissipated as thermal energy.

The nature of the radiant energy source is also very important and on substituting short-wavelength UV (i.e. at 254 nm) for near-UV and visible radiation, reduction of the dye by the polymer is more likely.

Egerton thought that fading is determined by treatments which are regarded as a normal part of the particular dyeing process, or by aftertreatments such as coating with resins designed to improve the physical properties of the dye material. For vat dyes the final treatment of boiling the dyed material with soap or detergent solution involves aggregation and crystallization of the dye and sometimes an alteration of polymorphic form. Improvements in lightfastness can result from a variety of causes, of which an increase in particle size is only one. Attachment such as may be between the dye and polymer molecules can be modified. Every dye/polymer combination should be thought of as being unique.

For after-treatment with resin, the interaction of the dye and the resin may favour reduction of the dye on irradiation. This can then be accompanied by air autoxidation with the consequent creation of a peroxy compound of the resin and subsequently more rapid fading of the dye.

When oxidative fading is obviously evident but there is no evidence of prior dye reduction, reaction of the excited dye molecule with oxygen would appear to be involved. Although direct oxidation in this way cannot be eliminated in every instance, it would appear to follow that fading of the dye in dry atmospheres is likelier to be a consequence of the effect of excited singlet oxygen on the dye molecule.

The singlet oxygen is probably formed by the reaction:[132]

$$^3D + {}^3O_2 \rightarrow {}^1D + {}^1O_2$$

Water vapour in conjunction with oxygen comprises fading of dyes by light. Normally dyed polymer materials are subject to constant changes in both temperature and moisture content throughout exposure to actinic radiation. The consequences of these two factors on fading are interlinked in practice, although studies of their separate effects which have been made under conditions in which the temperature of the dyed polymer materials is kept the same indicated that augmenting the relative humidity of the atmosphere increases the dye fading rate. Sometimes a linear relation between the rate of fading and the moisture content of the polymer material exists. Relative humidity effects vary considerably, both with the type of dye and the nature of the polymer.

On maintaining the relative humidity of the air constant, then an increase in the temperature of the polymer increases the rate of fading of the dye significantly ($\sim 10\,°C$ rise causes $\sim 10\%$ increase in rate of fading). The critical factor in the rate of fading is the relative humidity of the air in immediate contact with the exposed surface of the material and this may be substantially lower than that of the surrounding air.

The photochemical process can now be described in more detail.[133] As a consequence of the absorption of energy of the dye, D,

i.e. $$D \xrightarrow{h\nu} D^*$$

one of the electrons in the dye molecule is promoted to a higher energy level.

This is followed by one of the following processes:

1. The energy acquired by the dye molecule can be rapidly degraded to heat (by internal vibration) with the return of the excited molecule to the ground state. The excited dye molecule will be unable to transfer its energy to colliding molecules because of the shortness of its lifetime, and hence will be an inefficient photo-sensitizer.
2. The excited molecule may return to the ground state with the emission of light as fluorescence. The lifetime of the excited state will be comparatively long and

the absorbed energy is not converted into heat. Many dyes that sensitize the degradation of materials are fluorescent.
3. The excited dye molecule may return to the normal state as a consequence of a collision transferring energy to the colliding molecule.

i.e. $$D^* + M \rightarrow D + M^*$$

4. The excited dye molecule may enter into a chemical reaction with another molecule. The activated electron either leaves the dye molecule and enters the other molecule, resulting in oxidation of the dye.

$$D^* + M \rightarrow D^+ + e^- + M$$

or

$$D^* + M \rightarrow D^+ + M^-$$

or the energy level vacated by the excited electron is filled by an electron from the colliding molecule:

$$D^* + M \rightarrow D^- + M^+$$

In this case the dye is reduced.[134]

5. The activated dye species reacts preferentially with oxygen instead of with the substrate. The following reactions can be postulated:[135–136]

Because of the high ionisation potential of oxygen, which is greater than 12 eV, the reaction

$$D^* + O_2 \rightarrow D^- + O_2^+$$

appears to be energetically unlikely with visible and near-UV radiation.

The probability that the reaction

$$D^* + O_2 \rightarrow D^+ + O_2^-$$

can occur when an oxygen molecule collides with an activated dye molecule will be governed by the dye's ionization potential and the electron affinity of oxygen. This ionization potential should not be greater than the electron affinity. The more energetic the activated state of the dye molecule, the higher is the reaction probability. The value for the electron affinity of the oxygen molecule is dubious, as values in the range 3–62 kcal/mole have citings in the literature. On the assumption that the lower values are more probable, then reaction is likely to occur only when energy differences between the excited dye molecule and that necessary to cause ionization of the unexcited molecule are minimal. These conditions are likely with dyes in the highly excited state as is produced by UV radiation.

An important factor for the fading process may be the formation of hydrogen peroxide, possibly via the mechanism below:[137]

$$^3D^* + {}^3O_2 \rightarrow D^+ + O_2^-$$
$$O_2^- + H^+ \rightarrow HO_2\cdot$$
$$2HO_2\cdot \rightarrow H_2O_2 + {}^3O_2$$

Singlet oxygen would be produced in the absence of water vapour.

Apart from dye oxidation, hydrogen peroxide and associated products are likely to result in polymer degradation.

3.4.8 Pigment Photo-sensitization

Many pigments used in surface coating products can act as photosensitizers as they are able to absorb UV light and consequently transfer in some cases the energy to the surrounding medium. In particular, the white pigment oxides of zinc, titanium, and antimony are photosensitive, as are some chrome pigments.

Some of the effects produced by UV light on these compounds involve the phenomena of:

1. Photoconductivity.
2. Phosphorescence or fluorescence.
3. Phototropy.
4. Photolysis.

The three white oxides have high reflecting power for visible light. Titanium dioxide in particular is used as a filler or extender in paints, printing inks, paper, and sometimes in textiles.

The reflectance curves[138-140] of these oxides are shown in Figure 54. Zinc oxide in the region 400–450 nm reflects almost 100 per cent of the blue and violet light. The reflectance drops sharply below 400 nm and at 370 nm the absorption is complete. Zinc oxide becomes bright yellow when heated to a few hundred degrees centigrade and this may be interpreted as resulting from lattice disturbances with a consequent shift of the absorption curve into the blue and violet spectral regions. On cooling, zinc oxide resumes its white colour.

Most of these oxides have semiconductor properties. This means that under normal conditions they are poor electrical conductors but become much better conductors under irradiation with UV light or when heated. Zinc oxide has a region of UV light absorption corresponding with the photoconductivity region and this may be interpreted as the UV light causing electronic excitation.[141] The electron can then be mobile from atom to atom within the crystal and produce an electric current when the crystal is placed in an external electric field. Photoconductivity is also shown by titanium dioxide.[142]

Stoichiometric zinc oxide normally shows only weak fluorescence but if it has a deffiiency of oxygen, or contains excess interstitial zinc, it becomes brilliantly fluorescent, emitting blue-green light when irradiated with non-visible UV light at 365 nm. The excited electrons are thought to be trapped at defects in the crystal lattice, and eventually drop back to their normal sites, with partial loss of the exciting energy as visible light. These crystal defects may be slight deviations from stoichiometry or the presence of small amounts of foreign ions in the lattice.

Ultraviolet light is therefore likely to have the effect of transferring an oxide ion electron to a nearby metal ion.

The structure is thought to be not simply ionic in nature, but possessing large polarization energies. The outer electrons are strongly distorted within the crystal structure and the binding can be viewed as a compromise between ionic and covalent compounds.

Also, the crystal surface atoms are subjected to uneven force fields as the electrons are more powerfully attracted by the internal atoms of the crystal than towards the atmosphere on the crystal boundary side. Here, in the surface, electron transfer occurs with even lower energy than that necessary within the crystal. The absorption 'tail' in the reflectance curves of Figure 54 can be interpreted as being caused by the more active surface units, otherwise it would be expected to drop down more sharply.

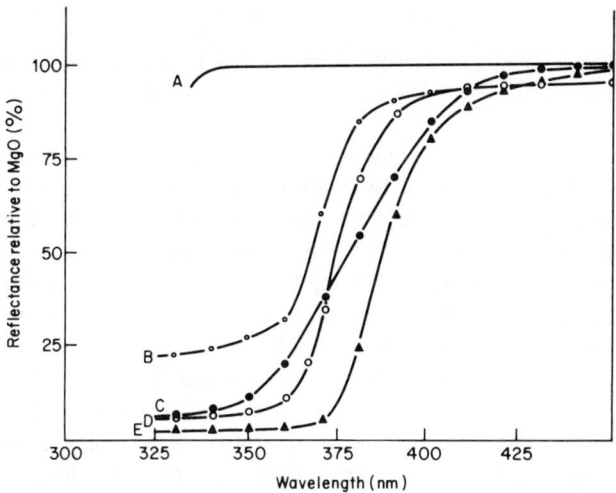

Figure 54   Reflectance curves for photochemically active oxides, relative to 100% reflectance for MgO obtained with reflectance attachment to Beckman DU spectrometer. A, MgO; B, Sb_2O_3 (c.p.); C, TiO_2; D, Sb_2O_3 (pr.); E, ZnO.[138] (Reproduced by permission of the New England Association of Chemistry Teachers.)

The gradient of the absorption curve is particularly noticeable for titanium dioxide.

Phototropy is shown by titanium dioxide and antimony trioxide in specific crystal forms. Phototropy is the change of colour that occurs on UV irradiation, reversibly returning to the normal white colour after the removal of the radiation.

The intensity of colour change is augmented when the crystals are placed in contact with organic substances during irradiation. In admixture with glycerol under UV irradiation, some types of titanium dioxide turn blue-grey, with very slow reversal back to white when the light source is removed. Titanium dioxide in non-stoichiometric form with oxide ion deficiencies yields the same blue-grey colour as also do Ti^{3+} ions in certain solvents. For a crystal with oxygen defficiency, two neighbouring (adjacent) titanium ions must have a $+3$ charge instead of the normal $+4$. The 'additional' electrons of these Ti^{3+} ions will tend to take the site of the missing oxygen to yield strangely distorted electron orbits that will absorb visible light, determining the blue-grey colour. On irradiation in the presence of oxidizable organic compounds, stoichiometric titanium dioxide undergoes electron transfer from the UV light to cause reduced Ti^{3+} and oxidized organic substances. Photolysis is the name allocated to this type of phenomenon, that is, the decomposition of the oxide by light. This type of reaction can play an important role as a cause in the problem of 'chalking', where a white removable powder is often observed in paint films containing extenders.

A suspension of zinc oxide in water[143] exposed to actinic radiation in the presence of oxygen or air produces hydrogen peroxide. The oxide does not appear to be destroyed or permanently changed in the process, and appears to be functioning as a photo-sensitizer or photocatalyst for the reaction

$$2H_2O + O_2 \rightarrow 2H_2O_2$$

This constitutes a means of storing solar energy in a chemical system, as zinc oxide absorbs throughout the UV region, which constitutes some 4 per cent of the total energy of sunlight.

The rate of formation of hydrogen peroxide and the total quantity yield may be augmented several magnitudes by adding even minute amounts of organic compounds. It is thought that direct participation of the organic material in the initial chemical reaction leads to peroxide formation.

The electron transferred by the UV light from an oxide ion to a zinc ion would transiently appear to form Zn^{+1} and then may be picked up by oxygen on the surface of the crystal, resulting in the reduction of oxygen and the creation of peroxide ion. The zinc oxide crystal would then remain with a positive charge. Organic materials present may provide electrons to the zinc oxide, which then gains its neutrality. The total result is the oxidation of the organic substance and the reduction of oxygen from the air. Possibly the reactions are:

$$ZnO \rightarrow Zn^{+1}O;$$
$$Zn^{+1} + O_2 \rightarrow Zn^{+2} + O_2^-$$
$$(ZnO)^+ + RH \rightarrow ZnO + RH^+ \text{ (oxidized organic compound)}$$

If this is what happens, zinc oxide cannot be a converter of solar energy to chemical energy, as the oxidation of these organic substances is an exothermic process, but would become important as a novel kind of photocatalyst. The zinc oxide and light are simply providing the activation energy to initiate the reaction as any thermal catalyst would do. Zinc oxide is a known thermal catalyst for the oxidation and dehydration of alcohols.

Zinc oxide will catalyse reactions in anaerobic conditions; for example, the polymerization of vinyl monomers in dilute aqueous solution.

Antimony trioxide appears to combine the possibility of photolysis characteristics of titanium dioxide with the photocatalytic properties of zinc oxide, the governing factor being probably the crystal form.

Goodeve[144] showed that antimony trioxide acts in a similar manner to titanium dioxide, absorbing light in the UV region of the spectrum and turning dark brown. The brown colour slowly fades in the dark. The rate and extent of darkening can be augmented by adding organic compounds, especially glycerol. Goodeve made his photochemically active antimony trioxide by precipitating it from antimony trichloride with concentrated ammonia, and washing the sample free of chloride ions. The crystal form was found to be prismatic by X-ray diffraction. Goodeve also noted an inert form of antimony trioxide with a rhombic crystal pattern which failed to darken under UV light irradiation.

The prismatic antimony trioxide does darken and the addition of glycerol enhances the darkening. For the former, antimony is probably both oxidized and reduced in the absence of organic matter:

$$5Sb_2O_3 \rightarrow 3Sb_2O_5 + 4Sb$$

Organic matter causes antimony to be produced, but oxidized organic substances probably result, instead of Sb_2O_5. It is unlikely that hydrogen peroxide is formed during the reaction.

Commercial antimony trioxide behaves like zinc oxide. It does not turn dark; hydrogen peroxide is found, and it oxidizes glycerol. Addition of chloride ion or ammonium ion, which could probably be impurities in the prismatic antimony trioxide owing to its method preparation, has no effect. This appears to be a case where a substance can either undergo photolysis or act as a catalyst for oxidation, depending only on the crystal form. These two oxides had slightly different absorption curves. Peroxide formation may readily occur if either water or oxygen (or both) are present on the crystal surface in a specific manner that will permit the transfer of the electron from the reduced metal to them. The metal is reoxidized, hydrogen peroxide accumulates, and the organic substance keeps on being oxidized. For a surface condition that stops one of the necessary reactants from being absorbed in a strategic position, the metal is not reoxidized, the oxide darkens owing to reduction to the metal, and the organic material is oxidized only to a specific limit and subsequently the reaction terminates.

Titanium dioxide is probably the most widely used of the white pigments. The rutile grade absorbs at wavelengths below about 420 nm,[145] and anatase much lower (see Figure 55).

Research has shown that when oxygen is present, isopropanol is oxidized to acetone under the influence of actinic radiation as is reported for zinc oxide.[146] In

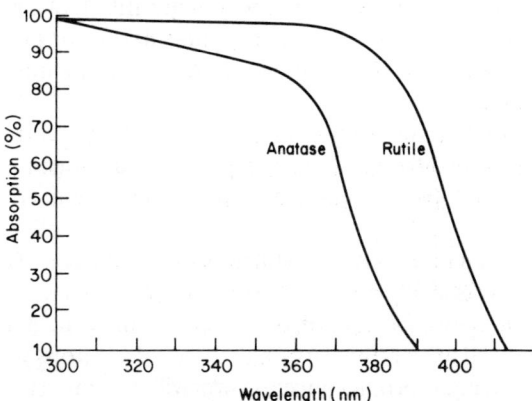

Figure 55 Absorption spectra of the rutile and anatase forms of titanium dioxide.[139] (Reproduced by permission of the Federation of Societies for Coatings Technology.)

nitrogen-purged systems, no reaction occurs, but oxidation of isopropanol will occur in the presence of nitrous oxide and hydrogen peroxide. Oxygen, hydrogen peroxide, and nitrous oxide are molecules of high electron affinity and, consequently, a scheme for radical generation could be:[143]

$$TiO_2 \xrightarrow{hv} [(h^+ - \bar{e})] + TiO_2$$
$$\text{exciton}$$

Dissociation of the exciton occurs and the 'positive hole' and electron become trapped at separate sites on the pigment surface.

$$h^+ + OH^- \text{ (surface)} \rightarrow OH \text{ (surface)}$$
$$\bar{e} + O_2 \rightarrow O_2^-$$
$$\bar{e} + N_2O \rightarrow O^- + N_2 \uparrow$$
$$\bar{e} + H_2O_2 \rightarrow OH^- + OH$$

In the case of oxygen, the effect of this upon isopropanol is:

$$O_2^- + (CH_3)_2CHOH \rightarrow HO_2 \cdot + (CH_3)_2CHO^-$$
$$(CH_3)_2CHO^- + OH \text{ (surface)} \rightarrow (CH_3)_2CHO \cdot + OH^- \text{ (surface)}$$
$$(CH_3)_2CHO \cdot + (CH_3)_2CHOH \rightarrow (CH_3)_2CHOH + \dot{C}(CH_3)_2OH$$
$$2\dot{C}(CH_3)_2OH \rightarrow (CH_3)_2CO + (CH_3)_2CHOH.$$

Similar reaction schemes can be written for the other additives.

Radicals generated at the surface of TiO_2 particles can then bring about photopolymerization or crosslinking.

Singlet oxygen may be involved. This species could arise from either energy transfer or decomposition of the superoxide ion.[148]

The role of diluents such as TiO_2 in accelerating fading of films has been recognized for many years. In general the lightfastness of pigments, especially red and yellow azo pigments, in tints with TiO_2 is worse than that in full-strength colours. Anatase TiO_2 results in faster fading than rutile TiO_2, possibly owing to differences in crystal structure, the number and nature of lattice defects, and consequently certain crystal surface differences, such as the number of hydroxyl groups per unit area.

Current ideas concerning the photochemistry of TiO_2 centre on the exciton theory. There are many different ways of depicting and understanding this 'solid' photochemical species phenomenon and it will be discussed in more detail below.[149-153]

Excitons are created in the TiO_2 crystal under the influence of actinic radiation and can be written as $(Ti^{3+} - O^-)$. These may transfer energy with adjacent atoms and can therefore be regarded as free to move in a random fashion throughout the crystal lattice. In an irradiated crystal, excitons will be created at a constant rate and an equilibrium concentration will become established when the formation rate equals the disappearance rate.

Excitons arriving at the crystal surface in the neighbourhood of a site carrying a hydroxyl group will form a trivalent titanium ion linked to an uncharged hydroxyl radical:

$$Ti^{4+}-OH^- + exciton \rightarrow Ti^{3+}-OH$$

The presence on the surface of the particle of the $(Ti^{3+}-OH)$ species produced by UV absorption in the crystal permits two possible reactions to occur:

(a) In the presence of oxygen, but with no oxidizable species available the irradiated pigment chemisorbs oxygen molecules to form a surface layer of hydroperoxide:

$$O_2 + 2(Ti^{3+}-OH) \rightarrow 2(Ti^{4+}-O-OH^-)$$

a typical yellow colour is developed.

(b) If oxygen is excluded, the exciton may dissociate to form free hydroxyl radicals:

$$(Ti^{3+}-OH) \rightarrow Ti^{3+} + OH\cdot$$

On the other hand, the hydroxyl radicals can initiate polymerizing chain reactions or oxidize in a variety of well known reactions. In such cases the surface titanium ions remain in the reduced state and a blue-grey colour is developed on the pigment. This colour can be discharged subsequently, if oxygen is allowed free access.

Returning back to the former example of irradiation in the presence of oxygen, the peroxide group in the presence of oxidizable material may form a free oxygen atom.

The yellow colour will then be discharged and the surface can return to its original unexcited state without showing any features of reduction:

$$(Ti^{4+}-OOH^-) \rightarrow (Ti^{4+}-OH^-) + O$$

The excited hydroxylated TiO_2 surface therefore provides one of two alternative free-radical species depending on whether conditions are aerobic or anaerobic. For the former case the atomic oxygen species $O\cdot$ is generated, the surface continuously reverts to the fully oxidized state, and the activity remains constant. In the latter case the hydroxyl radical species OH is generated, the surface becomes reduced and discoloured, and the generating process slows down as the hydroxyl population is depleted.

For effective photo-oxidation potential the ultraviolet excitation must cause an electron from a surface hydroxyl to move from the surface to a Ti^{4+} ion. Only in this manner may the OOH group be created from oxygen or the hydroxyl group become a free radical.

In order to overcome this detrimental effect to the lightfastness of films, a strong electron acceptor needs to be attached to the surface hydroxyl, and this can be obtained by heating the pigment surface with another material, for example, hydrated alumina. The electronic structure of a hypothetical $-Ti-OH-Al(OH)_3$ surface group is depicted below.[154]

The counter-effect of the donated lone pair from hydroxyl reduces the probability of the valency electron moving to Ti^{4+} by quantum absorption or by transfer of energy from an exciton.

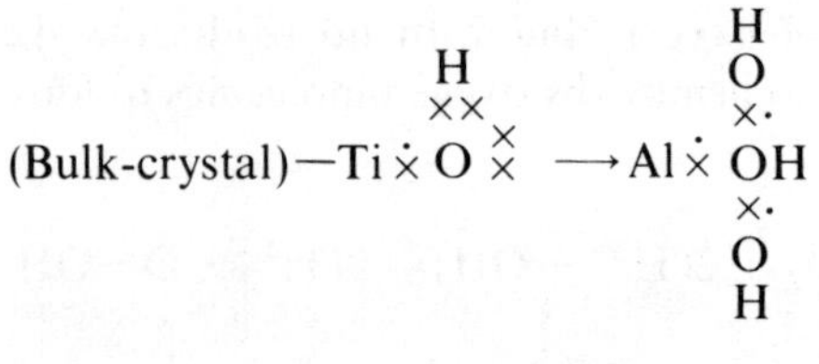

$$(\text{Bulk-crystal})-Ti \overset{\times\times}{\underset{}{\times}} \overset{H}{\underset{}{O}} \overset{\times}{\underset{}{\times}} \longrightarrow Al \overset{\overset{H}{\underset{}{O}}\cdot}{\underset{\underset{H}{O}}{\overset{\times}{\underset{\times\cdot}{\times}}}} OH$$

The lone-pair acceptance of aluminium becomes stronger if the aluminium is bonded to silica through oxygen, in a further hypothetical surface group such as:

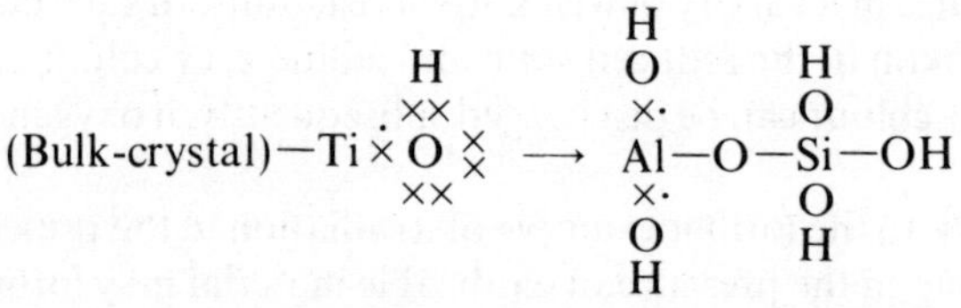

$$(\text{Bulk-crystal})-Ti \overset{\times\times}{\underset{\times\times}{\cdot\cdot}} O \overset{\times}{\underset{}{\times}} \longrightarrow \overset{\overset{H}{\underset{}{O}}\cdot}{\underset{\underset{H}{O}\cdot}{Al}}-O-\overset{\overset{H}{\underset{}{O}}}{\underset{\underset{H}{O}}{Si}}-OH$$

The previously discussed mechanisms refer to anhydrous conditions, but practical conditions usually involve exposure to both oxygen and water. In the presence of a photo-sensitizer such as TiO_2, the formation of H_2O_2 is probably on irradiation by the simple mechanism (Egerton[155]) shown below:

$$TiO_2 \overset{h\nu}{\rightarrow} TiO_2^*$$
$$TiO_2^* + O_2 \rightarrow TiO_2 + O_2^*$$
$$O_2^* + H_2O \rightarrow H_2O_2 + O\cdot$$

and this hydrogen peroxide is available for oxidation of compounds such as dyes. Egerton demonstrated that on irradiation of a suspension of TiO_2 in water through which oxygen was bubbled, peroxide was produced, although he was not able to demonstrate conclusively that this was hydrogen peroxide.

Another, but not dissimilar, mechanism has been proposed by Voltz et al.,[156] who stress the important role determined by water in chalking. They advocated that electron transfer results in the formation of a charged absorbed water molecule which decomposes to produce a proton, and on further irradiation the surface complex produces a hydroxyl radical, and this is the effective oxidizing agent. These workers appear to dispute the possibility of hydrogen peroxide being formed.

The Voltz et al. mechanism can be depicted:

$$(Ti^{4+}) + H_2O \rightarrow (Ti^{4+}\ ^{\dagger}H_2O)$$

$$(Ti^{4+}\ ^{\dagger}H_2O) \xrightarrow{h\nu} (Ti^{3+}\ ^{\dagger}H_2O^+)$$

$$(Ti^{3+}\ ^{\dagger}H_2O^+) \rightarrow (Ti^{3+}\ ^{\dagger}OH) + H^+$$

$$(Ti^{3+}\ ^{\dagger}OH) \xrightarrow{h\nu} (Ti^{3+}) + OH$$

Voltz et al. demonstrated the formation of the hydroxyl free radical in a TiO_2/water system on irradiation by electron spin resonance spectroscopy.

This free radical is likely to have a short life, though it is highly energetic. It is possible that it may continuously reform and propagate in a chain reaction and so reach a dye molecule. The hydrogen peroxide molecule would be expected to be more stable and may move by activated diffusion throughout the surface-coating film.

On this basis the promotion of fading (or chalking) by TiO_2 is either a complete combustion of the pigment to CO_2, H_2O, and other elementary volatile products, or alternatively an oxidation to less highly coloured intermediates.

3.4.9 Organic Peroxides

The peroxo group absorbs strongly at 254 nm but does not respond to radiation above 320 nm. It is therefore transparent to the 365 nm wavelength emitted from the medium pressure UV mercury lamp and remains inert to this line.

The absorption of the peroxo group leads to an electronically excited state. The electronic transition is probably one involving the transfer of a non-bonding electron (n) on one of the oxygen atoms to the anti-bonding sigma (σ^*) orbital localized in the O—O region. Such an n → σ^* transition would lead to a weakening and possible rupture of the O—O bond. The energy required to cause the dissociation of this bond is likely to be similar to that required to dissociate hydrogen peroxide into two hydroxyl radicals, which is about 48 kcal mole^{-1}. The energy of a 254 nm quantum is about 112 kcal mole^{-1}, which is in excess of that required for the dissociation. Quenching of the excited state,

$$(R-O-O-R_1)^* \rightarrow R-O-O-R_1$$

by collision with surrounding solvent or medium molecules is another possible fate. This is only possible if the process of collision is rapid compared with the time of vibration of the O—O bond. If it is not rapid then photodissociation is the more probable fate of $(R-O-O-R_1)^*$. The asterisk denotes excitation energy.

The likely photodissociation products are the free radicals RO· and $R_1O·$ by homolytic cleavage. The dot after the free radical atomic or molecular formula signifies paramagnetism due to an unpaired electron, which is a free radical characteristic.

The excited $(R-O-O-R_1)^*$ molecule would be produced in a solvent or medium cage (denoted by brackets). Once, the two free radicals have been produced in the solvent cage, they can undergo any combination of the following three processes:

1. Recombination in the cage as the activation energy to do so is low.
2. Diffusive transfer to the bulk of the system, each radical then possessing its own solvent cage.
3. Reaction of the two radicals with the solvent cage.

These processes described can be summarized as:

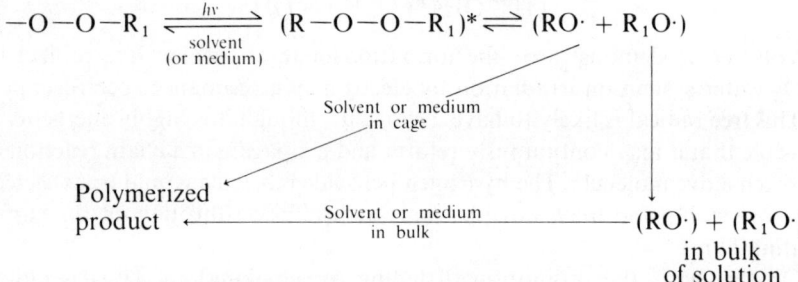

There are many common organic peroxides that have been used in two pack curing systems owing to their thermal instability, including:

1. *Di-tert-butyl peroxide*
   (activation energy for dissociation is about 30 kcal mole^{-1})

$$(CH_3)_3C-O-O-C(CH_3)_3$$

2. *Diacetyl peroxide*

$$CH_3-\overset{O}{\overset{\|}{C}}-O-O-\overset{O}{\overset{\|}{C}}-CH_3$$

3. *Dibenzoyl peroxide*

$$C_6H_5\overset{O}{\overset{\|}{C}}-O-O-\overset{O}{\overset{\|}{C}}C_6H_5$$

4. *Cumene hydroperoxide*

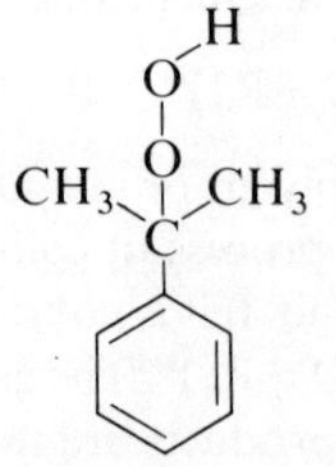

$$CH_3-\underset{\underset{Ph}{|}}{\overset{\overset{O-H}{|}}{\underset{|}{C}}}-CH_3$$

Dibenzoyl peroxide initially fragments as[157–160]

$$Ph\cdot\overset{O}{\overset{\|}{C}}O-O\overset{O}{\overset{\|}{C}}\cdot Ph \xrightarrow{h\nu} Ph\cdot\overset{O}{\overset{\|}{C}}-O\cdot + Ph\cdot + CO_2$$

Di-tertbutyl peroxide yields different products at radiation less than 300 nm[161] from those that occur at 313 nm.[162] All product radical fragments have the capability of polymerization initiation.

Inorganic peroxides such as hydrogen peroxide, H_2O_2, and peroxodisulphates have application for photopolymerization reactions.[163]

3.4.10 Organic Sulphur Compounds

Numerous sulphur compounds have been used as photo-initiators/sensitizers. Some of these include:[20]

Alkyl disulphides	such as di-*n*-butyl disulphide
Aralkyl disulphides	such as dibenzyl disulphide
Aryl disulphides	such as diphenyl disulphide
Aroyl disulphides	such as dibenzoyldisulphide
Acyl disulphides	such as diacetyl disulphide
Cyclo alkyl disulphides	such as dibornyl disulphide

Simple disulphides, like dimethyl disulphide, possess relatively strong S—S bonds, needing 72 kcal per mole to undergo dissociation.[164] Those with more of a complicated nature such as diaryl disulphides have less stability and generate free radicals when irradiated with actinic radiation in the wavelength range 280–400 nm.[165]

The S—S bond cleavage often follows the pattern:[166]

$$C_4H_9S-SC_4H_9 \xrightarrow{h\nu} 2C_4H_9S\cdot$$

The type of fission in the disulphide linkage can be of paramount importance. Asymmetric rupture of the C—S linkage may provide the inhibiting radical $C_6H_5SS\cdot$ for diphenyl disulphide, resulting in a retarding effect on the system in contrast to the formation of the faster reacting $C_6H_5S\cdot$ free radicals. Further examples of these are discussed later in conjunction with mono and polysulphides.

Some other important sulphur-containing photo-initiators include:

(a) *Mercaptans*[167]

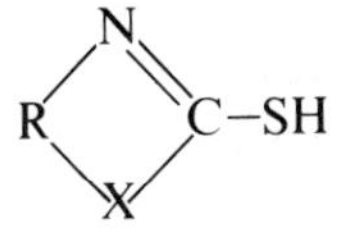

e.g. 2-mercaptobenzothiazole

R = *ortho*-arylene radical
X = S, O, or NH

(b) *Thiols*[168]
R—SH   e.g. thiophenol
R = aryl or substituted aryl

Photoinitiators possessing the S—H linkage create via photo-induced bond rupture the radicals RS· and H·, which initiate the subsequent chain reaction.

(c) *Sulphonyl chlorides.* These possess the general structure RSO_2Cl. As the organic radical R changes from alkyl to aryl, the photoactivity markedly increases.

Some particularly active types for the photopolymerization of methyl methacrylate are: 1 and 2 naphthalene sulphonyl chlorides and 1, 5 naphthalene disulphonyl chloride.

(d) *Chlorosulphonamides.* Dichloramine-T compounds are reported as active photo-sensitizers as they possess high chain transfer characteristics.[169]

An example of these compounds is sodium *p*-toluene chloramide, $CH_3C_6H_4SO_2NClNa$.

(e) *O-Alkyl xanthate esters.* These possess the grouping,

$$-S-\overset{S}{\underset{\|}{C}}-OR.$$

Examples of these compounds are 2-oxopropylene *bis*(methyl xanthene) and carbethoxymethylene *bis*(ethyl xanthate), the latter having the structure[170]

$$C_2H_5OOC-CH[S-C(=S)O-C_2H_5]_2$$

Xanthates such as $R-S-CS-O-C_2H_5$, where $R = C_6H_5\,(CH_2)_n$, $C_2H_5OOC(CH_2)_n$; $n = 0$ to 2, are reported as photo-sensitizers for methyl methacrylate photopolymerization.[171]

(f) *Metal mercaptides.*[172–173] Mercuric phenyl mercaptide is a typical example of this class of compound which has the general formula, $(RS)_xM$. R is an organic radical, M is a metal such as Hg, Pb, Ag, Zn, with a valency x.

(g) *Mono-, di- and polysulphides.* Diaryl and diaroyl disulphide derivatives with the absence of nitro, hydroxyl, and primary amino groups on the aromatic carbon have been found to give the best results.

(h) *Sulphenates.*[200] These have the general structural formula:

$$R-O-S-R'$$

Typical examples of these are ethyl trichloromethane sulphenate[201] and 2,4,5 trichlorophenyl trichloromethane sulphenate.

These kind of photosensitizers tend to yield polymers with relatively low molecular weight. The presence of many chlorine atoms yield poor thermal stability products.

(i) *Thiocyanate Radical Formation.* A molecule that homolytically dissociates is thiocyanogen, $(SCN)_2$, to give the thiocyanate radical. A source of these radicals is the trithiocyanatotripyridine iron[100] complex $(Fe(SCN)_3Py_3)$. Pyridinium thiocyanate also gives these radicals. In all cases, monomers such as methyl methacrylate and the additives seem to form a photochemically active adduct that efficiently fragments to initiating radicals.[202–203]

(j) *β-Keto Sulphides and Sodium Benzyl thiosulphate.*[204–206] These are found to be, in decreasing order of effectiveness for polymerization of methyl methacrylate, acrylonitrile, styrene, and vinyl acetate: phenyl phenacyl sulphide > phenyl acetonyl sulphide > ethyl acetonyl sulphide > ethyl phenacyl sulphide.

Sodium benzyl thiosulphate polymerizes acrylamide and methyl methacrylate (the energy of activation for the latter being 5.1 kcal mole^{-1}).

There are many monosulphides, disulphides and polysulphides that have shown notable photoactivity. Some examples of these are shown below:[174–176]

1. $C_6H_5-S_n-C_6H_5 \quad n = 2-4$
   diphenyl polysulphide
2. $C_6H_5-CH_2-S_n-CH_2C_6H_5 \quad n = 2-4$
   dibenzyl polysulphide
3. $C_6H_5-CO-S-S-CO-C_6H_5$
   dibenzoyl disulphide
4. 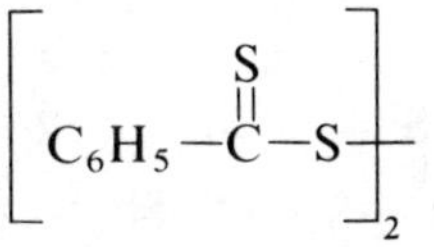
$$\left[C_6H_5-\overset{S}{\underset{\|}{C}}-S \right]_2$$
   dithiobenzoyl disulphide

   dithiobenzoyl disulphide

5. 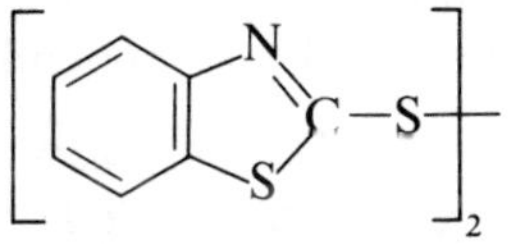
$$\left[\underset{S}{\underset{|}{\overset{N}{\overset{\|}{\diagup}}}}C-S \right]_2$$
   dibenzothiazoyl disulphide

   dibenzothiazoyl disulphide

6. 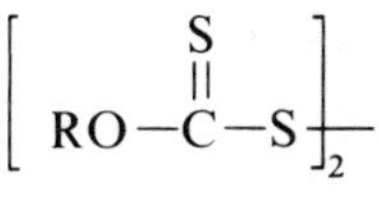
$$\left[RO-\overset{S}{\underset{\|}{C}}-S \right]_2 \quad \text{where } R = CH_3, C_2H_5, C_4H_9$$

   dialkyl xanthogene disulphide

7. *Dithiocarbamates*[177]
   These have the general formula

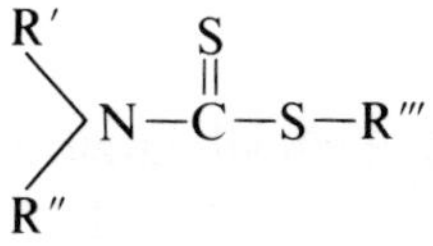

$$\underset{R''}{\overset{R'}{\diagup}}N-\overset{S}{\underset{\|}{C}}-S-R'''$$

Examples are; N, N', dimethyldithiocarbamate,

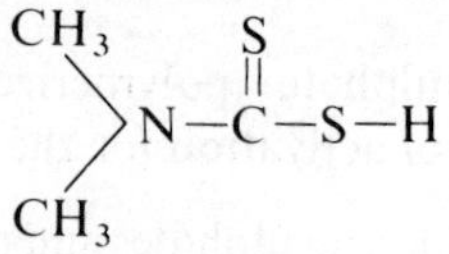

$$\begin{array}{c}CH_3\\ \\ CH_3\end{array}\!\!\!>\!\!N-\underset{\|}{\overset{S}{C}}-S-H$$

and methyl diethyl carbamate,

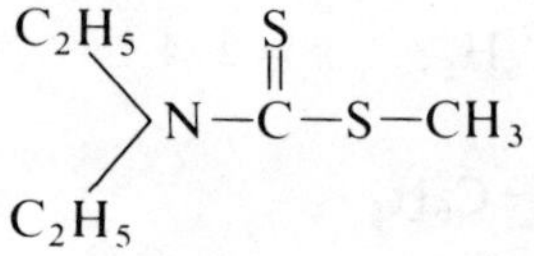

$$\begin{array}{c}C_2H_5\\ \\ C_2H_5\end{array}\!\!\!>\!\!N-\underset{\|}{\overset{S}{C}}-S-CH_3$$

8. *Thiuram derivatives*[178-181]

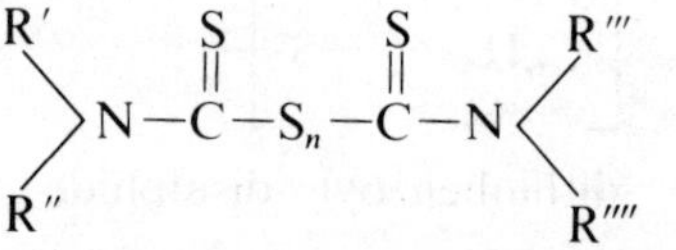

$$\begin{array}{c}R'\\ \\ R''\end{array}\!\!\!>\!\!N-\underset{\|}{\overset{S}{C}}-S_n-\underset{\|}{\overset{S}{C}}-N\!\!<\!\!\!\begin{array}{c}R'''\\ \\ R''''\end{array}$$

R', R'', R''', R'''', may be alkyl, phenyl, or pentamethylene groups. n is 1–2.

An example of these compounds is tetramethylthiuram monosulphide, where R' = R'' = R''' = R'''' = CH$_3$, and $n = 1$.

A possible mechanism of action for this compound is:

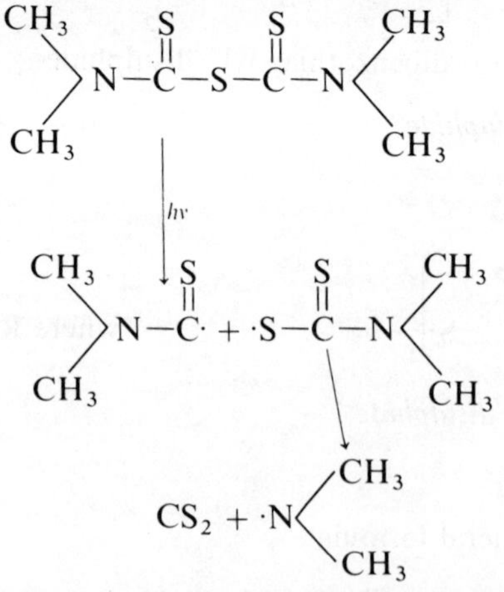

$$\begin{array}{c}CH_3\\ \\ CH_3\end{array}\!\!\!>\!\!N-\underset{\|}{\overset{S}{C}}-S-\underset{\|}{\overset{S}{C}}-N\!\!<\!\!\!\begin{array}{c}CH_3\\ \\ CH_3\end{array}$$

$\Big\downarrow h\nu$

$$\begin{array}{c}CH_3\\ \\ CH_3\end{array}\!\!\!>\!\!N-\underset{\|}{\overset{S}{C}}\cdot\; +\;\cdot S-\underset{\|}{\overset{S}{C}}-N\!\!<\!\!\!\begin{array}{c}CH_3\\ \\ CH_3\end{array}$$

$$CS_2\;+\;\cdot N\!\!<\!\!\!\begin{array}{c}CH_3\\ CH_3\end{array}$$

The photodissociation of tetramethylthiuram disulphides may be depicted in a similar manner as:[182]

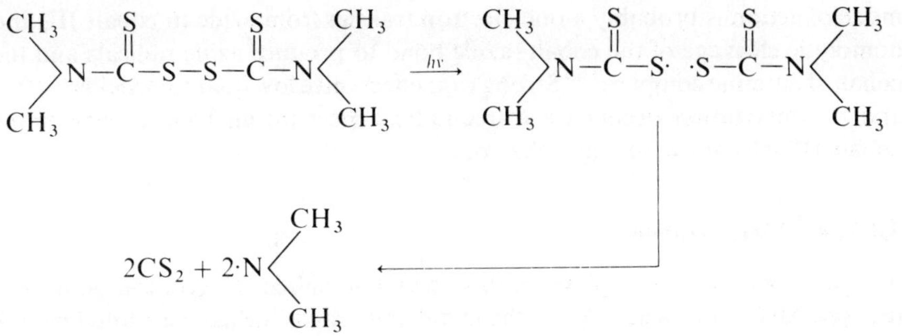

$$\underset{CH_3}{\overset{CH_3}{>}}N-\overset{S}{\overset{\|}{C}}-S-S-\overset{S}{\overset{\|}{C}}-N\underset{CH_3}{\overset{CH_3}{<}} \xrightarrow{h\nu} \underset{CH_3}{\overset{CH_3}{>}}N-\overset{S}{\overset{\|}{C}}-S\cdot \quad \cdot S-\overset{S}{\overset{\|}{C}}-N\underset{CH_3}{\overset{CH_3}{<}}$$

$$2CS_2 + 2\cdot N\underset{CH_3}{\overset{CH_3}{<}} \longleftarrow$$

3.4.11 Metal Compounds and Ions

The mode of photo-initiation/sensitization of metal compounds and ions often occurs by energy transfer or electron transfer mechanisms, with some exceptions such as the iron complex (previously mentioned) that yields thiocyanate radicals upon irradiation. Some useful systems are outlined below.

3.4.11.1 Uranyl Salts

Uranyl salts are photo-sensitizers. The uranyl ion absorbs the incident radiation and the subsequent formation of the initiating species appears to be due to energy transfer to a monomer followed by a monomer–monomer electron transfer reaction. The uranyl ion is a true sensitizer as it is not reduced or oxidized. UO^{2+} readily sensitizes monomers such as acrylamide,[183] methacrylamide,[184] and hydroxyalkyl acrylates.[185] This behaviour may be depicted as:[110]

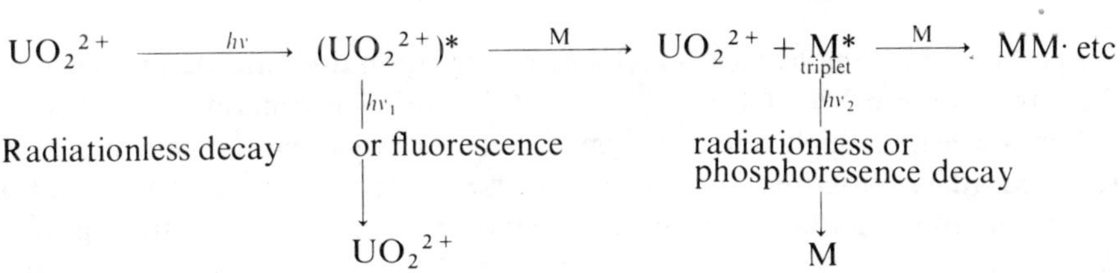

$$UO_2^{2+} \xrightarrow{h\nu} (UO_2^{2+})^* \xrightarrow{M} UO_2^{2+} + \underset{\text{triplet}}{M^*} \xrightarrow{M} MM\cdot \text{ etc}$$

Radiationless decay | $h\nu_1$ or fluorescence | $h\nu_2$ radiationless or phosphoresence decay

UO_2^{2+} | M

3.4.11.2 Gold Salts

Aqueous photopolymerization of acrylamide can occur with sodium chloroaurate ($NaAuCl_4 2H_2O$). The absorption of 365 nm radiation is probably due to excitation of the charge transfer to metal band of Au(111) to Au(0) by a one-electron transfer process.[186,110]

3.4.11.3 Cobalt Salts

Azidoammine cobalt III salts photopolymerize acrylamide, methylacrylamide, methyl methacrylate, acrylonitrile, and N-vinyl pyrrolidone.[110,187–189] The mode of action is probably a one-electron transfer from azide to cobalt III, and homolytic cleavage of the cobalt–azide bond to produce azide radicals and the cobalt II ammine complex.[190] Strong cage effects give low quantum yields (<0.2) and polymerization occurs via a free-radical mechanism. Diazidotetrammine cobalt III salts show similar behaviour.[191]

3.4.11.4 Metal Halides

Acetylacetone forms complexes with some metal halides to give compounds of the type $Me(acac)_n$, where Me is the metal and acac is the acetoacetonyl ligand. $Co(acac)_3$ is very effective,[192] as is $Mn(acac)_3$.

Initiation is by acetylacetonyl radicals from dissociation of the Co or Mn chelate, free-radical propagation ensuing.[110]

3.4.11.5 Metal Carbonyl Compounds[110]

Metal carbonyl compounds such as rhenium carbonyl, $Re_2(CO)_{10}$, and manganese carbonyl, $Mn_2(CO)_{10}$, cause photopolymerization at 365 nm and 436 nm respectively.[193] Irradiation in the presence of monomer results in polymerization slowly with $Re_2(CO)_{10}$, possibly by hydrogen abstraction. $Mn_2(CO)_{10}$ is ineffective under similar conditions.

When specific chlorinated or brominated compounds are present both carbonyl complexes initiate with high efficiency ($\phi \sim 1$). The primary radicals are probably produced by:

$$Me_2(CO)_{10} \xrightarrow{h\nu} Me(CO)_4 + Me(CO)_6$$
$$Me(CO)_4 + CCl_4 \rightarrow Me(CO)_4Cl + \cdot CCl_3$$

For Me = Mn, the hexacarbonyl species, initiation does not occur, and it is thought to reform $Mn_2(CO)_{10}$ by loss of CO and recombination reactions.

For Me = Re, a high rate of polymerization occurs when the light source is removed (dark reaction), probably because of very slow scission ($-CO$) recombination reactions, permitting a build-up of potentially active species. Additives such as acetylacetone in the $Mn_2(CO)_{10}$ system show this dark polymerization, probably owing to trapping of a species such as $Mn(CO)_6$ which subsequently acts as an initiator. The Mn system with such an additive reacts similarly to the Re system without additive.

Efficient photopolymerization of tetrafluorethylene without an added halide occurs with $Mn_2(CO)_{10}$, $Os_3(CO)_{12}$, $Ru_3(CO)_{12}$, $Re_2(CO)_{10}$ with a mechanism comparable to the above. The initiating species is thought to be $(CO)_nMeCF_2CF_2$.[193-194]

3.4.11.6 Photolysis of Dichromates

Ammonium alkali metal and organic quaternary ammonium dichromates are readily reduced to Cr^{3+} in the presence of oxidizable organic materials. The reaction is accelerated by light, but the exact mechanism is not clear. Several system types are shown below:

1. Inorganic dichromate/water-soluble polymer, such as gelatin polyvinyl alcohol and polysaccharides.[195] These find application in the photo-engraving process.
2. Inorganic dichromate/water-insoluble polymer, such as polyvinyl butyral in ethanol,[196,197] used as metal plate resist coatings.
3. Organic dichromate/water-insoluble polymer, such as trimethyl lauryl ammonium dichromate $[C_{12}H_{25}N(CH_3)_3]_2Cr_2O_7$, which is insoluble in water but soluble in some organic solvents.[198]
4. Dye-sensitized dichromate photolysis. This type of system is depicted by methylene blue + chelating agent/dichromate–gelatin system such as EDTA.

The chelating agent functions as an electron donor for the light-excited dye molecule but is inert to dichromate (does not reduce it) in the dark.[199] The application of dichromated systems is mainly in the photoresist area and is discussed more fully in Chapter 6.

3.4.12 Organic Phosphorus-containing Compounds[207]

Electron-rich compounds that function as Lewis bases to form complexes with electron acceptors (Lewis acids) are often useful additives in photo-polymerization formulations and are in this sense used as synergistic agents.[110] Two common types are:

1. Organic phosphines, R_3P:
2. Organic phosphites, $(RO)_3P$:

where R is alkyl or aryl in nature.

Triphenylphosphine (TPP) undergoes charge-transfer complex formation with unsaturated acceptor molecules. This complex absorbs UV light to initiate polymerization via a free-radical route.[208]

Tri-orthotolylphosphine (TOTP) has a faster cure. The mechanism would seem to be very concentration-dependent. For less than 10^{-4} M, a complex is created that absorbs and dissociates to free radicals. For concentrations greater than 10^{-2} M, the primary radical species seem to be generated by direct homolytic fission of a phosphorus–carbon bond. Steric interference of resonance of TOTP is thought to be due to the greater activity of TOTP compared with TPP. The plane of the ring has to be perpendicular to the phosphorus orbital possessing the lone-pair electrons for delocalization into the π system of the aromatic ring. The electrons are localized on the phosphorus atom and are more available to an electron acceptor if this is impossible.[209]

Charge-transfer complexes can be created between triphenylphosphite and an appropriate unsaturated acceptor. Acrylonitrile will form a UV-sensitive complex with triphenylphosphite. Under irradiation conditions, where acrylonitrile does not polymerize alone, it has been found that in the presence of triphenyl phosphite it will form an UV sensitive complex that absorbs the energy, forms radical fragments, and induces polymerization.[210]

3.4.13 Chlorosilanes

These act as true photo-initiators.[110,211-212] Cleavage generally occurs as depicted below for trimethylchlorosilane:

$$(CH_3)_3SiCl \xrightarrow{h\nu} (CH_3)_3Si\cdot + Cl\cdot$$

3.4.14 Azo Compounds

This class of compounds containing the $-N=N-$ grouping, causing them to be sensitive to actinic radiation especially in the 300–400 nm range, is enormous. Apart from those used for photo-ionic curing previously described in section 3.4.1, there are many others for different applications that are well documented and are not described here as they are too numerous. Some azo compounds of other interest are described briefly below.[110]

If the carbon atom adjacent to the azo group also possesses a nitrile substituent, the azo compound is a particularly efficient photo-initiator.

Azo-*bis* (isobutyronitrile) (AIBN) is a typical example of an azo compound photo-initiator.[213-214]

Another is diazirine;[215]

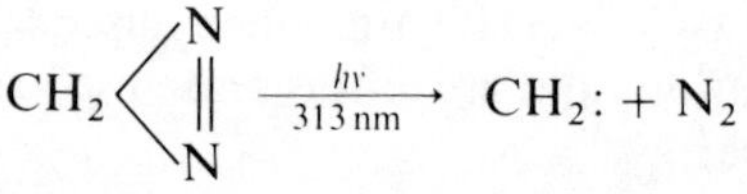

$$CH_2\begin{matrix}N\\\|\\N\end{matrix} \xrightarrow{h\nu}_{313\,nm} CH_2: + N_2$$

Cyclic azo compounds such as 3-acetoxy-3,5,5-trimethyl-1-pyrazoline are used in photopolymerization of vinyl monomers.[216]

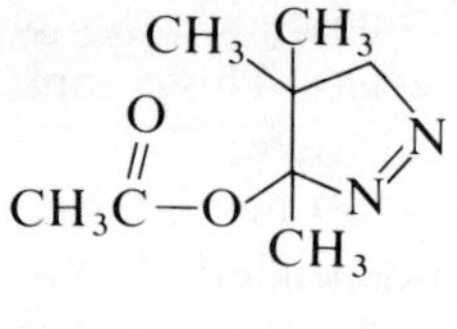

$$CH_3C(=O)-O-\underset{CH_3}{\underset{|}{C}}(CH_3)(CH_3)-N=N-$$

Pyrazolines can be used as precursors to cyclopropanes,[217] and pyrazoles may be employed as precursors to cyclopropenes.[213]

pyrazoline $\begin{bmatrix}N\\ \|\\ N\end{bmatrix} \xrightarrow{h\nu} \triangle + N_2\uparrow$ cyclopropane formation

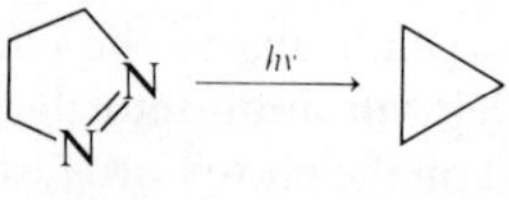

pyrazole $\begin{bmatrix}N\\ \|\\ N\end{bmatrix} \xrightarrow{h\nu} \triangle + N_2\uparrow$ cyclopropene formation

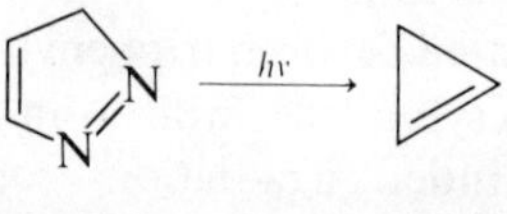

Dimethylamino radicals result on photolysis of tetramethyltetrazene:[219]

$$(CH_3)_2N-N=N-N(CH_3)_2 \xrightarrow{h\nu} 2(CH_3)_2N\cdot + N_2\uparrow$$

Probably one of the most useful compounds is α-*bis*-1-cyclohexane-carbonitrile. It absorbs in the UV region strongly and does not decompose thermally at moderate temperatures.[220–221]

Azides containing the grouping N_3- form nitrenes by nitrogen loss when irradiated. These are thought to have a triplet ground state.[222] Rearrangement and abstraction mechanisms are the mode of action for these compounds as depicted below:[223–225]

3.5 AIR INHIBITION OF PHOTOPOLYMERIZATION

Most photopolymerization reactions are performed in air and this can prove detrimental to cure performance. The oxygen molecule may be regarded as a free radical. The ground electronic configuration is a triplet energy state. Competition may occur between the photo-initiator radicals and oxygen for reactive monomer sites. The competition rate is largely governed by the concentrations of the components. At higher oxygen levels ($\sim 10^{-3}$ M) the reaction of the photo-initiator radicals with oxygen predominates that with the monomer. The oxygen effect may be limited by decreasing the amount of oxygen in the film or conversely increasing the monomer reactivity.

The energy level diagram for oxygen is shown in Figure 56,[226–227] in comparison with benzophenone and naphthalene.

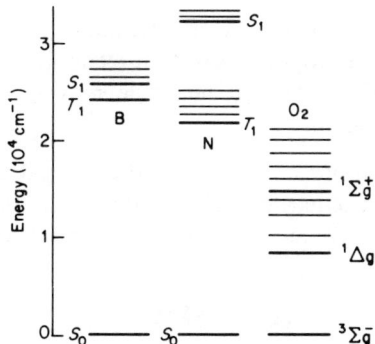

Figure 56 Energy levels of the oxygen molecule in comparison to benzophenone (B) and naphthalene (N).[9] (Reproduced by permission of the Oil & Colour Chemists' Association.)

The photochemistry of the oxygen molecule involves several of the states from above.

Some absorption bands can be seen in sunlight. These are called the Fraunhofer bands and lie as very weak absorption at 759–765 nm and 687–692 nm. They exist owing to a forbidden transition involving changes in both multiplicity and symmetry:

$$O_2(^3\Sigma_g^-) \xrightarrow{h\nu} O_2(^1\Sigma_g^+)$$

Another transition occurs about 245 nm:

$$O_2(^3\Sigma_g^-) \xrightarrow{h\nu} O_2(^1\Delta_g) \text{ and } O_2(^1\Sigma_u^+)$$

and is again weak owing to rule violation.

In the short UV range the Schumann–Runge region, at 176–195 nm, the mechanism below is noted:

$$O_2(^3\Sigma_g^-) \xrightarrow{h\nu} O_2(^3\Sigma_u^-) \to O_2(^3\Pi_u) \to 2O(^3P)$$

In the Schumann–Runge region the bands converge to a continuum at 176 nm where $O(^1D)$ and ground $O(^3P)$ atoms are formed.

$$O_2(^3\Sigma_g^-) \xrightarrow{h\nu} [O_2(^3\Sigma_u^-)] \to O(^1D) + O(^3P)$$

It is in this region that oxygen absorbs the 184.9 nm radiation from the mercury arc lamp to form ozone gas.

At wavelengths below 134 nm,

$$O_2(^3\Sigma_g^-) \xrightarrow{h\nu} O(^2P) + O(^1S)$$

Below 92.3 nm,

$$O_2(^3\Sigma_g^-) \xrightarrow{h\nu} 2O(^1S)$$

For photo-oxidation reactions, the metastable singlet state of molecular oxygen $O_2(^1\Sigma_g^+)$ might be involved, although in other cases the less energetic $(^1\Delta_g)$ state may be active.[228]

Below 184.9 nm the absorption characteristics of oxygen are of no major significance to current photopolymerization technology.

The ground-state triplet oxygen molecule may interact in photopolymer systems as shown below:[229]

(a) It forms charge-transfer complexes in the ground state, which on photo-excitation create excited complexes that are able to initiate photopolymerization. Styrene is probably photopolymerized in this manner.[230–231]

$$S_t + O_2 \to (S_t\text{---}O_2) \xrightarrow{h\nu} (S_t\text{---}O_2)^* \xrightarrow{S_t} \text{polystyrene}$$

(b) 'Oxyplex' formation. These are complexes formed with the excited state of a sensitizer or chromophore in the system. Within the lifetime of the complex, processes such as those below usually happen, so that frequently the final result is quenching of the excited state and suppression of photopolymerization. In the vinyl carbazole/carbon tetrabromide system, in which photopolymerization occurs via a simultaneous cationic polymerization of VCZ via the VCZ cation radical formed from a VCZ–CBr$_4$ exciplex and radical addition polymerization, the presence of oxygen completely suppresses polymerization, resulting in cyclodimer formation as suggested below[9] in polar solvents:[230,232–233]

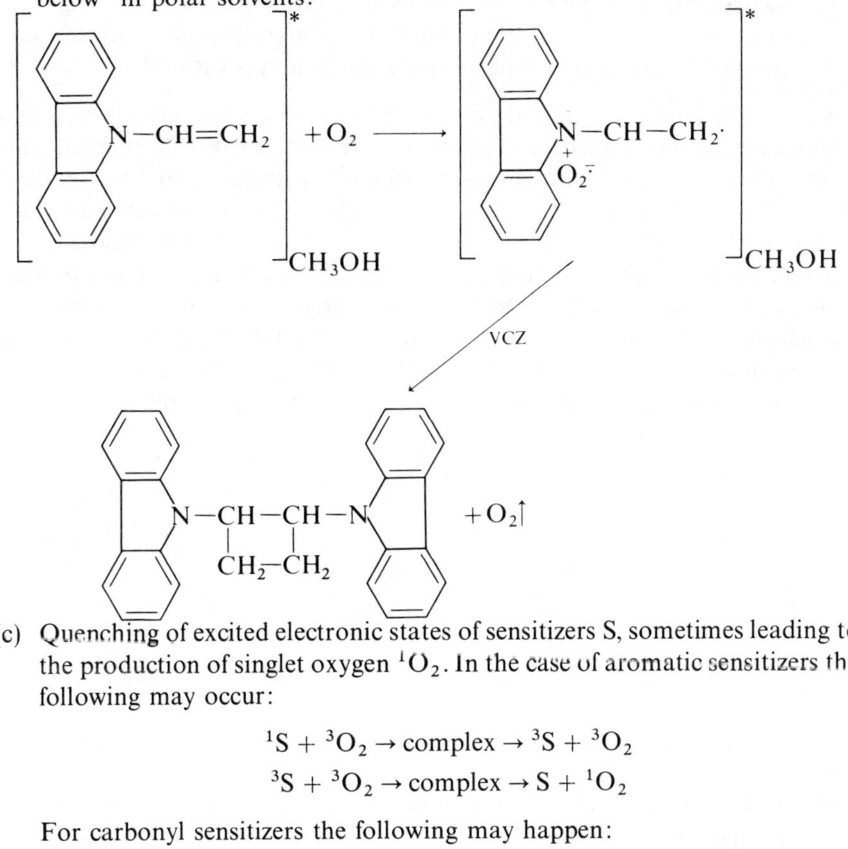

(c) Quenching of excited electronic states of sensitizers S, sometimes leading to the production of singlet oxygen $^1O_2$. In the case of aromatic sensitizers the following may occur:

$$^1S + {}^3O_2 \rightarrow \text{complex} \rightarrow {}^3S + {}^3O_2$$
$$^3S + {}^3O_2 \rightarrow \text{complex} \rightarrow S + {}^1O_2$$

For carbonyl sensitizers the following may happen:

$$^3S + {}^3O_2 \rightarrow \text{complex} \rightarrow S + {}^1O_2$$

Other reactions are possible, all subject to the Wigner spin-conservation law.
(d) Scavenging the radical species in radical polymerization, and altering the polymerization mechanism. This may be depicted as:

$$R\cdot + O_2 \rightarrow ROO\cdot$$
$$ROO\cdot + RH \rightarrow ROOH + R\cdot \quad \text{etc.}$$

There are several methods of achieving a reduction of oxygen inhibition, which include:

1. Performing the photopolymerization under inert blanketing that is an inert atmosphere (nitrogen, argon, etc.).

2. Application of a surface/air barrier layer (paraffin or filler materials).
3. Using resins possessing alkyl groups. These react rapidly with oxygen. Oxygen then has less probability of reacting with the photo-initiator.
4. Using more powerful-intensity lamps.
5. Incorporating faster-reacting monomers.
6. Improving the degree of surface cure:
   (a) incorporate more photo-initiator to alter surface/through-cure balance;
   (b) incorporate synergists or co-initiators to accelerate the surface cure. Amine/benzophenone is particularly useful in this case.[233]

As acrylates have fast cure response, the difference between through-cure and surface cure resulting from oxygen inhibition is notable. In order to eliminate this adverse effect of oxygen, the photo-initiator concentration and hence radical density should be increased so as not to lose the bonus of fast cure rate. This oxygen inhibition is discussed by Berner et al.[234] and is outlined below.

Benzophenone/amine combination allows good surface cure relative to bulk through-cure. Aliphatic amines tend to lower the film hardness, probably as a plasticizing effect results, or they function as chain transfer agents, causing a decrease in average chain length and hence a decrease in the hardness.

A possible mechanism for amines in supressing oxygen inhibition is:

$$R\cdot + -\overset{H}{\underset{|}{C}}-NR_2 \longrightarrow RH + \cdot\overset{|}{\underset{|}{C}}-NR_2$$

$$\cdot\overset{|}{\underset{|}{C}}-NR_2 + O_2 \longrightarrow \overset{\cdot O_2}{\underset{\wedge}{C}}-NR_2$$

$$-\overset{O_2\cdot}{\underset{|}{C}}-NR_2 + -\overset{H}{\underset{|}{C}}-NR_2 \longrightarrow -\overset{O_2H}{\underset{|}{C}}-NR_2 + -\overset{\cdot}{\underset{|}{C}}-NR_2$$

The role of amines appears to be a reduction of dissolved oxygen by a radical chain process.

Owing to the reaction of oxygen with propagating radicals, to produce peroxy radicals, oxygen concentration is the highest at the coating/air surface. Probably, oxygen concentration is inversely related to double-bond conversion. An increase in photo-initiator concentration increases double-bond conversion beneath the surface but decreases substrate/coating interface double-bond conversion.

3.6 REFERENCES

1. Ledwith, A., *J.O.C.C.A.*, **59**, 157–165, (1976).
2. Heine, H. G., Rosenkranz, H. J., and Rudolph, H., *Angew. Chem. Internat. Edit.*, **Vol. 11** (11), 974–978, (1972).
3. Tazuke, S., and Okamura, S., *J. Polym. Sci.*, Part B., **6**, 173, and Part A-1, **6**, 2907, (1968).
4. Irie, M., Yamamoto, Y., and Hayashi, K., *J. Macromol. Sci. Chem.* **A9**(5), 817, (1975).
5. Tamura, H., Sakaue, K., Tanaka, M., and Murato, N., *J. Chem. Soc. Japan, Ind. Chem. Sect.*, **72**, 304, (1969).

6. Kodaini, T., and Hayashi, K., *J. Polym. Sci., Part* , (*Polym. Letters*), **9**, 907, (1971).
7. Schlessinger, S. I., *U.S. Pat. 3, 708, 296*, (1973).
8. Ledwith, A., *The Exciplex*, (Eds. Gordon, M., & Ware, W. R.), London: Academic Press, (1975).
9. Phillips, D., *J.O.C.C.A.*, **59**, 202–207, 1976.
10. Davis, G. A., Carapellucci, D. A., Szoc, K., and Gresser, J. D., *J. Amer. Chem. Soc.*, **91**, 2264, (1969).
11. Cohen, S. G., and Parsons, G. H., *J. Amer. Chem. Soc.*, **92**, 7603, (1970).
12. Sander, M. R., Osborn, C. L., and Trecker, D. J., *J. Polymer Sci. Part A-1, Polymer. Chem.* **10**, 3173, (1972).
13. Cohen, S. G., Parola, A., and Parsons, G. H., *Chem. Rev.* **73**, 141, (1973).
14. Sandner, M. R., Osborn, C. L., and Trecker, D. J., *J. Polymer Sci.*, A-1, **10**, 3173, (1972).
15. Caldwell, R. A., and Gajewski, R. P., *J. Amer. Chem. Soc.*, **93**, 532, (1971).
16. Cundall, R. B., and Gilbert, A., *Photochemistry*, Thomas Nelson & Sons Ltd., London, (1970).
17. Chimmayanandam, R. B., and Melville, H. W., *Trans. Faraday Soc.*, **50**, 73, (1954).
18. Crivello, J. V., *UV Curing: Science & Technology*, (Ed. S. Peter Pappas), pp. 24–77, Technology Marketing Corporation, (1978).
19. Delzenne, G. A., *Revs. in Polym. Technology*, **Vol. 1**, p. 185, Marcel Dekker, New York, (1972).
20. Kosar, J., *Light Sensitive Systems*, John Wiley, (1965).
21. Bamford, C. H., and Dewar, M. J. S., *Proc. Roy. Soc. (London)*, **A.192**, 309, (1948); *Faraday Soc. Discussions* **2**, 310, (1947).
22. Osborn, C. L., *J. Radiation Curing*, 3 (3), 2, (1976).
23. Gaylor, N. G., *Soc. Plast. Eng., Regional Tech. Conf., Ellenville*, New York, Nov. 6–7, p. 32, (1967).
24. Nickerson, R., Sr., *Industrial Finishing*, p. 10, (Feb. 1974).
25. Tazuki, S., Asai, M., Ikeda, S., and Okamura, S., *J. Polym. Sci.*, Pt. C, 453, (1967).
26. Asai, M., and Tazuki, S., *Macromolecules*, **6**(6), 818, (1973).
27. Asai, M., Tazuki, S., and Okamura, S., *J. Polym. Sci.*, **12**(45), 56 (1974).
28. Asai, M., Takeda, Y., Tazuki, S., and Okamura, S., *Polymer. J.* (Japan), **1**(3), 359 (1975).
29. Kaeriyama, K., and Shimura, Y., *J. Polym. Sci.*, **10**, 2833, (1972).
30. Marek, M., and Toman, L, *J. Polymer Sci.*, Symposium No. 42, 339, (1973).
31. Toman, L., Marek, M., and Joki, J., *J. Polymer Sci.*, **12**, 1897, (1974).
32. Irie, M., Yamamoto, and Hayashi, K., *J. Macromol. Sci. Chem.*, **A9**(5), 817, (1975).
33. Tazuki, S., Asai, M., Ikada, S., and Okamura, S., *J. Polymer Sci.*, Pt. B, **5**, 453, (1967).
34. Sakamoto, M., Hayashi, K., and Okamura, S., *J. Polym. Sci.*, Pt. B, **3**, 205, (1965).
35. Roe, A., *Org. Reactions*, **5**, 193–228, (1949).
36. Saunders, K. H., *The Aromatic Diazo Compounds*, (2nd Edn,) E. Arnold & Co., London, (1949).
37. Calvert, J., and Pitts, J. N., *Photochemistry*, John Wiley & Sons, Inc., New York, p. 473, (1966).
38. Gamble, A. A., *J.O.C.C.A.*, **59**, 240–244, (1976).
39. Bikales, N. M., *Encyc. of Polym. Sci & Tech.*, Vol. 6, p. 224, John Wiley & Sons, New York, (1967).
40. Schlessinger, S. I., *Photogr. Sci., & Eng.*, **18**(4), 387, (1974).
41. Schlessinger, S. I., *Polym. Eng., & Sci.*, **14**(7), 513 (1974).
42. Casserio, M. C., Glusker, D. L., and Roberts, J. D., *J. Am. Chem. Soc.*, **81**, 336, (1959).
43. Crivello, J. V., and Lam, J. H. W., *4th International Symposium on Cationic Polym.*, Akron, Ohio, June 24, 1976; *J. Polymer Sci.*, Symposium No. 56, 1–11, (1976).
44. Crivello, J. V., and Lam, J. H. W., *Macromolecules*, **10**(6), 1307–1315, (1977).
45. Crivello, J. V., *U.S. Patent 3,981,897*, 21 Sept., (1976).

46. Smith, G. H., *Belg., Patent 828, 841*, (1975).
47. I.C.I. – *Belgium Patent 837,782*, 22 June, (1976).
48. May, C. B., and Tanaka, Y., *Epoxy Resin Chemistry and Technology*, Marcel Dekker, Inc., New York, pp. 199–205, (1975).
49. Billmeyer, F. W., Jr., *Textbook of Polymer Science*, Interscience Pub., New York, p. 294, (1964).
50. Crivello, J. V., Lam, J. H. W., and Volante, C. N., *J. Rad. Curing*, **4**(3), 2, (1977).
51. Odian, G., *Principles of Polymerization*, McGraw-Hill Pub. Corp., New York, p. 321, (1970).
52. Crivello, J. V., and Schroeter, S. H., *U.S. Patent 4,026,705*, 31 May, (1977).
53. Crivello, J. V., Lam, J. H. W., and Volante, C. N., *ACS Mtg; Chicago III, Ctgs., & Plast., Preprints*, **37**(2), 4, 28 Aug.–2 Sept., (1977).
54. Sandin, R. B., and Hay, A. S., *J. Am. Chem. Soc.*, **74**, 274, (1952).
55. Crivello, J. V., *Belg. Patent, 828,670*, 2 May, (1974).
56. I.C.I., *Belgium Patent 833,472*, 16 March, (1976).
57. Murov, S. L., *Handbook of Photochemistry*, Marcel Dekker, Inc., New York, p. 19, (1973).
58. Leicester, H. M., and Bergstrom, F. W., *J. Am., Chem. Soc.*, **51**, 3587, (1929).
59. Closs, G. L., and Paulson, D. R., *J. Amer. Chem. Soc.*, **92**, 7229, (1970).
60. Ledwith, A., Russell, P. J., and Sutcliffe, L. H., *J. Chem. Soc.*, Perkin II, (1925, 1972).
61. Heine, H. G., Rosenkranz, H. J., and Radolph, H., *Angew, Chem., Int. Edit.*, **11**, 974, (1972).
62. Heine, H. G., *Tetrahedron Lett.*, 4755, (1972).
63. Heine, H. G., and Traenckner, H. J., *Progr. Org. Coatings*, **3**(2), 115, (1975).
64. Pappas, S. P., and Chattopadhyay, *J. Amer. Chem. Soc.*, **95**, 6484, (1973).
65. Solly, R. K., and Benson, S. W., *J. Amer. Chem. Soc.*, **93**, 1592, (1971).
66. Pappas, S. P., and Chattopadhyay, A. K., *J. Polymer Sci., Polymer Letters* Edition, **13**, 483, (1975).
67. Ogata, Y., and Sawaki, Y., *J. Org. Chem.*, **41**, 373, (1976).
68. Sandner, M. R., and Osborn, C. L., *Tetrahedron Letters*, 415, (1974).
69. Brueniskolz, J., and Kirchmayr, R., *Ger. Offen. 2,337,813*, (1974); *Brit. Pat. 1,390,006* (1976).
70. Berner, G., Kirchmayr, R., and Rist, G., *J.O.C.C.A.*, **61**, 105–113, (1978).
71. Hamity, M., and Scaiano, J. C., *J. Photochem.*, **4**, 229, (1975).
72. Russell, K. E., and Toboloski, A. V., *J. Amer. Chem. Soc.*, **76**, 395, (1954).
73. Osborn, C. L., and Watson, S. L., *Abstracts, 9th Central Regional Meeting of the Amer. Chem. Soc.*, Charleston, West Virginia, Poly 21, p. 75, 12–14 Oct., (1977).
74. *Ger. 1922627; 2317846; U.S. 382 7596-60*.
75. Braun, D., and Becker, K. H., *Makromolekulare Chem.*, **147**, 91, (1971).
76. Neckars, D. C, *Mechanistic Organic Photochemistry*, New York; Reinhold, (1967).
77. Porter, G., and Suppan, P., *IUPAC Symposium on Organic Photochemistry*, Strassbourg, 499, (1964).
78. Gibian, M. J., *Tetrahedron Lett.*, 5331, (1967).
79. Filipescu, N., and Minn, F. L., *J. Amer. Chem. Soc.*, **90**, 1544, (1968).
80. Weiner, S. A., *J. Amer. Chem. Soc.*, **93**, 425, (1971).
81. Pitts, J. N., Letsinger, R. L., Taylor, R. P., Patterson, J. M., Recktenwald, G., and Martin, R. B., *J. Amer. Chem. Soc.*, **81**, 1068, (1959).
82. Wamser, C. C., Hammond, G. S., Chang, C. T., and Bayler, C., Jr., *J. Amer. Chem. Soc.*, **92**, 6362, (1970).
83. Sander, M. R., Osborn, C. L., and Trecker, D. J., *J. Polymer. Sci.*, Part A-1, *Polym. Chem.* **10**, 3173, (1972).
84. Schuster, D. I., and Goldstein, M. D., *J. Amer. Chem. Soc.*, **95**, 986, (1973).
85. Suppan, P., and Porter, G., *Trans. Faraday Soc.*, **61**, 1664, (1965).
86. Koch, T. H., and Jones, A. H., *J. Amer. Chem. Soc.*, **92**, 7503, (1970).
87. McGinniss, V. D., and Dusek, D. M., *Polymer Preprints.*, **15**, 480, (1974).

88. Delzenne, G. A., Lariden, U., and Peters, H., *Europ. Polym. J.*, **6**, 933, (1970).
89. Heine, H. G., and Traenckner, H. J., *Progr. Org. Coatings*, **3(2)**, 115, (1975).
90. Ward Blenkinsop & Co., Ltd., *Technical Bulletin*.
91. Davis, M. J., Doherty, J., Godfrey, A. A., Green, P. N., Young, J. R. A., and Parrish, M. A., *J.O.C.C.A.*, **61**, 256–263, (1978).
92. Yoshitiara, K., and Kearns, D. R., *J. Chem. Phy.*, **45**, 1991, (1965).
93. Singer, L. A., *Tetrahedron Lett.*, 923, (1969).
94. Davidson, R. S., and Lambeth, P. F., *Chem. Commun.*, 1098, (1969).
95. Guttenplan, J. B., and Cohen, S. G., *Tetrahedron Lett.*, 2125, (1969).
96. Davis, G. A., Carapellucci, D. A., Szoc, K., and Gresser, J. D., *J. Amer. Chem. Soc.*, **91**, 2264, (1969).
97. Cohen, S. G., and Parsons, G. H., *J. Amer. Chem. Soc.*, **92**, 7603, (1970).
98. Caldwell, R. A., and Gajewski, R. P., *J. Amer. Chem. Soc.*, **93**, 532, (1971).
99. Ledwith, A., Bosley, J. A., and Parbrick, M. D., *J.O.C.C.A.*, **61**, 95–104, (1978).
100. Wilkinson, F., *J. Phys. Chem.*, **66**, 2569, (1962).
101. Tickle, K., and Wilkinson, F., *Trans. Faraday Soc.*, **61**, 1981, (1965).
102. Neely, W. C., and Dearman, H. H., *J. Chem. Phys.*, **44**, 1302, (1966).
103. Ledwith, A., Ndaalio, G., and Taylor, A. R., *Macromolecules*, **8**, 1, (1975).
104. Rubin, M. B., and Zwitkowits, P., *J. Org. Chem.*, **29**, 2362, (1964); *Tetrahedron Lett.*, 2453, (1965).
105. Clark, K. P., and Stonehill, H. I., *Journal of the Chem. Soc., Faraday Transactions I*, **68**, 577–590, and 1676–1686, (1972).
106. Anwarrudin, Q., and Santappa, M., *J. Polymer. Sci.*, Part A-1, **7**, 1315, (1969).
107. Morton, T. H., *J. Soc. Dyers & Col.*, **65**, 597, (1949).
108. Oster, G. K., Oster, G., and Prati, G., *J. Amer. Chem. Soc.*, **79**, 595, (1957).
109. Meier, H., *The Chemistry of Synthetic Dyes*, (Ed. K. V. Venkataraman), Vol. 4, New York Academic Press, p. 389, (1971).
110. Kinstle, J. F., *Journal Radiation Curing*, **1**(2), April, (1974).
111. Nagabhushanam, T., and Santappa, M., *J. Polym. Sci.*, Part A-1, **10**, 1511, (1972).
112. Rust, J. B., Miller, L. J., and Margerum, J. D., *Polym. Eng. Sci.*, **9**, 40, (1969).
113. Schenk, G. O., *Liebigs, Annalen*, **584**, 125, 156, 177, 199, 221, (1953).
114. Harper, D. J., and McKellar, J. F., *Chem. & Ind.*, 848, (1972).
115. Carlsson, D. J., Sproule, D. E, and Wiles, D. M., *Macromolecules*, **5**, 569, (1972).
116. Foote, C. S., *Accts. Chem. Res.*, **1**, 104, (1968).
117. Roffey, C. G., *The Fading of Coloured Paint Films*, Paint Research Association, RS/T/70/71, (Oct. 1971).
118. Gates, A. P., and Patterson, D., *J. Oil Col. Chem. Assocn.*, **50**, 1008–1022, (1967).
119. Pappas, S. P., and Kühhirt, J., *Paint Technology*, **47**,(610), 42, (1975).
120. McIntyre, N. S., Thompson, K. R., & Weltner, W., Jr., *J. Phys. Chem.*, **75**, 3243, (1971).
121. Addis, R.,R., Jr., Ghoch, A. K., and Wakim, F. G., *Appl. Phys. Lett.*, **12**, 397, (1968).
122. Allen, N. S., McKellar, J. F., Phillips, G. O., and Wood, D. G. M., *J. Polymer Sci. Polymer Lett. Ed.*, **12**, 241, (1974).
123. Wells, A. F., *Structural Inorganic Chemistry*, Oxford University Press, (1950).
124. Morton, T. H., *J. Soc. Dyers & Col.*, **65**, 597, (1949).
125. Vesce, V. C., *Official Digest*, **31**, No. 419, Pt. 2, 1–143, (1959).
126. Giles, C. H., and McKay, R., *Textile Research, J.*, **33**, 527, (1963).
127. Maikowski, M. A., *Farbe und Lack*, **77**(7), 640, (1971).
128. ICI Dyestuffs Ltd., *Pigments for Paint*, (1963).
129. Desai, M. F., and Giles, C. H., *J. Soc. Dyers & Col.*, **65**, 639.
130. Atherton, E., and Peters, R. H., *J. Soc. Dyers & Col.*, **68**, 64, (1952).
131. Egerton, G. S., *Br. Polymer J.*, **3**, (1971).
132. Kawaoka, K., Khan, A. U., and Kearns, D. R., *J. Chem. Phys.*, **46**, 1842, (1967).
133. Egerton, G. S., and Morgan, A. G., *J. Soc. Dyers & Col.*, 268–277, August (1971).
134. Egerton, G. S., and Asaad, *N.E.N.*, **86**, 203, (1970).

135. Weiss, *Naturwiss*, **35**, 610, (1935).
136. Mullikan, and Stevens, *Phys. Rev.*, **44**, 720, (1933).
137. Uri, N., *Chem. Rev.*, **50**, 375, (1952).
138. Markham, Sister Maria Clare, Paper presented at the 280th Meeting of the *NEACT*, Mt. Holyoke College, 23 April, (1955). (Later published in *J. Chem. Education*, 540–543, Oct. (1955).
139. Hird, M. J., *J. Coatings Technology*, **48**(620), (1976).
140. Hulme, B. E., *Paint Manufacture*, 12–16, May (1975).
141. Schwarz, E., *Proc. Phys. Soc.*, **A62**, 530, (1949).
142. Earle, M. D., *Phys. Revs.*, **61**, 56, (1942).
143. Chari and Qureshi, *J. Indian Chem. Soc.*, **21**, 97, (1944).
144. Goodeve, C. F., and Cohn, G., *Trans. Faraday Soc.*, **36**, 433–40, (1940).
145. Hird, M. J., *J. Coatings Technology*, **48**(620), 75 (1976).
146. Kuriacose, J. C., and Markham, M. C., *J. Catalysis*, **1**, 498, (1962).
147. Cundall, R. B., *J.O.C.C.A.*, **59**, 95–101, (1976).
148. Pappas, S. P., and Fischer, R. M., *J. Paint Tech.* **46**, 65, (1974).
149. Egerton, T. A., and King, C. J., *J.O.C.C.A.*, **62**, 386–391, (1979).
150. Cundall, R. B., Hulme, B., Rudham, R., and Salim, M. S., *J.O.C.C.A.*, **61**, 351.
151. Cundall, R. B., Rudham, R., and Salim, M. S., *J.C.S. Faraday I.*, **72**, 1642, (1976).
152. Bickley, R. I., and Stone, F. S., *J. Catalysis*, **31**, 389, (1973).
153. Hughes, W. H., *Xth FATIPEC Congress*, Verlag Chemie, Weinheim, 67, (1970).
154. Kämpf, G., and Völz, H. G., *Farbe und Lack*, **74**, 37, (1968).
155. Egerton, G. S., *J. Soc. Dyers & Col.*, **65**, 764, (1949).
156. Völtz, H. G., Kämpf, G., and Fitzky, H. G., *Xth FATIPEC Congress Book*, pp. 107–113, (1970).
157. Bevington, J. C., *Proc. Roy. Soc.*, **A239**, 420, (1957).
158. Bevington, J. C., and Lewis, T. D., *Trans. Farad. Soc.* **54**, 1340, (1958).
159. Bevington, J. C., and Brooks, C. S., *J. Polymer Sci.*, **22**, 257, (1958).
160. SenGupta, P. K., and Bevington, J. C., *Polymer*, **14**, 527, (1973).
161. Martin, J. T., and Norrish, R. G., *Proc. Roy. Soc.*, **A220**, 322, (1953).
162. Frey, H. M., *Proc. Chem. Soc.*, 385, (1959).
163. Roffey, C. G., Ph.D. Thesis, Chelsea College, London University, (1970).
164. Franklin, J. L., and Lumpkin, H. E., *J. Am. Chem. Soc.*, **74**, 1023–1026, (1952).
165. Richards, L. M., *U.S. Pat. 2,460,105*, (1949).
166. Russel, K. E., and Tokolsky, A. V., *J. Am. Chem. Soc.*, **76**, 395–399, (1954).
167. Kern, R. J., *U.S. Pat. 2,773,822*, (1956).
168. Kern, R. J., *U.S. Pat. 2,861,934*, (1958).
169. McCloskey, C. M., and Bond, J., *Ind. Eng. Chem.*, **47**(2), 2125, (1955).
170. Du Pont de Nemours & Co., *U.S. Patent. 2,716,633*.
171. Okawara, M., and Nakai, T., *Internat. Symposium on Macrol. Chem.* Tokyo, Preprint IV-16, (1966).
172. Kern, R. J., *U.S. Pat. 2,738,319*, (1956).
173. Engelhard, V. A., and Peterson, M. L., *U.S. Pat. 2,716,633*, (1955).
174. Du Pont de Nemours & Co., *U.S. Patent 2,460,105*.
175. Otsu, T., *Macromol. Chem.* **27**, 1/2, 142, (1958).
176. Otsu, T., *J. Polymer Sci.*, **21**, 559, (1956).
177. Richards, L. M., *U.S. Pat. 2,423,520*, (1947).
178. Kern, R. J., *U.S. Pat. 2,861,933* (1958).
179. Gerhart, H. L., *U.S. Pat. 2,673,151*, (1954).
180. Kern, R. J., *J. Am. Chem. Soc.*, **77**, 1382–1383, (1955).
181. Ferington, T. E., and Tobolsky, A. V., *J. Am. Chem. Soc.*, **77**, 4510–4512, (1955).
182. Roth, C. B., *U.S. Pat. 3,147,116*, (1964).
183. Venkatarao, K., and Santappa, M., *J. Polym. Sci.*, Part A-1, **8**, 3429, (1970).
184. Venkatarao, K., and Santappa, M., *J. Polym. Sci.*, Part A-1, **5**, 637, (1967).
185. Higgins, C. E., and Baldwin, W. H., *J. Appl. Polym. Sci.*, **12**, 1471, (1968).

186. Imamura, K., Assai, M., Tazuke, S., and Okamura, S., *Makromol. Chem.* **174**, 91, (1973).
187. Natarajan, L. V., and Santappa, M., *J. Polym. Sci.*, Part A-1, **6**, 3245, (1968).
188. Natarajan, L. V., and Santappa, M., *J. Polym. Sci.* Part B-5, 357, (1967).
189. Kothandaramaun, H., Srinivasan, K. S. V., and Santappa, M., *J. Polym. Sci.*, Part A-1, **10**, 3685, (1972).
190. Penkett, S. A., and Adamson, A. W., *J. Amer. Chem. Soc.*, **87**, 2514, (1965).
191. Kothandaraman, H., and Santappa, M., *J. Polym. Sci.*, Part A-1, **9**, 1351, (1971).
192. Kaeriyama, K., and Shimura, Y., *Makromol. Chem.*, **167**, 129, (1973).
193. Bamford, C. H., and Mullik, S. U., *Polymer*, **14**, 38, (1973).
194. Bamford, C. H., and Mullik, S. U., *J. Chem. Soc., Faraday Transactions I*, **71**, 625–636, (1975).
195. Duncan, B., and Dunn, A. S., *J. Appl. Polymer Sci.*, **8**(4), 1763, (1964).
196. Philips, N. V., Gloeilampenfabrieken, *German Patent, 1,222,372*.
197. Powers, Chemico Inc., *Brit. Pat. 1,058,690*.
198. Philips, N. V., Gloeilampenfabrieken, *Fr. Pat. 1,450,338*.
199. Oster, G., *J. Polymer Sci.*, **48**, 321, (1960).
200. Schenek, G. O., *Angew Chem.*, **73**, 578, (1961).
201. Monsanto Chemical Co., *U.S. Pat. 2,769,777*.
202. Robert, G., and Jeffrey, T. J., *Chem. Ind.* (London), **47**, 1499, (1970).
203. Nishihara, K., and Sakota, N., *Makromol. Chem.* **165**, 105, (1973).
204. Tsunooka, M., Araki, S., Tanake, M., and Murata, N., *J. Chem. Soc. Japan, Ind. Chem. Sect.*, **72**, 284, (1969).
205. Tsunooka, M., Tanake, M., Kusube, M., and Murata, N., *J. Chem. Soc., Japan, Ind. Chem. Sect.*, **72**, 287, (1969).
206. Tsunooka, M., Fujii, M., Kusube, M., Tanake, M., and Murata, N., *J. Chem. Soc., Japan, Ind. Chem. Sect.*, **72**, 292, (1969).
207. Takeshi, Ogawa, and Takumi Taninaka, *J. Polym. Science*, Part A-1, **10**, 2005–2012, (1972).
208. Mao, T. G., and Eldred, R. J., *J. Polym. Sci.*, Part A-1, **5**, 1741, (1967).
209. Eldred, R. J., *J. Polym. Sci.*, Part A-1, **7**, 265, (1969).
210. Ogawa, T., and Taninaka, T., *J. Polym. Sci.*, Part A-1, **10**, 2005, (1972).
211. Minoura, Y., and Toshima, H., *J. Polym. Sci.*, Part A-1, **7**, 2837, (1969).
212. Minoura, Y., and Toshima, H., *J. Polym. Sci.*, Part A-1, **8**, 273, (1970).
213. Oster, G., and Yang, N., *Chem. Rev.*, **68**, 125, (1968).
214. Miyami, H., Harumiya, N., and Takeda, A., *J. Polym. Sci.*, Part A-1, **10**, 943, (1972).
215. Frey, H. M., *Adv. Photochem.*, **4**, 225, (1966).
216. Nakaya, T., Ikeda, H., and Imoto, M., *Makromol. Chem.*, **161**, 241, (1972).
217. Jacobs, T. L., in *Heterocyclic Compounds*, Vol. 5, (Ed. Elderfield, R.), Wiley, New York, (1957).
218. Closs, G. L., Boll, W. A., Heyn, H., and Dev, V., *J. Amer. Chem. Soc.*, **90**, 173, (1968).
219. Sugiyama, K., Nakaya, T., and Imoto, M., *J. Polym. Sci.*, Part A-1, **10**, 205, (1972).
220. Miyami, H., Harumiya, N., and Takedo, A., *J. Polym. Sci.*, Part A-1, **10**, 154, (1972).
221. Ogo, Y., Yokawa, M., and Imoto, T., *Makromol Chem.*, **171**, 123, (1973).
222. Hafner, K., Kaiser, W., and Puttner, *Tetrahedron Letter*, 3953, (1964).
223. Barton, D. H. R., and Morgan, L. R., Jr., *J. Chem. Soc.*, 622, (1962).
224. Barton, D. H. R., and Starrat, A. N., *J. Chem. Soc.*, 2444, (1965).
225. Moriarty, R. M., and Raliman, M., *Tetrahedron*, **21**, 2877, (1965).
226. Foote, Wexler, Ando, and Higgins, *J. Amer. Chem. Soc.*, **90**, 975, (1968).
227. Cundall, R. B., and Gilbert, A., *Photochemistry*, Thomas Nelson & Sons Ltd., London, (1970).
228. Kautsky, H., *Trans. Faraday Soc.*, **35**, 216, (1939).
229. Phillips, D., *J.O.C.C.A.*, **59**, 202–7, (1976).
230. Kodaira, T., Hayashi, K., and Ohnishi, T., *Polym. J.* **4**, 1, (1973).
231. Kodaira, T., and Hayashi, K., *J. Polym. Sci.*, Part B, *Polymer Letters*, **9**, 907, (1971).

232. Olaf, O. F., Breitenbath, T. W., and Kaufman, H. F., *J. Polym. Sci.*, Part B, *Polymer Letters*, **9**, 877, (1971).
233. Bartholemew, R. F., and Davidson, R. S., *J. Chem. Soc.*, (C), 2342, (1971).
234. Berner, G., Kirchmayr, R., and Rist, G., *J.O.C.C.A.*, **61**, 105–113, (1978).

4

Photopolymerizable film-forming materials

4.1 PHOTOPOLYMERIZATION

Photopolymerization can occur by two chief reaction paths involving either addition or condensation reactions:[1]

1. Photocondensation polymerization – a step growth mechanism.
2. Radical or ionic addition polymerization – a chain growth mechanism.

For 1, each step in the growth of a polymer is a photochemical reaction.

In 2, the absorption of photochemical energy and creation of a reactive species occurs in the initiation step. Polymer formation itself occurs by non-photochemical chain propagation reactions of free radicals, cations, or anions. Mechanism 2 is generally the more efficient and occurs mainly by free radicals as the most likely species of propagation.

For photo-induced chain polymerization the photochemistry occurs in the initiation step. A component of the irradiated mixture has to absorb actinic radiation energy in the region given out by the light source and consequently interaction or reaction of the absorbing molecule must lead to the formation of a primary reactive species. At some instance in this sequence a chemical bond must be severed and therefore the absorbed actinic radiation must supply adequate energy for the bond-breaking process.

Bond scission alone does not ensure that efficient initiation of polymerization will occur. The primary reactive species must react within a monomer to form a new reactive species capable of chain propagation. For efficient chain polymerization to occur, the primary reactive species has to be sufficiently stable relative to the initial reactants to allow its formation and also sufficiently unstable so that it reacts rapidly with a monomer to yield the chain propagating species. In photopolymerization reactions, the initiation step can proceed by one of several different mechanisms, including the two important ones below:

(a) Direct absorption followed by dissociation of a monomer is the simplest process. However, most monomers do not have significant absorption characteristics to match the light irradiation sources commonly in use. They are generally transparent to wavelengths longer than 254 nm.
(b) The presence of another vinyl or a carbonyl group, an aromatic ring or a vinylic bromide atom can shift the UV absorption into a usable range. These substituted monomers may function as photo-sensitizers or photo-initiators.

A survey of the literature[2] reveals that photo-initiated/sensitized photopolymerization reactions can fall into several main addition and condensation categories which are discussed in the following sections.

4.1.1 Photopolymerization of Compositions Possessing Vinyl Unsaturation

4.1.1.1 Mono- and Multifunctional Unsaturated Monomers

See section 4.3.3.

4.1.1.2 Polymers with Unsaturation

These include:

1. Unsaturated polyesters.[3]
2. Unsaturated polyvinyl alcohol derivatives[4-5] such as:
   (a) Polyvinylesters containing the grouping

$$\sim\sim\sim-\underset{\underset{H}{|}}{\overset{\overset{H}{|}}{C}}-\underset{\underset{O-\overset{\overset{}{\underset{\|}{C}}-R}{\|}}{|}}{\overset{\overset{H}{|}}{C}}-\sim\sim\sim$$

   $$O$$

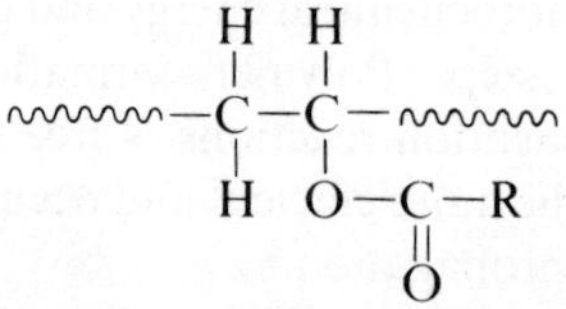

   (b) Polyvinyl acetals containing the grouping

$$-CH_2-CH_2-CH_2-CH-O-CH(R)-O$$

   or mixed ester-acetals with non-terminal polymerizable side groups.
   (c) Unsaturated polyamide type[6-7]

$$-CO-N(CH_2-O-CO-C(R')=CH_2)R''$$

   (d) Salt-forming polymers containing unsaturation with complementary polymerizable salt-forming monomers.[8] These are essentially polymerizable polymeric salts which on polymerization become resistant to non-polar solvents. Generally they are vinyl or diene addition polymers possessing either acid groups such as carboxyl, sulphonic, sulphate, phosphate, etc., or primary, secondary, or tertiary amine groups with the complementary photopolymerizable monomer components having the respective balancing amino or acids groups.

(e) Epoxide/acrylic or methacrylic acid reaction products.[9] They may be epoxides from soya bean oil, corn oil, or esters of tall oil fatty acids reacted with an ethylenically unsaturated carboxylic acid such as acrylic, methacrylic, α-halogenated acrylics, α-aryl acrylic, etc.

4.1.2 Photo-crosslinking and Photopolymerization of Saturated Polymers

4.1.2.1 Vinyl Polymerization of Monomers with Simultaneous Chain-transfer Reaction with Saturated Polymer

Abstraction of hydrogen or halogen atoms by growing chains of monomer units involving chain transfer with the saturated polymer creates active sites that can initiate block copolymerization by transfer in the chain.[10-11]

4.1.2.2 Purpose-modified Saturated Polymers

Polymers may be prepared containing functional groups that are particularly susceptible to actinic radiation and which may be used to form graft and block copolymers. There are several important types, which include:

1. *Halogenated polymers possessing carbon–halogen bonds.* Polystyrene chains with terminal Br groups on irradiation with suitable actinic radiation in the presence of vinyl monomers produces Br. free radicals which initiate homopolymerization whilst the polymeric radicals form linear block polymers. CCl_3 groups on the polystyrene do not photolyse in this environment, whereas CBr_3 groups undergo further photolysis causing branched chains.

   Metal carbonyls such as $Mo(CO)_6$ and $Mn_2(CO)_{10}$ in the presence of suitable organic halides such as CCl_4, $CHCl_3$, CH_2Cl_2, CCl_3COOH, CCl_3CN, or bromine components are useful photoactivators for these systems.[13-14] $Mn_2(CO)_{10}$ is the preferred carbonyl. The initiating reaction is thought to involve the abstraction of a halogen atom as an ion from the halide molecule CX_4 and causes an increase in the oxidation number of the metal atom by unity.[15] The CX_3 radicals are then the initiating species. Polymer chains are formed with terminal halide groups proceeding from the growing monomer radical with a CX_3 end group,

$$CX_4 + (n+1)M \xrightarrow[\text{carbonyl}]{\text{Metal}} CX_3(M)_n M \cdot (X = Br \text{ or } Cl) + X \cdot$$

$$2CX_3(M)_n M \cdot \longrightarrow CX_3(M)_{2n+2}CX_3$$

   where X is a halogen atom. These polymers can be used for block copolymer preparation.[16]

   Prepolymers containing halide groups such as polyvinyl trichloroacetate $(Cl_3CCO-O-CH=CH_2)_n$ can be used to prepare graft polymers. Initiation of polymerization occurs via a radical from the chlorinated polymer and it is likely that substantially no homopolymer is formed.

When a prepolymer molecule such as polyvinyl trichloroacetate contains more than two halide groups, networks may be formed. This network formation is depicted below:

$$RCCl_3 \xrightarrow{\text{Monomer + }}_{\text{Metal carbonyl}} RCCl_2M_n$$

$$RCCl_2M_n + RCCl_2M_m \longrightarrow RCCl_2-M_{n+m}-Cl_2CR$$

CCl_2, CBr_3, and CBr_2 are other suitable halide groups. If the termination reaction is completely disproportionate, crosslinking will not occur and the final polymer will be graft.[17]

2. *Sulphur-containing polymers*. Photo-responsive sulphide end groups can be incorporated into polymers by polymerizing a monomer in the presence of certain sulphides such as methyl dithiocarbamate, *tetra*-ethylthiuram disulphide.[18-21]

3. *Keto-polymers*. Actinic irradiation of polymers and copolymers based on vinyl ketones gives free-radical grafting sites in the polymer chain.[22] Alternatively, when the ketone groups are incorporated as a part of the main chain of the polymer, photolytic cleavage gives radicals which form block copolymers by initiating the polymerization of a vinyl monomer in the absence of extensive chain transfer.[23] A typical mechanism could be:

$$R-O-\overset{O}{\underset{\|}{C}}-CH_2CH_2COCH_2CH_2-\overset{O}{\underset{\|}{C}}-O-R \quad \text{(keto polymer)}$$

$$\downarrow h\nu(x > 300\,nm)$$

$$R-O-\overset{O}{\underset{\|}{C}}-CH_2-CH_2\cdot \quad + \quad \cdot CCH_2CH_2-\overset{O}{\underset{\|}{C}}-O-R$$

4. *Polymers possessing linearly-linked iso- or hetero-cyclic groups*. Groups used include anthracene, benzanthrene, acridine, phenazine linked to a polyvinyl alcohol to produce a photo-responsive acetal. These type of polymers have been used as photo-sensitizers in the polymer printing plate industry.[24]

5. *Polymers incorporating dye groups*. Dyes such as eosin, safranine, acridine can become incorporated into the polymer as a leuco derivative.[25-27]

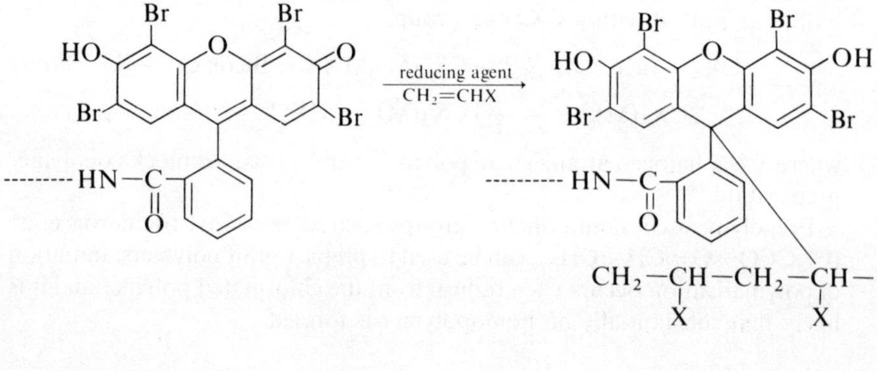

6. *Polymers possessing 2,2-dimethyl-1,3-dioxolane groups.* Ultraviolet radiation crosslinks polymers of this type as below:[28-30]

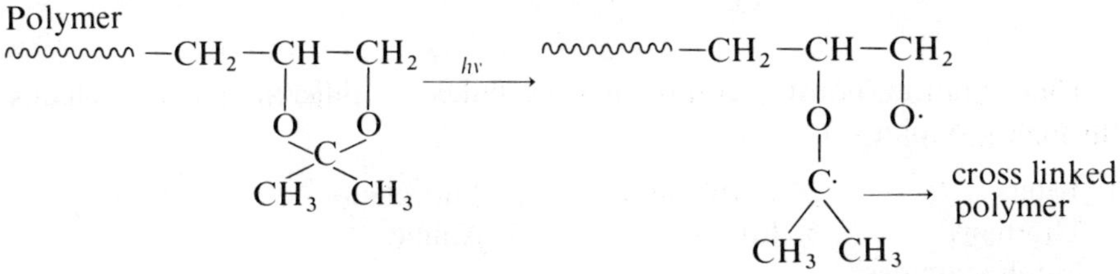

Polymer

$$\sim\sim\sim-CH_2-CH-CH_2 \xrightarrow{h\nu} \sim\sim\sim-CH_2-CH-CH_2$$

with dioxolane ring (O-C(CH$_3$)$_2$-O) opening to give O· and O-C·(CH$_3$)$_2$ radicals → cross linked polymer

4.1.2.3 Photopolymerization and Crosslinking Induced by Photo-sensitizers

See Chapter 3 concerning commercial photo-initiators/sensitizers.

4.1.3 Alternative Photo-induced Crosslinking Reactions to Vinyl Addition Polymerization

Light-sensitive photopolymers can be macromolecules which may become crosslinked by the formation of interchain bonds under the influence of actinic radiation.

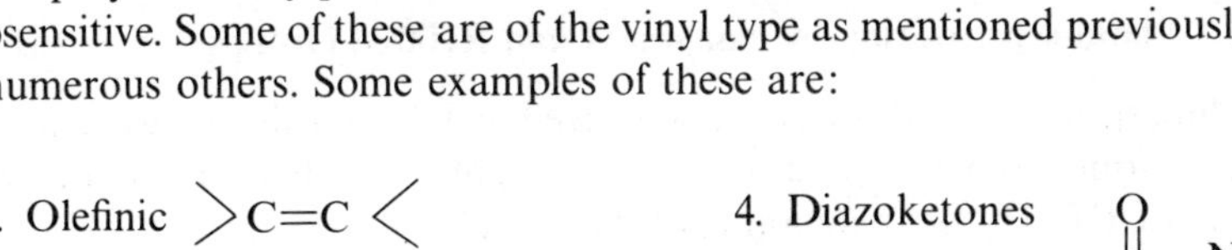

$$\sim\sim=M-(M)_n-M=\sim\sim \xrightarrow{h\nu} \sim\sim-M-(M)_n-M-\sim\sim$$
$$\sim\sim=M-(M)_m-M=\sim\sim \phantom{\xrightarrow{h\nu}} \sim\simM-(M)_m-M-\sim\sim$$

Photopolymers may possess in their chain certain functional groups that are photosensitive. Some of these are of the vinyl type as mentioned previously and also numerous others. Some examples of these are:

1. Olefinic $>C=C<$

2. α-β unsaturated ketone $>C=C<_{C=O}$

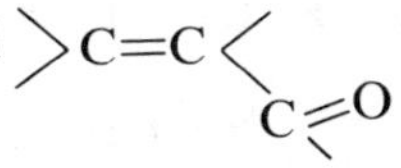

3. α-β unsaturated esters $\overset{R}{>}C=C<_{\underset{R'}{C=O}}^{}$

4. Diazoketones (cyclohexanone with =N$_2$ group)

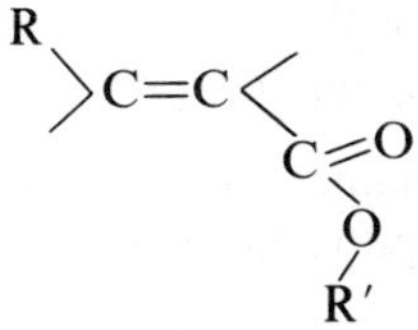

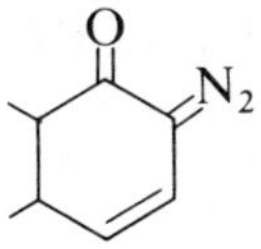

5. Azide —N$_3$

6. Carboazide —C(=O)—N$_3$

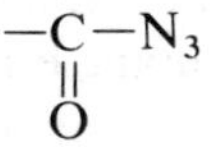

7. Sulphone azide

$$-\underset{\underset{O}{\|}}{\overset{\overset{O}{\|}}{S}}-N_3$$

8. Diazonium salt

$$R-N{\equiv}N\}^+X^-$$

These photoresponsive groups may be linked to different polymer chains through groupings such as:

1. Esters
2. Carbonates
3. Phosphates
4. Urethans
5. Ethers
6. Amides
7. Sulphonamides

Two types of particular interest are:

4.1.3.1 Photolysis of Azido Groups

(a) Di- and poly-functional azides.
Polycarboxylic acid azides[31–32] $R(CON_3)_x$

$$RCON_3 \xrightarrow{h\nu} RNCO + N_2 \uparrow$$

(b) Polymers containing azide groups.[33–35]
A typical polymer of this group could be:[36]

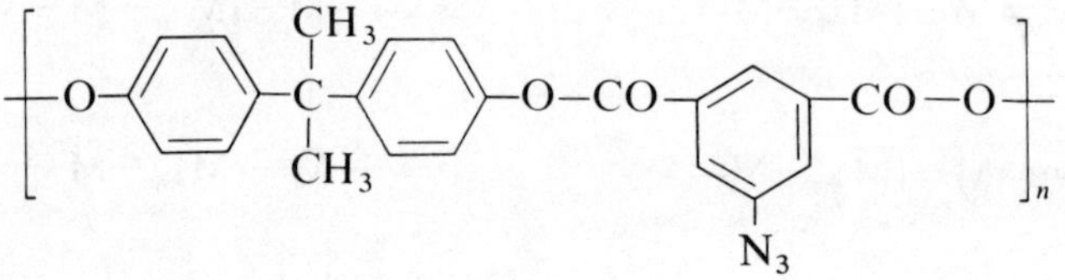

$$\left[-O-\underset{}{\bigcirc}-\underset{\underset{CH_3}{|}}{\overset{\overset{CH_3}{|}}{C}}-\underset{}{\bigcirc}-O-CO-\underset{\underset{N_3}{|}}{\bigcirc}-CO-O- \right]_n$$

4.1.3.2 Photo-sensitization by cinnamoyl and Related Groups

Photodimerization reactions can be sensitized by UV (not short-wave length UV as bonds formed may be broken) and visible light in the 350–450 nm range. For $R = OH$ or OC_2H_5 the reaction is of the form[37–40]

$$2R-CO-CH{=}CH-Ph \xrightarrow{h\nu} \begin{array}{c} RCO-CH-CH-Ph \\ \quad\quad\quad | \quad\quad | \\ Ph-CH-CH-CO-R \end{array}$$

A typical ester is polyvinyl cinnamate.[41]

There are two opposing hypothesis for this photo-crosslinking mechanism:

1. Photocyclodimerization of cinnamic groups with the formation of cyclobutane bridges between the polymer chains.[42]
   i.e.

```
∼∼—CH₂—CH∼∼         Ph         ∼∼—CH₂—CH—∼∼
      |              |                |
      O              CH               O        Ph
      |              ||               |        |
      CO             CH       hν      CO       CH—
      |              |        ──→     |        |
      CH             CO               CH——————CH
      ||             |                |        |
      CH             O                CH——————CH
      |              |                |        |
      Ph             CH₂—CH—∼∼        Ph       CO
                         |                     |
                         ξ                     O
                                               |
                                       ∼∼—CH₂——CH—∼∼
```

2. The crosslinking is due to free radicals formed during the UV irradiation.[43]

```
∼∼—CH₂—CH∼∼         Ph         ∼∼—CH₂—CH—∼∼
      |              |                |
      O              CH               O        Ph
      |              ||               |        |
      CO             CH       hν      CO       CH—
      |              |        ──→     |        |
      CH             CO               CH——————CH
      ||             |                |        |
      CH             O                —CH      CO
      |              |                |        |
      Ph             CH₂—CH—∼∼        Ph       O
                         |                     |
                         ξ                 ∼∼—CH₂—CH—∼∼
```

Covalent crosslinks are formed. Photodimerization of this nature can be regarded as a topochemical reaction. Pairs of adjacent cinnamoyl groups can dimerize if incident light is absorbed by such a pair of groups. The reaction is not oxygen inhibited as in vinyl polymerization.

At most, the formation of one crosslink results from each absorbed photon.

Several similar types to the cinnamoyl groups are:

(a) Cinnamate (ester) type. Cinnamic acid esters of polyvinyl alcohol and cellulose.
(b) Cinnamide type[44-45]
(c) Coumarin type.[46-48]
    Polymers containing coumarin substituents.
(d) Styryl ketone ($C_6H_5CH=CH-CO-$) type.[47-48]
(e) Cinnamylidene type,[49] $-\overset{|}{C}=CH-CH=CH-Ph$
(f) Benzoyl acrylic acid type,[50] $-C_6H_4-CO-CH=CH-COOH$

Photo-crosslinking reactions are finding application to various methods of image and data storage and retrieval and of printing-plate and printed-circuit manufacture. Some of these are discussed in Chapter 6.

Film-forming materials for photopolymerizable systems are of two main kinds. These are:

1. Resins – these may be oligomers or prepolymers.
2. Diluents – These may be:
    (a) reactive monomers
    (b) unreactive compounds that plasticize the cured film.

The main areas of current commercial interest are described in more detail in sections 4.2 and 4.3.

4.2 Resins

For aqueous-based photoemulsion systems the resins used are generally of the polyvinyl alcohol or polyvinyl cinnamate derivatives. These are described in greater detail in Chapter 6.

The development of early non-aqueous solventless systems for ultraviolet curing was restricted to non- or light pigmentation. Ultraviolet radiation was unable to penetrate deep into pigmented layers and UV curing was restricted in the first instance to the wood-finishing industry for curing clear lacquers and chipboard fillers. These often involved saturated polyester resins dissolved in volatile monomers such as styrene or vinyl toluene and were photo-initiated with benzoin ethers.[51] Saturated polyesters were eventually replaced with unsaturated polyesters to confer faster photopolymerization response.

There were several disadvantages to these early systems,[52] which included such properties as slow cure, volatility of the monomers and associated toxicity, poor through cure due to the then inferior lamp technology and poor storage stability owing to the use of benzoin ethers as photo-initiators. A major adverse factor of these systems was that surface polymerization was air-inhibited. The addition of waxes helped to remedy this, by retarding oxygen diffusion, as the waxes floated to the surface/air boundary at the time of the first hardening stage. Unfortunately, this destroyed the film gloss characteristics and this could only be obtained by sanding and polishing the surface.

The present industrial needs of fast cure response and elimination of atmospheric pollution (pioneered in the USA by the Los Angeles 1966 Act), in several areas of the surface-coatings industry, have caused many novel polymer systems to be investigated, the most widely used systems being the acrylated resins. A photopolymerizable medium when polymerized by actinic radiation should give a hard but flexible film with good adhesion to various substrates. The polymer must have acceptable pigment wetting properties and rheology in order that the final surface coating has acceptable flow properties, and open time on the press in the case of inks.

Resin types currently found to meet these requirements include:

Unsaturated polyesters;
Acrylated polyesters;

Acrylated epoxy esters;
Acrylated isocyanates;
Acrylated triazines (melamines);
Acrylated polyethers;
Thiol/ene systems;
Cationic cured epoxy systems;
Aminoplasts cured by photoliberated acids.

Unsaturation may be typically introduced in the following manner to some of the base conventional resin groups:

1. Epoxy resins generally modified with acrylic or methacrylic acid.
2. Urethans with hydroxy alkyl acrylates.
3. Unsaturated polyesters are often formed using maleic/fumaric acids as the photoresponsive component.
4. Chain addition of thiols (mercaptans) can be achieved photochemically to allylic compounds.

Modifications of some conventional resins are discussed below to show how response to light is achieved and the desirable physical properties of the original resin retained as far as possible.

4.2.1 Epoxy Resins

Notable properties of epoxy resins[53] include: good adhesion; a high level of chemical resistance; non-yellowing colour; and flexibility. The functionality of an epoxy resin depends upon the number of epoxide groupings present. It can therefore be mono-, di-, tri-, or polyfunctional. Two epoxide rings occur at the ends of the molecule in a bisphenol A–derived epoxy resin. A number of hydroxyl groups along the chain will confer polarity to the resin and promote adhesion properties to polar or metallic substrate surfaces.

The polymer chain of an epoxy resin contains two prime linkages. These are carbon–carbon linkages and ether linkages. These are both stable, conferring upon epoxy resins the characteristic of good chemical resistance.

On their own, epoxy resins are rather poor film formers (somewhat rigid because of aromatic rings) but in conjunction with other resins crosslinking can take place in the reactive epoxide rings and hydroxyl groups (if present). These are well separated by several atoms giving a long backbone in crosslinked form and conferring flexibility upon the molecule. Flexibility is a characteristic of epoxy resins also because of free rotation of the two methyl groups sandwiched between the two aromatic rings, i.e.

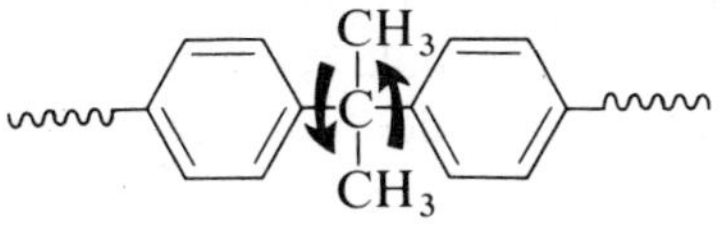

There are two stages in the preparation of epoxy resins:[54] a condensation reaction and an addition reaction.

1. *Condensation reaction.*

$$HO-\underset{bisphenol\ 'A'}{\underset{|}{\overset{|}{\underset{CH_3}{\overset{CH_3}{C}}}}}-OH + \underset{epichlorohydrin}{\overset{O}{CH_2-CH-CH_2Cl}} \xrightarrow{dilute\ NaOH}$$

$$\underset{the\ basic\ epoxy\ unit}{\overset{O}{CH_2-CH-CH_2}-O-\underset{CH_3}{\underset{|}{\overset{CH_3}{\overset{|}{C}}}}-OH} + NaCl$$

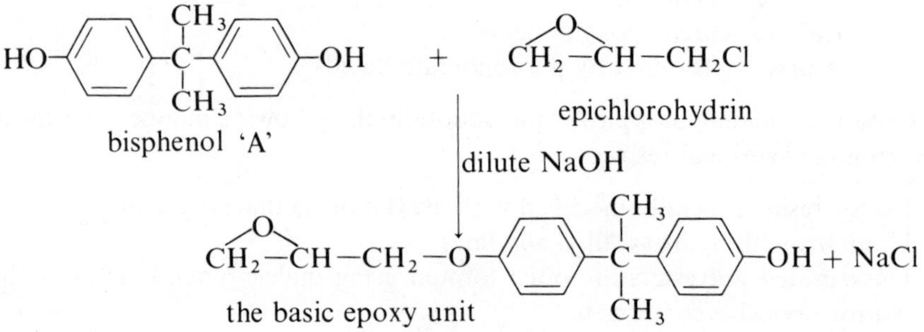

2. *Addition reaction.* Opening of the epoxide ring occurs, forming hydroxyl groups, (see page 147).

Some typical physico-chemical characteristics manifested by parts of the molecule are also depicted above.[55] The physical state of the unmodified resin, i.e., liquid or solid, is determined by the value of n in the general formula. For $n = 0$ or 1 a liquid results. For $n \geq 2$, the resins are solid in nature.

Modification of epoxy resins for UV curing by acrylation can occur by the routes shown below:[56-58]

1. Reaction of epoxy resin with an acrylic or methacrylic acid to give an epoxy acrylate or methacrylate, causing epoxy ring opening, i.e.,

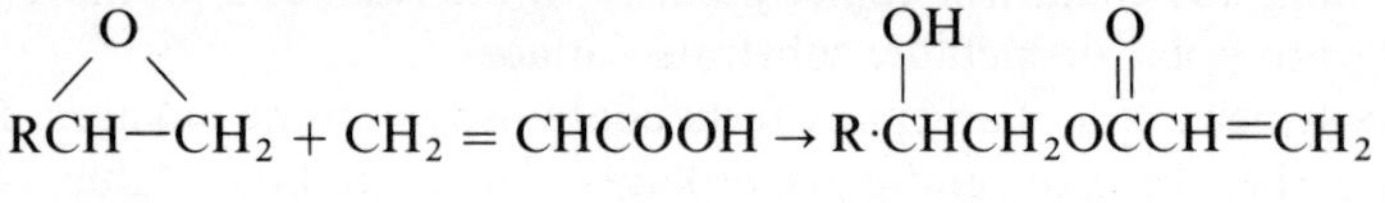

$$\overset{O}{RCH-CH_2} + CH_2=CHCOOH \rightarrow R\cdot \overset{OH}{\underset{|}{C}}HCH_2O\overset{O}{\overset{\|}{C}}CH=CH_2$$

Large structures may be formed in this manner, such as the one on page 148.

2. Epoxy resins can be reacted with the intermediate reactant of hydroxyalkyl acrylates and maleic or other anhydrides,[59] which increases the molecular weight of epoxy acrylates by forming acrylate terminated hydroxy functional polyesters. The unsaturated double bond thus introduced is then available for free-radical addition polymerization reactions.

The epoxy backbone provides the characteristics of toughness, flexibility, adhesion, and resistance to the cured film. A bifunctional molecule is to be avoided as it could cause the whole system to polymerize during synthesis. Further means of introducing double-bond functionality is by half-esters such as that of fumaric acid.

Epoxy resin polymer.

Acrylated Epoxy resin

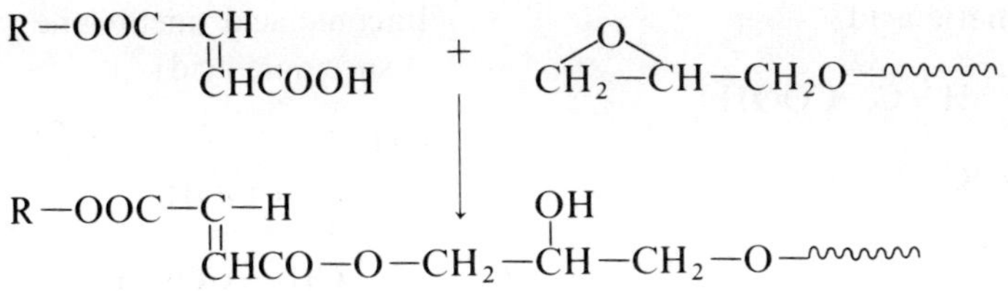

$$\text{R}-\text{OOC}-\overset{\displaystyle ||}{\underset{\displaystyle \text{CHCOOH}}{\text{CH}}} \quad + \quad \overset{\displaystyle O}{\overset{\displaystyle \frown}{\text{CH}_2-\text{CH}}}-\text{CH}_2\text{O}-\sim\!\sim\!\sim$$

$$\downarrow$$

$$\text{R}-\text{OOC}-\underset{\displaystyle ||}{\overset{\displaystyle \text{C}-\text{H}}{\underset{\displaystyle \text{CHCO}-\text{O}-\text{CH}_2-\overset{\displaystyle \text{OH}}{\text{CH}}-\text{CH}_2-\text{O}-\sim\!\sim\!\sim}{}}}$$

A hydroxyl group as well as an unsaturated bond is therefore introduced by this means. The molecule is therefore capable of photopolymerizing through the double bond, and also the presence of the hydroxyl group confers better adhesion and pigment wetting properties.

Other half-esters can be used, e.g., more complicated ones such as the half-ester of itaconic acid:

$$CH_2=C(COOR)CH_2-COOH$$

4.2.2 Unsaturated Polyester Systems

Early UV systems were based on styrene/polyester. Chief factors of consideration were:

1. Polymer structure and molecular weight.
2. Monomer/polymer ratio.
3. Monomer reactivity.
4. Inhibitor level.

Unsaturated polyester resins are based on components which introduce unsaturation directly into the polyester backbone, and this unsaturation is capable of direct addition copolymerization with vinyl monomers.

Unsaturated polyesters are the condensation products obtained by polycondensation from dicarboxylic acids or anhydrides with polyhydric alcohols.

A linear polymer can be formed from dibasic acids and dihydric alcohols, but the resin must include some unsaturated components. These are often acids or anhydrides, such as:[60]

Maleic anhydride   Citraconic anhydride (methyl maleic anhydride)

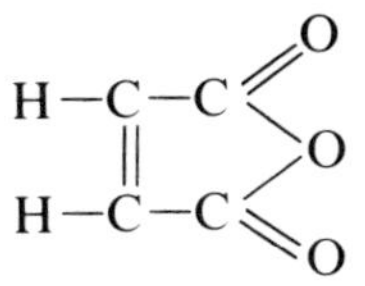

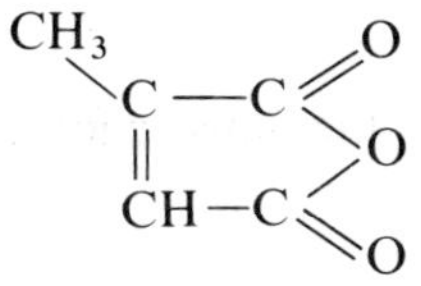

$$\begin{array}{c} \text{H}-\text{C}-\text{C} \overset{\displaystyle O}{\underset{\displaystyle O}{\diagdown}} \\ || \qquad \qquad \diagup \\ \text{H}-\text{C}-\text{C} \underset{\displaystyle O}{} \end{array} \qquad\qquad \begin{array}{c} \text{CH}_3 \\ \diagdown \text{C}-\text{C} \overset{\displaystyle O}{\underset{\displaystyle O}{\diagdown}} \\ || \qquad \qquad \diagup \\ \text{CH}-\text{C} \underset{\displaystyle O}{} \end{array}$$

Fumaric acid

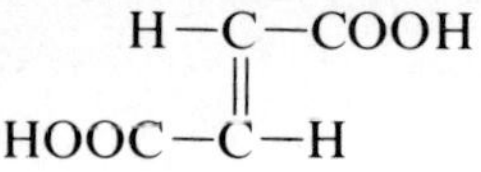

$$\begin{array}{c} H-C-COOH \\ \| \\ HOOC-C-H \end{array}$$

Itaconic acid (methylene succinnic acid)

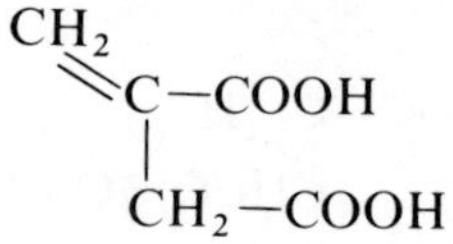

$$\begin{array}{c} CH_2 \\ \backslash\!\!\!= \\ C-COOH \\ | \\ CH_2-COOH \end{array}$$

Mesaconic acid (methyl fumaric acid)

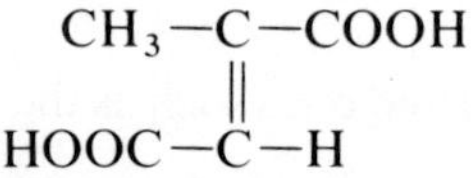

$$\begin{array}{c} CH_3-C-COOH \\ \| \\ HOOC-C-H \end{array}$$

Acenitic acid (propene-1,2,3-tricarboxylic acid)

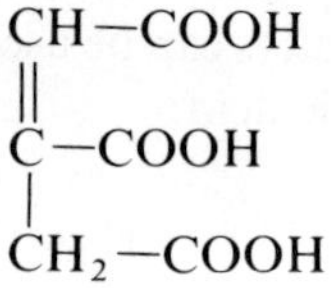

$$\begin{array}{c} CH-COOH \\ \| \\ C-COOH \\ | \\ CH_2-COOH \end{array}$$

Numerous polyhydric alcohols are available, such as:

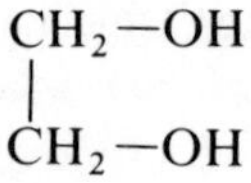

$$\begin{array}{c} CH_2-OH \\ | \\ CH_2-OH \end{array} \qquad \begin{array}{c} CH_2-OH \\ | \\ CH_3-C-H(OH) \end{array}$$

ethylene glycol 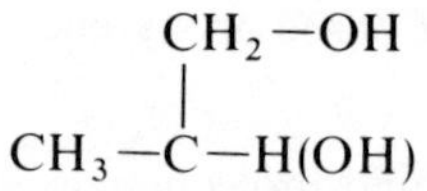   1,2-propylene glycol

A typical unsaturated polyester could be prepared from the following components:[61] 1,2-propylene glycol; phthalic anhydride; and maleic anhydride

An excess of hydroxyl content is generally used and a typical ratio of the above components could be of the order $2:1:1\frac{1}{2}$. Nitrogen is generally used as an inert gas blanket over the liquid mixture to prevent the unsaturation being destroyed. Standard alkyd preparation technique is used and as no monoglyceride stage occurs, the three reactants are present from the start of the reaction, although if iso-phthalic acid is incorporated, reaction with the diol is carried out prior to the addition of maleic anhydride to prevent any unreacted iso-phthalic acid remaining at the termination of the reaction.

At elevated temperatures, the maleic form of the rigid double bond,

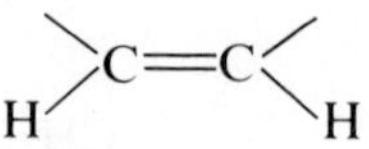

$$\underset{H}{\diagup}C=C\underset{H}{\diagdown}$$

largely converts to the fumaric form,[62]

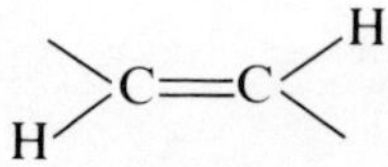

$$\underset{H}{\diagup}C=C\underset{}{\diagup}^{H}$$

The overall reaction can be summarised as probably being of the form:[53]

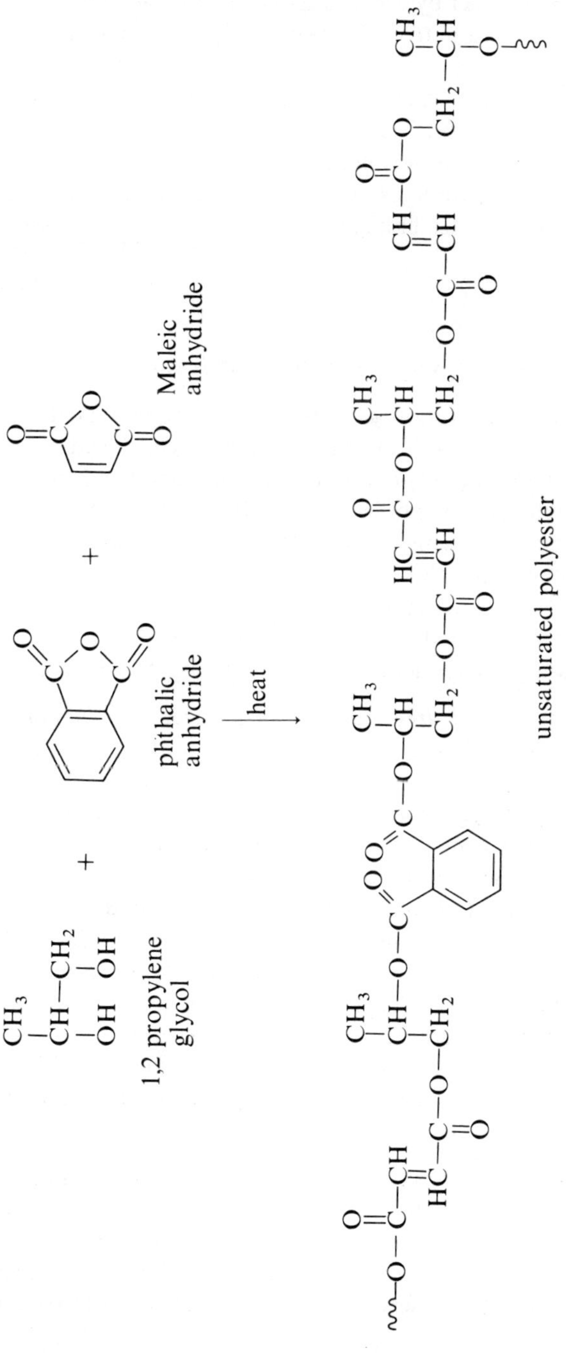

Polyester finishes are very hard,[63] tough, and solvent resistant. They are also resistant to fairly hot objects and therefore find application for furniture finishes.

Polyester films are not very flexible but this can be improved by spacing out the crosslinks in the polymer by decreasing the proportion of unsaturated acid in the polyester resin, or alternatively by using long-chain flexible saturated acids or alcohols.

Polyesters tend to undergo appreciable volume shrinkage on photopolymerization which may cause loss of adhesion, pulling away from the edges, and other defects. Adhesion is generally poor and polyesters are often used in conjunction with polyurethans. Alkali resistance is generally poor owing to the saponification of the ester linkage.

4.2.2.1 Preparation of Acrylic Functional Polyesters

The commercial preparation of these reactive resins via the esterification of polyhydric esters with acrylic acid has been restricted because of the low boiling point and thermal instability of the acrylic acid monomer. Esterification catalysts functional at temperatures of less than 100 °C are needed.

The polyester component is formed from the direct fusion of the intermediates under a nitrogen atmosphere.

A typical polyester composition with dibutyl tin oxide as catalyst is from trimethylol propane, triethylene glycol, 1,6-hexanediol, and adipic acid.[64]

Some necessary components for the acrylating process are:

1. *An efficient thermal free radical polymerization inhibitor.* Phenothiazine (60–75 ppm) is used as it is a more effective inhibitor in the absence of oxygen than the various ether derivatives of hydroquinone. Nitrobenzene can be included (up to 25 ppm) to prevent polymerization of the otherwise uninhibited acrylic acid that condenses overhead.
2. *An azeotroping solvent.* Benzene and toluene are used in combination with hydrocarbon cuts (a blend of aromatic and aliphatics).
3. *An appropriate low-temperature esterification catalyst.* Organometallics function best in the temperature range 150–250 °C and are not adequate at lower temperature conditions of esterification. The most active in the 80–100 °C temperature range include the strong acids such as sulphuric, methane sulphonic acid, and benzene sulphonic acid. These catalysts can eventually be removed from the finished product with an anion exchange resin if necessary (e.g. Amberlyst 15).

Low-temperature esterification is executed with acrylic acid and the polyhydric ester at stoichiometric equivalents of acrylic acid to hydroxyl. Reaction is of the order of 90 per cent conversion at 100 °C.

Unsaturated polyesters are often of high molecular weight and although a high molecular weight prepolymer will require fewer crosslinks to obtain a cured non-tacky state, the unsaturated polyester backbone has an intrinsically slow crosslinking rate. High molecular weight materials tend also to give viscosities, leading to bad rheology and application problems in the sense of poor flow and transfer properties. For these reasons, unsaturated polyesters are generally diluted with monomeric materials to give good application properties and increase the cure speed. The choice of reactive diluent is important. Styrene was originally the most widely used because of its high reactivity with maleic/fumaric structures.[65–66] Vinyltoluene was also popular. Substituted styrenes have also been used and generally it has been found that the reactivity is of the following order:[67]

$$\text{styrene} > \text{vinyl toluene} > \text{p-chlorostyrene} > \text{t-butylstyrene}$$

Response is probably independent of the styrene–maleic/fumaric component ratio, suggesting that styrene/styrene intrachain propagations control cure rate.

The polyester resin molecular weight is likely to be more important to cure response than the styrene level, and is limited by the application viscosity requirements. Properties such as scratch and chemical resistance are more related to resin and monomer types than to molecular weight or diluent level. Application problems can also occur with the more volatile monomers such as vinyl acetate. Acrylic monomers are now mainly used owing to their fast curing properties.

4.2.3 Polyurethans

These polymers are frequently incorrectly called polyurethanes. Standard chemical nomenclature requires that the term 'ane' be used for saturated hydrocarbons and specific fully saturated heterocyclic compounds.[68]

Urethans have the characteristic structure

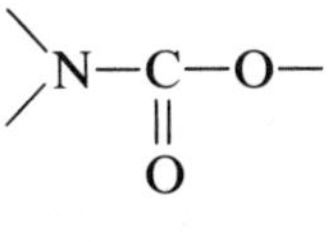

$$\begin{matrix} \diagdown \\ \diagup \end{matrix} N-\underset{\underset{O}{\|}}{C}-O-$$

and may be regarded as esters of the unstable carbamil acid or amide esters of carbonic acid.

A finally dried polyurethan film does not necessarily contain polyurethan resins and the urethan linkage is not always predominant in the final cured film. Polyurethan finishes do, however, contain isocyanates or the corresponding reaction product.

A major drawback to polyurethans is that they yellow in the presence of sunlight, probably owing to chromophoric groups resulting from nitrogen being present.

4.2.3.1 Preparation of a Urethan

Urethans may be prepared from reactions of isocyanates with compounds containing hydroxyl groups such as alcohols,

$$RNCO + R'OH \longrightarrow R-\underset{H}{N}-\overset{O}{\underset{\|}{C}}-OR'$$

Unsaturation may then be readily introduced when R' is acrylic or allylic or vinylic (but not vinyl alcohol which does not exist in nature).
e.g.

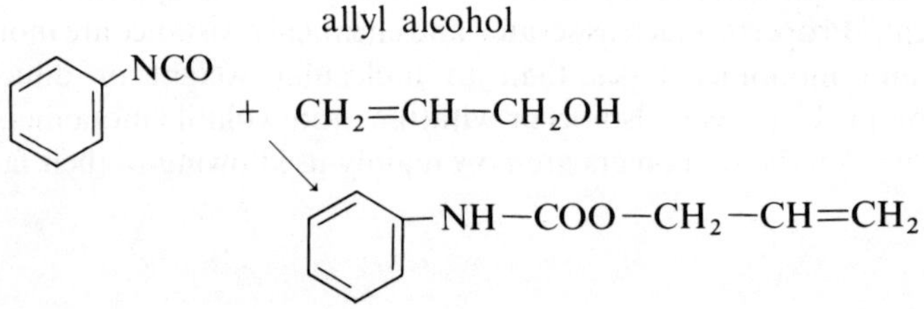

Ph-NCO + CH$_2$=CH-CH$_2$OH (allyl alcohol) → Ph-NH-COO-CH$_2$-CH=CH$_2$

More complicated urethans may be prepared by reacting:

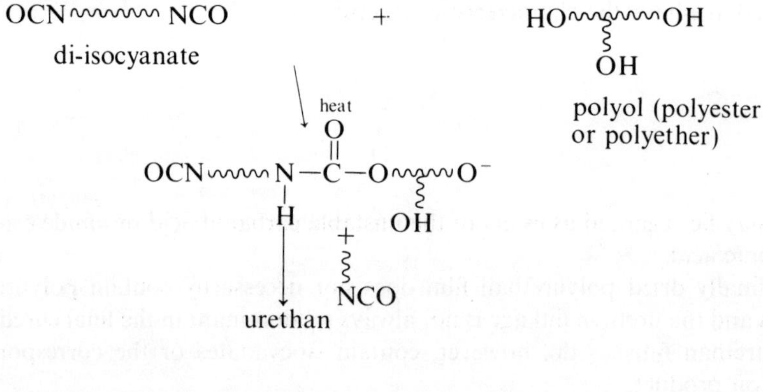

OCN~~~NCO + HO~~~OH
di-isocyanate |
                     OH
                polyol (polyester or polyether)

↓ heat

OCN~~~N-C(=O)-O~~~O⁻
     | |
     H   OH
                    +
                    ~~~
                    NCO
     urethan

Unsaturation may also be introduced by reacting a hydroxy-modified acrylic or methacrylic monomer with a polyisocyanate to obtain a urethan-type resin containing acrylic linkages which can then undergo photopolymerization with free radicals. The terminal groups can then be attached to introduce polymerization. The carbamate groups in these isocyanate-modified acrylic systems give a tough, flexible film with good adhesion, and abrasion resistance.

Di-isocyanates are frequently used, allowing large structures to be formed, especially when the chain is lengthened by ethylene oxide derivatives, amino alcohols, diols, polyesters, diamines, etc.[70-75]

e.g.

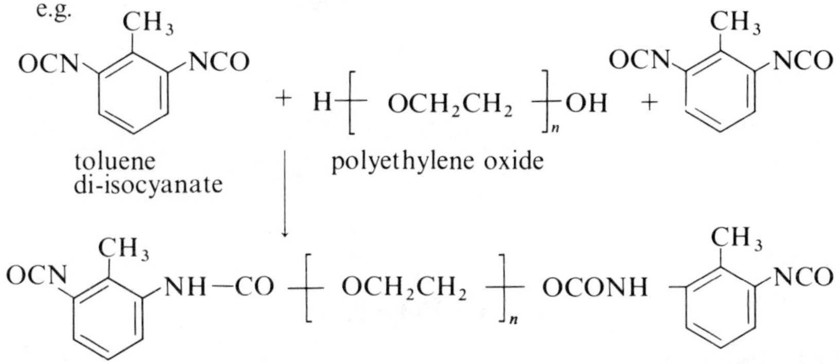

toluene di-isocyanate + polyethylene oxide

$$OCN\text{-}C_6H_3(CH_3)\text{-}NH\text{-}CO\text{-}[OCH_2CH_2]_n\text{-}OCONH\text{-}C_6H_3(CH_3)\text{-}NCO$$

Very large complicated urethan structures may be built up by combinations with acrylics and polyester/urethan complexes,[76]

e.g. a polyester/acrylate may be based upon adipic acid (AD) and hexanediol (HD), reacted through its terminal hydroxyl groups with acrylic acid to give a structure of the form:

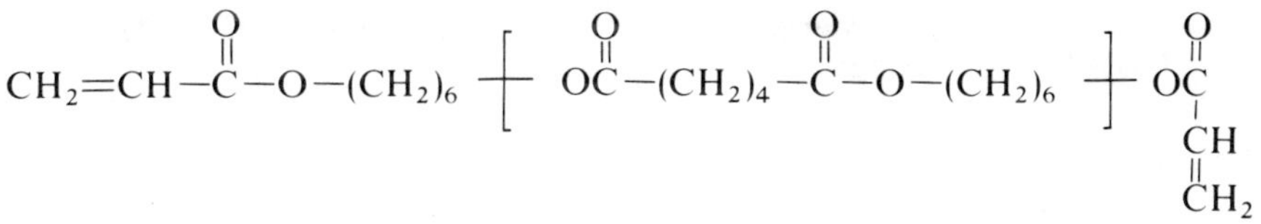

$$CH_2{=}CH{-}\overset{O}{\underset{\|}{C}}{-}O{-}(CH_2)_6 \left[\overset{O}{\underset{\|}{OC}}{-}(CH_2)_4{-}\overset{O}{\underset{\|}{C}}{-}O{-}(CH_2)_6 \right] \overset{O}{\underset{\|}{OC}}{-}\underset{\|}{\overset{CH}{CH_2}}$$

and this same polyester acrylate may then be reacted with toluene di-isocyanate (TDI),

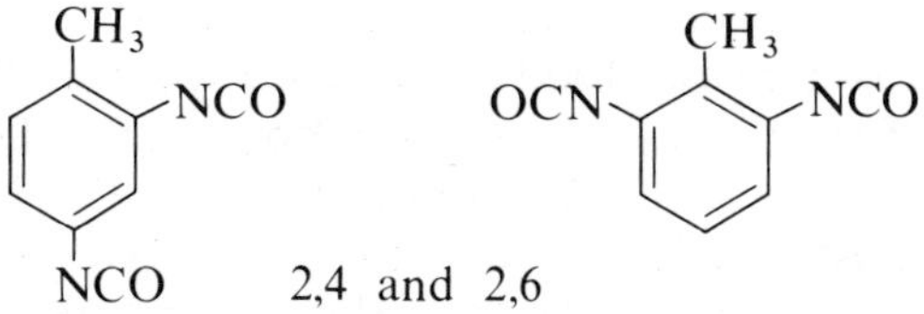

2,4 and 2,6 toluene di-isocyanates

followed by further reaction with hydroxyethyl acrylate to give a possible structure of the form:[77-78]

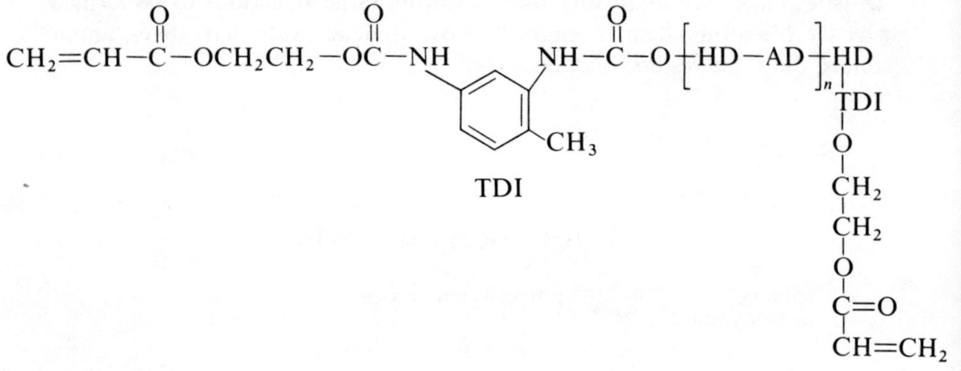

$$CH_2=CH-\overset{O}{\underset{\|}{C}}-OCH_2CH_2-O\overset{O}{\underset{\|}{C}}-NH-\underset{\underset{CH_3}{\text{TDI}}}{\bigcirc}-NH-\overset{O}{\underset{\|}{C}}-O\left[HD-AD\right]_n HD$$

with side chain: TDI—O—CH$_2$—CH$_2$—O—C(=O)—CH=CH$_2$

4.2.4 Polyethers

4.2.4.1 Preparation

Polyethers are prepared by reacting ethylene oxide or propylene oxide and a polyol with acid or basic catalysts such as BF_3 or $NaOH$ respectively.

A typical example is[53]

$$\begin{array}{c} CH_2OH \\ | \\ CHOH \\ | \\ (CH_2)_3 \\ | \\ CH_2OH \end{array} + 3n\,CH_2\!\!-\!\!\overset{O}{\underset{\diagdown}{}}\!\!-\!\!CH\!-\!CH_3 \xrightarrow[\text{(3) catalyst}]{\text{(1) pressure, (2) heat}} \begin{array}{c} CH_3 \\ | \\ CH_2(OCHCH_2)_nOH \\ | \\ CH_3 \\ | \\ CH(OCHCH_2)_nOH \\ | \\ (CH_2)_3 \\ | \quad CH_3 \\ | \;\;/ \\ CH_2(OCHCH_2)_nOH \end{array}$$

1,2,6-hexane triol   propylene oxide   polyether

Direct esterification of polyhydric ethers with acrylic acid may be possible but many polyethers will degrade under the acidic conditions of esterification. Transesterification is quite useful and will not adversely effect the ether linkages. One of the most effective catalysts is tetraisopropyl titanate. Ethyl acrylate also appears to be a most useful acrylate source because of its boiling point and azeotrope characteristics. A binary ethanol/ethyl acrylate azeotrope is used.[79] The use of a more highly functional crosslinking acrylate decreases the coating flexibility while increasing hardness and chemical resistance.

Polyethers are generally low-viscosity resins. The lower internal binding as opposed to polyesters of similar molecular weights and structure results in even lower viscosities. They are also quite cheap.

Common polyethers that have been converted to acrylic functional forms include ethoxylates and propoxylates of trimethylolpropane, and pentaerythritol and polyethers of 1,4-butanediol.

4.2.5 Thiol/ene System

Mercaptan olefin copolymerization has been investigated in depth by W. R. Grace and Co. Ltd.[80–81] for the preparation of photopolymer relief printing plates, and the process is known as Letterflex for the printing of newspapers.

The overall reaction is a free-radical addition of the thiol to an olefinic or allylic double bond:

$$RSH + CH_2=CHR \xrightarrow{hv} RSCH_2CH_2R$$

Several other workers have studied this type of reaction.[82–84]

A general mechanism is thought to be of the form:[85]

[1] $R_2C=O \xrightarrow{hv} (R_2C=O)^1$ (n,π*) (singlet)

[2] $(R_2C=O)^1$ (n,π*) $\longrightarrow (R_2C=O)^3$ (n,π*) (triplet)

[3] $(R_2C=O)^3$ (n,π*) + R'SH $\longrightarrow R_2\dot{C}OH$ + R'S·
                                                           ketyl   thiyl
                                                           radical radical

[4] $2R_2\dot{C}OH \longrightarrow R_2C(OH)\,C(OH)R_2$
                                     pinacol

[5] R'S· + $CH_2=CHCH_2R'' \rightleftarrows R'SCH_2\dot{C}H\,CH_2R''$

[6] $R'SCH_2\dot{C}HCH_2R'' + R'SH \longrightarrow R'SCH_2CH_2CH_2R'' + R'S·$
                            chain reaction

[7] $R'SCH_2\dot{C}HCH_2R'' + nCH_2=CHCH_2R''$

$$\longrightarrow R'SCH_2CH{-}\!\!\left[CH_2{-}CH{-}\!\right.\!\!\underset{n-1}{\left.\vphantom{|}\right]}\!CH_2CH_2CH_2R''$$
$$\quad\quad\quad\quad\quad |\quad\quad\quad\quad\quad |$$
$$\quad\quad\quad\quad\quad CH_2R''\quad\quad CH_2R''$$

The mechanism is excitation of the sensitizer by the absorption of light to the singlet state, followed by intersystem crossing, giving the photochemically reactive (n, π^*) triplet which abstracts a hydrogen atom from the thiol to form ketyl and thiyl radicals. The ketyl radicals dimerize to pinacol and the thiyl radicals add to the polyene by a chain mechanism.

Evidence for the above mechanism is:

1. Popular agreement indicates that the excited state responsible for the photochemistry of carbonyl compounds in solution is usually the triplet state.
2. The reaction yields 95 per cent and higher anti-Markovnikov products as exhibited by NMR.
3. The quantum yield is thought to be of the order 400 moles/einstein.
4. Free-radical inhibitors such as hydroquinone greatly reduce the rate.
5. For C^{14}-labelled benzophenone as the sensitizer, inverse isotope-dilution analysis on the solution from a solvent-extracted cured polymer indicates the presence of benzopinacol.

In reaction [3], the ketyl radicals produced may also initiate and terminate chains either by direct reaction [8] and [9] or by hydrogen transfer [6] and [11].

[8] $R_2\dot{C}OH + CH_2=CH-CH_2R'' \rightarrow R_2COHCH_2\dot{C}HCH_2R''$
[9] $R_2\dot{C}OH + R'SCH_2\dot{C}HCH_2R'' \rightarrow R'SCH_2(R_2COH)CHCH_2R''$
[10] $R_2\dot{C}OH + CH_2=CHCH_2R'' \rightarrow R_2C=O + CH_3-\dot{C}H-CH_2R''$
[11] $R_2\dot{C}OH + R'SCH_2\dot{C}HCH_2R'' \rightarrow R_2C=O + R'SCH_2CH_2CH_2R''$

There appears to be no real evidence for the hydrogen transfer reactions [10] and [11]. Proportional counting of extracted cured films indicates that no more than 3 per cent of the initial C^{14} labelled benzophenone is incorporated in the polymer in the time required for curing. There are many possible paths for incorporation of the sensitizer in these polymers. This data indicates only that reactions [8] and [9] are occurring.

The addition of phosphines will eliminate any inhibition that occurs.

Under favourable conditions, when the thiol/ene radical chain transfer addition reaction predominates via hydrogen abstraction, end-to-end addition may result, such as:

$$HS-R-SH + nCH_2=CH\sim\sim\sim\sim CH=CH_2$$
$$\downarrow h\nu \; | \; Ph_2CO$$
$$HS-[RSCH_2CH_2 \sim\sim\sim\sim CH_2CH_2S]_n-R-SH$$

Enormous structures may be built up by using an admixture of a polyene and polythiol, which causes curing to occur by simultaneous chain-extending and cross-linking reactions[85,86]

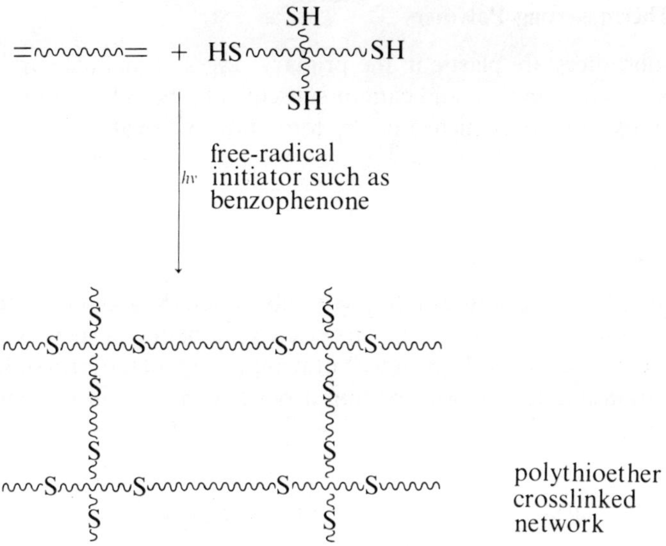

$$\equiv\!\!\sim\!\!\sim\!\!\sim\!\!\equiv \;+\; HS\!\!\sim\!\!\sim\!\!\underset{SH}{\overset{SH}{\sim\!\!\sim\!\!\sim}}\!\!\sim\!\!\sim\!\!SH$$

$\Big\downarrow h\nu$ free-radical initiator such as benzophenone

polythioether crosslinked network

As the properties of the cured material depend primarily on the prepolymer structure, the one end group and functionality of the polyene can be varied, but the backbone can be modified to include different groups such as urethans, esters, polyethers, imides amides, etc. Also the backbone and functionality of the polythiol can be varied. The properties of the cured material can be tailored almost as desired with so many degrees of freedom available. For example, these polymers may be rendered thermoplastic or thermosetting in character. If urethans are incorporated, thermoplastic systems result, but if trifunctional olefins or trimercaptans are incorporated into the backbone instead, thermosetting properties can be achieved. The terms 'thermoplastic' and 'thermosetting' are described below.

4.2.5.1 Thermoplastic Polymers

During vinyl polymerization, a mobile liquid monomer will change from a free-flowing state due to weak Van der Waals' forces operating between the relatively small molecules, to a relatively viscous liquid when the molecules reach a size where the Van der Waals' forces become effective. When polymerization approaches completion, the molecules become very extended. The sum of Van der Waals' forces operating between adjacent molecules then become considerable and the material so viscous that it may be termed a solid. It will not flow unless heated, which causes an increase in the energy of the molecules so that they are able to overcome the Van der Waals' forces acting between them. A substance exhibiting this behaviour is termed thermoplastic as it may be softened repeatedly by the application of heat.

4.2.5.2 Thermosetting Polymers

These substances are plastic in the primary stages of manufacture but once moulded to shape they 'set' and cannot subsequently be softened by reheating. Thermosetting is a consequence of the formation of covalent bonds between chain molecules.

4.2.5 (cntd.)

It is reported that the nature of the olefin substituent (X) grouping is important on reactivity rate, as substituents having electron-donating properties accelerate the cure rate, whereas with electron-withdrawing groups of the form shown below may be responsible for this by creating a positive charge on the unsaturated component.[87]

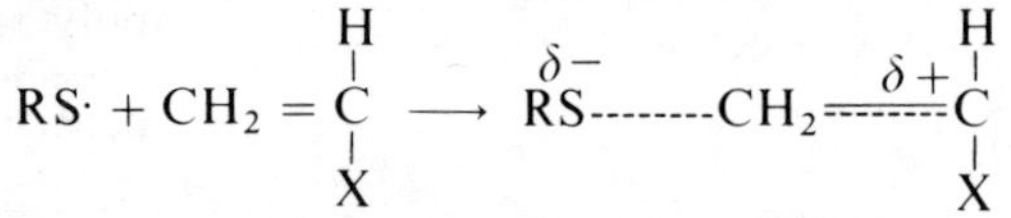

$$RS\cdot + CH_2 = \underset{X}{\overset{H}{C}} \longrightarrow RS\text{--------}CH_2\overset{\delta-}{\text{=======}}\underset{X}{\overset{\overset{H}{|}\delta+}{C}}$$

There are two notable advantages to the thiol/ene system. In contrast to the majority of radiation-curable systems, air inhibition does not occur to any extent and inert blanketing gas systems are therefore not necessary. This is possibly because the addition of the thiyl radical to the allylic bond or the more stable rearranged radical to form the initial radical can rapidly scavenge oxygen. Hydrogen abstraction from the thiol by the newly created peroxy radical then initiates a new chain reaction.

This may be depicted as:[88]

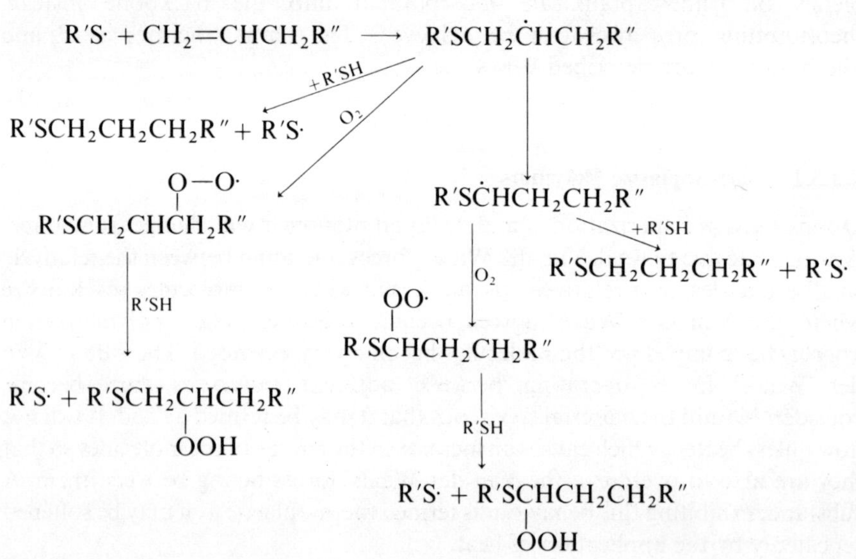

$$R'S\cdot + CH_2=CHCH_2R'' \longrightarrow R'SCH_2\dot{C}HCH_2R''$$

$$R'SCH_2CH_2CH_2R'' + R'S\cdot \xleftarrow{+R'SH} \quad \xrightarrow{O_2} \quad R'\dot{S}CHCH_2CH_2R''$$

$$\underset{R'SCH_2CHCH_2R''}{\overset{O-O\cdot}{|}}$$

$$\downarrow R'SH$$

$$R'S\cdot + \underset{OOH}{\overset{R'SCH_2CHCH_2R''}{|}}$$

$$\underset{R'SCHCH_2CH_2R''}{\overset{OO\cdot}{|}} \xrightarrow{+R'SH} R'SCH_2CH_2CH_2R'' + R'S\cdot$$

$$\downarrow R'SH$$

$$R'S\cdot + \underset{OOH}{\overset{R'SCHCH_2CH_2R''}{|}}$$

The second advantage is that by careful selection of the structure of the thiol and polyene constituents in the thiol/ene systems, a broad range of properties can be obtained. Predominantly, flexibility may be achieved on curing formulations composed from relatively low-viscosity materials, eliminating the need of low molecular weight diluents. This increased flexibility may occur because the introduced thioether linkages have low hindrance to bond rotation in the crosslinked network.

These systems, however, tend to be somewhat expensive in some applications owing to the use of polythiols.

Hybrid systems have been developed by W. Grace & Co. Ltd. because of this. Small quantities of polythiols are employed to modify lower-cost acrylate systems, retaining many advantages of the more costly system.

Two types of polythiol structures used in this chemistry are:

1. Pentaerythritol tetrakis (thioglycolate)

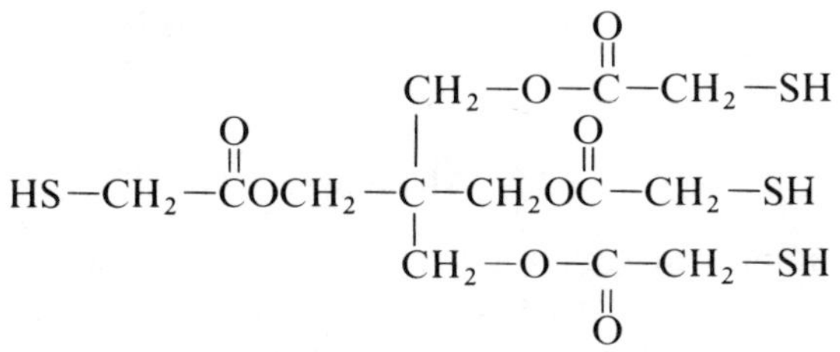

$$HS-CH_2-\overset{O}{\underset{\|}{C}}OCH_2-\underset{\underset{\underset{O}{\|}}{CH_2-O-C-CH_2-SH}}{\overset{\overset{\overset{O}{\|}}{CH_2-O-C-CH_2-SH}}{C}}-CH_2O\overset{O}{\underset{\|}{C}}-CH_2-SH$$

2. Trimethylol propanetris (β-mercaptopropionate)

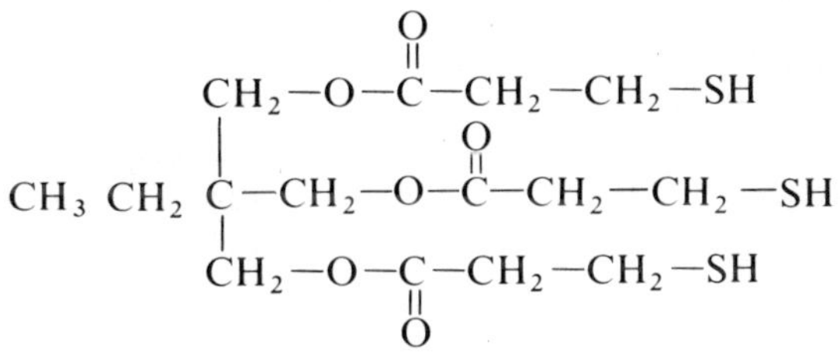

$$CH_3\ CH_2\ \underset{\underset{\underset{O}{\|}}{CH_2-O-C-CH_2-CH_2-SH}}{\overset{\overset{\overset{O}{\|}}{CH_2-O-C-CH_2-CH_2-SH}}{C}}-CH_2-O-\overset{O}{\underset{\|}{C}}-CH_2-CH_2-SH$$

The thiol/ene photosensitizers such as benzophenone are less effective for the hybrid thiol/acrylate systems. Benzoin ethers have been found to be particularly effective for the latter.

These novel systems of W. Grace have found application in the flooring and paper coating industries.

4.3 DILUENTS

Viscosity reduction is achieved by three common methods:[89]

1. The use of suitable wetting agents in pigmented coating formulations.
2. Preparation of low-viscosity resins.
3. Selection of appropriate monomeric diluents.

4.3.1 Wetting Agents

These should be incorporated into the diluent monomers prior to pigment addition or they will often result in preferential absorption of the diluent by the pigment and consequently increase the paste viscosity.

Some wetting agents used are:

> Witconal NP-40 (Witco Chem)
> Witconal NP-100 (Witco Chem)
> Igepal CO-430 (GAF Corp.)
> Triton X-100 (Rohm and Haas Co.)

4.3.2 Low-viscosity Resins

Most acrylic functional resins are extremely viscous due to the urethan or epichlorohydrin-bisphenol A epoxy backbones. For low-viscosity application, acrylic functional alkyds, polyesters, or polyethers having lower viscosities can be utilized.

4.3.3 Monomers

Viscosity can be monitored in photopolymerizable systems by thin monomers, in particular monoacrylates. Alternatively, low-viscosity multifunctional acrylates such as TMPTA (trimethylol propane triacrylate) are employed as reactive diluents.

The monomer selected for a system depends upon its:

1. Photoresponse.
2. Contribution to the photopolymerized film properties.
3. Relative volatility.
4. Odour and toxicity.
5. Solvating efficiency.
6. Cost.

The molar concentration of monomers in a system is important as the molar concentrations of the monomer and oligomer are then reflected in their contributions to properties such as photoresponse, film hardness, chemical resistance, flexibility, etc. Crosslinking mechanisms are intricate and depend on many interwoven factors such as the molecular weight, degree of unsaturation, and resin/monomer molecular architecture.

The molarity of monomer influences significantly the chemical resistance and adhesion of the film. Monomers can be mixed to obtain a compromise in properties if the monomers needed are unobtainable. For any resin system there is an optimum resin/monomer ratio for a requested quantity of radiation needed for hardening, degree of crosslinking, and film characteristics. Too high a monomer concentration often produces a lower gel fraction due to a decrease in viscosity, an increased rate of bimolecular termination (preferential homopolymerization of the monomer), and a higher radiation dose is necessary. Too low monomer content permits the resin alone to predominate in the crosslinking mode and this also results in a lowering of the overall gel fraction.

In the case of a need for highly polar solvents for solvency of the resin system, these polar attributes can destroy resistance to compounds such as water, ethanol, etc. The choice of reactive diluent also affects the hardness and flexibility.

The multifunctional acrylates of the types such as pentaerythritol triacrylate and trimethylolpropane triacrylate improve significantly the hardness. The high crosslink density of these triacrylates however, creates an upper limit on concentration to prevent polymer shrinkage, which consequently can reduce adhesion and flexibility on the non-porous substrates.

Monomers perform five chief functions in a system:

1. As a solvent for the prepolymer, especially where the latter is a solid, but this affects pigment dispersion and final film properties.
2. As a rheological (viscosity and tack) control agent. Monofunctional monomers are particularly important for this role.
3. As crosslinking agents. Polyfunctional monomers are used for this.
4. They determine the photoresponse speed in conjunction with the photoinitiator/sensitizer.
5. They determine the nature of the surface of a coating. This arises because many monomers are relatively volatile. Their use can lead to a disturbed surface. Acrylic acid esters of the higher alcohols (higher boiling points) can be used to surmount this.

Diluents are of two classes:

1. Those that are reactive monomers.
2. Non-photopolymerizable compounds that effectively plasticize the final cured film.

Monomers fall into three main chemical categories such as acrylics and derivatives, vinyls, and allylics. A general order of cure speed for these is:[90]

$$\text{acrylic} > \text{methacrylic} > \text{vinyl} > \text{allylic}$$

The *functionality* of a monomer can be defined as the number of reactive double bonds per unit, and it has a definite effect on cure response. A prepolymer having as little as two reactive groups per molecule may be quite sensitive to UV radiation in spite of being dissolved in a monofunctional monomer if the prepolymer molecular weight is low enough. Difunctional or tri-, tetra-, and penta-functional monomers are themselves reactive. For these, molecular weight or functionality of the prepolymer is not so significant. Owing to the fact that acrylic systems are highly responsive, they are employed for those applications requiring fast speed and should not be combined with slower responsive groups such as vinyl, methacrylates, allyl, etc.

Some monofunctional diluents in use are:[91]

4.3.3.1 Vinyls

These diluents possess vinyl unsaturation.

1. Styrene and derivatives such as α-methyl styrene and vinyl toluene.

styrene — C_6H_5–CH=CH$_2$

vinyl toluene isomers — ortho, meta, para (methyl-substituted styrenes with CH=CH$_2$)

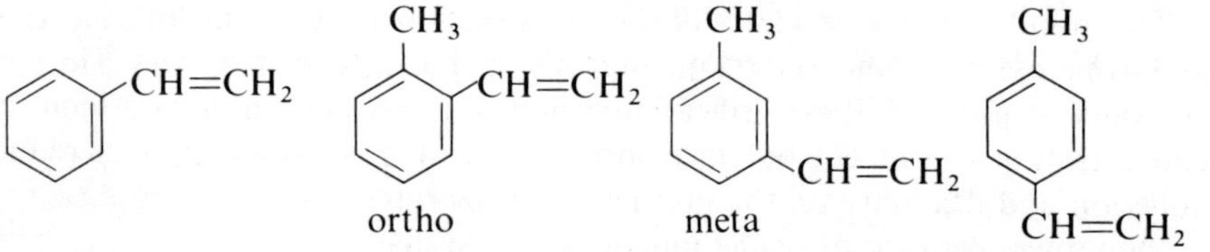

These have low cost, give hardness, slow cure and are volatile, but have good copolymerization characteristics with many resins and other monomers.

2. Vinyl acetate

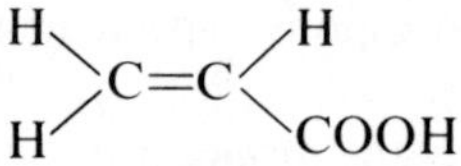

$$\text{H}_2\text{C=CH–COOH}$$

This is an excellent reducer but is volatile and inflammable. Coatings formulated from this monomer tend to have poor weatherability and water resistance.

3. *N*-Vinyl pyrrolidone

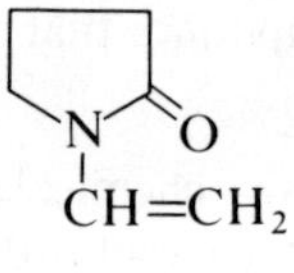

(pyrrolidone ring with N–CH=CH$_2$)

This is an excellent reducer with low reported toxicity. It has no acrylate functionality but copolymerizes with acrylates when used at the optimum mole ratio. Also, it is reported to improve flexibility of the cured film.

4.3.3.2 Acrylics

These diluents possess acrylic functionality[92] in the classes mono-, di-, tri- and tetra-acrylic and the corresponding methacrylates. Examples of these classes are listed below.

(i) Monoacrylates

n-Butyl acrylate. This has slow cure and poor solvent resistance but is a good viscosity reducer and has good flexing action. It is volatile and mainly used in wood coating. Isobutyl acrylate may also be used but again is very volatile.

2-Ethylhexylacrylate (EHA))

$$CH_2=CH-\underset{\underset{O}{\|}}{C}-O-CH_2-\underset{\underset{C_2H_5}{|}}{CH}-(CH_2)_3-CH_3$$

This is probably the most widely used diluent. It is volatile and has odour but is slowish in cure and gives poor solvent resistance. It has very good flexing action in films.

Isodecyl acrylate. This is a good viscosity reducer, being less volatile than EHA. It is reported to increase flexibility owing to long aliphatic chains.

Iso-bornyl acrylate

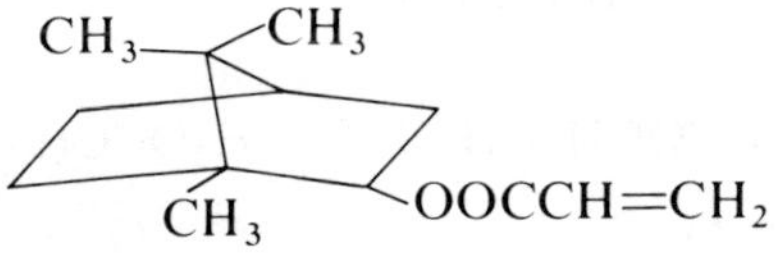

A strong odour is the main disadvantage of this monomer, which has low toxicity and volatility, imparts a hardness comparable to methyl methacrylate, but with the fast cure rate of acrylates, and has a low shrinkage rate.

Phenoxyethylacrylate. This is an excellent viscosity reducer.

Tetrahydrofurylacrylate. This is an excellent viscosity reducer but has a strong lingering odour and reduces the shelf-life of many systems.

2-Hydroxyethylacrylate.

$$HO-CH_2-CH_2-O-\underset{\underset{O}{\|}}{C}-CH=CH_2$$

2-Hydroxypropylacrylate

$$HO-CH_2-CH_2-CH_2-O-\underset{\underset{O}{\|}}{C}-CH=CH_2$$

These are good reducers but highly toxic.

Cyclohexyl acrylate.

3-Butoxy-2-hydroxypropyl acrylate.

(ii) Diacrylates

1,4-Butane-diol diacrylate (BDDA)

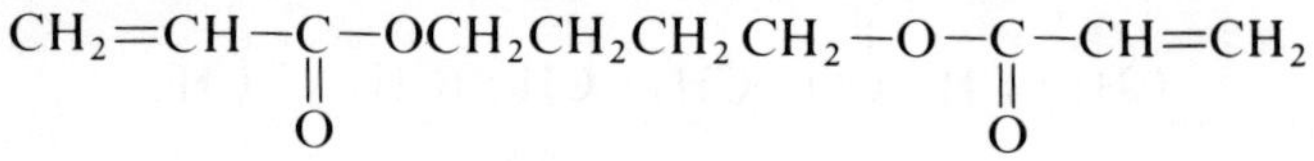

$$CH_2=CH-\underset{O}{\overset{\parallel}{C}}-OCH_2CH_2CH_2CH_2-O-\underset{O}{\overset{\parallel}{C}}-CH=CH_2$$

Neopentylglycol diacrylate (NPGDA)

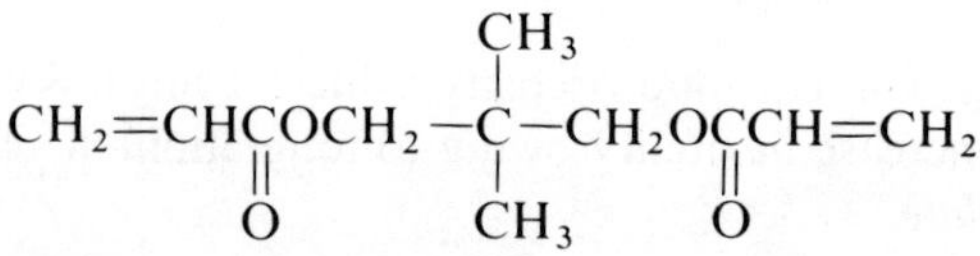

$$CH_2=CHC(O)OCH_2-\underset{CH_3}{\overset{CH_3}{C}}-CH_2OC(O)CH=CH_2$$

Diethylene glycol diacrylate (DEGDA)

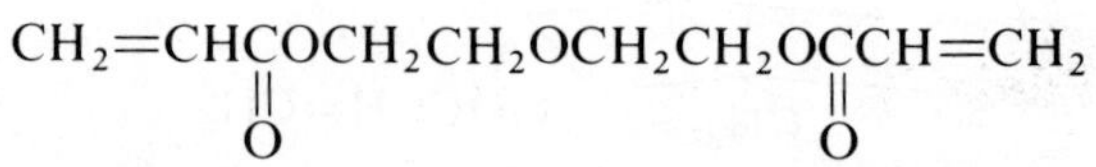

$$CH_2=CHC(O)OCH_2CH_2OCH_2CH_2OC(O)CH=CH_2$$

1,6-Hexanediol diacrylate

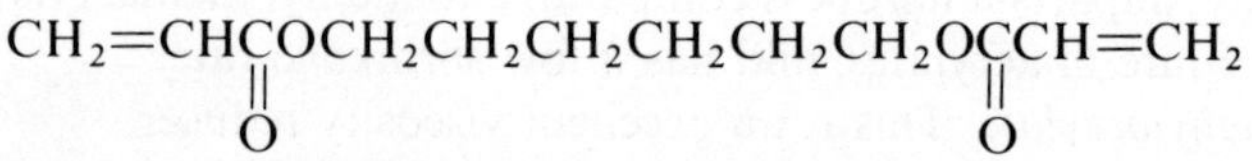

$$CH_2=CHC(O)OCH_2CH_2CH_2CH_2CH_2CH_2OC(O)CH=CH_2$$

(iii) Triacrylates

Trimethylol propane triacrylate

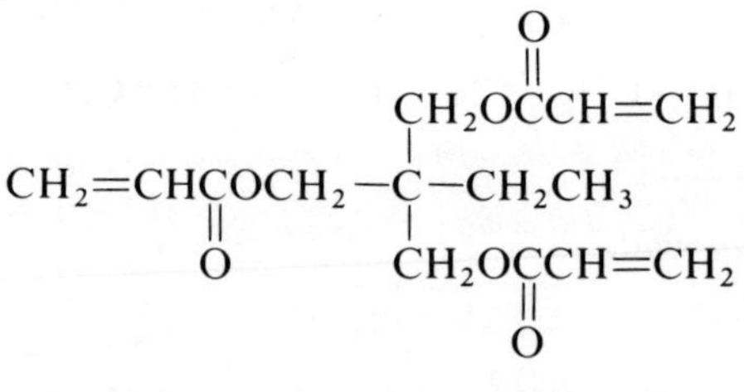

$$CH_2=CHC(O)OCH_2-\underset{CH_2OC(O)CH=CH_2}{\overset{CH_2OC(O)CH=CH_2}{C}}-CH_2CH_3$$

Pentaerythritol triacrylate

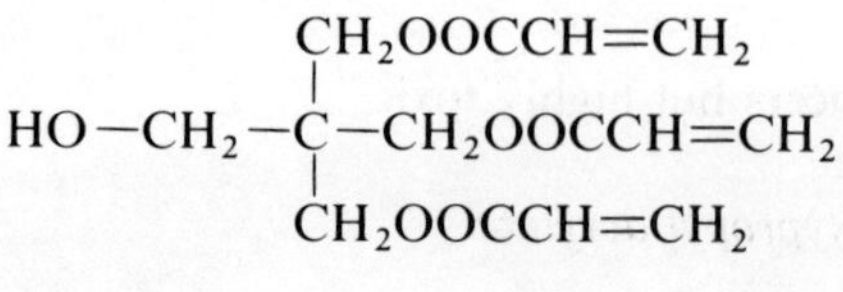

$$HO-CH_2-\underset{CH_2OOCCH=CH_2}{\overset{CH_2OOCCH=CH_2}{C}}-CH_2OOCCH=CH_2$$

(iv) Tetra-acrylates

Pentaerythritol tetra-acrylate

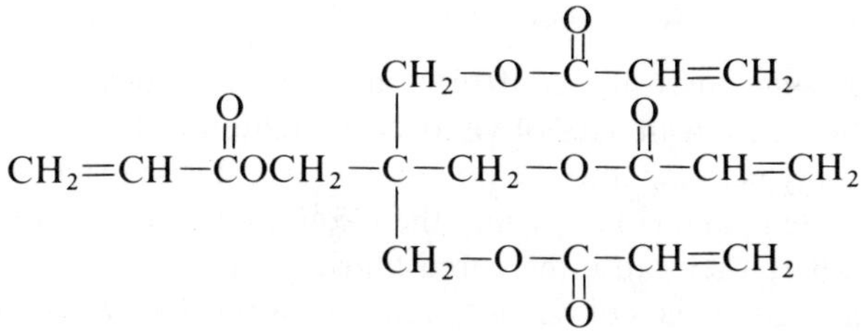

$$\begin{array}{c} CH_2-O-\overset{O}{\underset{\|}{C}}-CH=CH_2 \\ | \\ CH_2=CH-\overset{O}{\underset{\|}{C}}OCH_2-\overset{|}{\underset{|}{C}}-CH_2-O-\overset{O}{\underset{\|}{C}}-CH=CH_2 \\ CH_2-O-\overset{}{\underset{\underset{O}{\|}}{C}}-CH=CH_2 \end{array}$$

(v) Penta-acrylates

Dipentaerythrital (mono-hydroxy) penta-acrylate

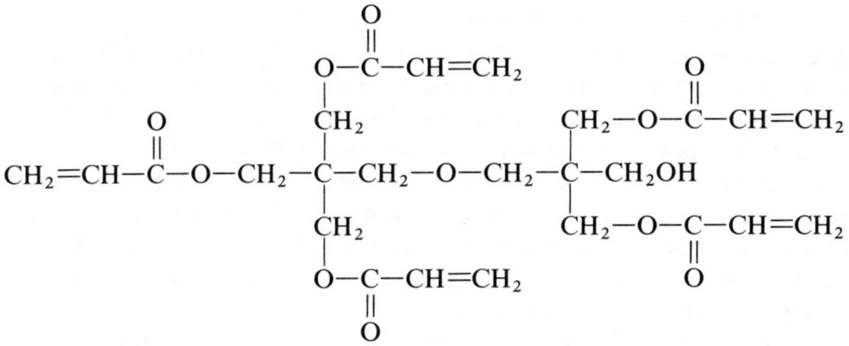

Many novel monomers and polymers have now been developed, such as the acrylic derivatives of:

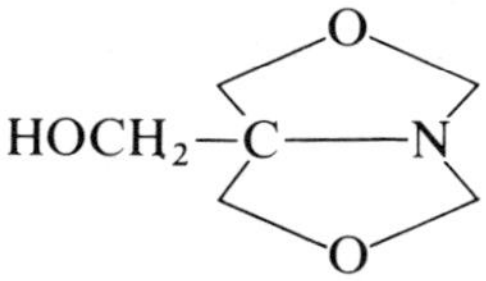

$$HOCH_2-C\overbrace{}^{O}N\underbrace{}_{O}$$

This is 1-aza-5-hydroxymethyl-3,7-dioxabicyclo (3.3.0) octane.[93-94] This has several advantages compared with available commercial monomers:

1. Simple preparation
2. Easy molecular modification to obtain optimum properties.
3. It has photopolymerizable activity even when polymerized to augment potential use.

The acrylic monomer has reactivity of the following groups:
1. A cyclic ether group.
2. Acrylic ester containing a cyclic ether group.
3. A tertiary amino group.
4. Acrylic ester of an amino alcohol.

For formulating, viscosity control is a critical problem. A few acrylate monomers offer the advantages of viscosity reduction to fit coating equipment and/or photopolymerizing into the polymer matrix. These reactive diluents stay in the cured system, and consequently their concentrations must not be greater than a limit where they cause the cured film to soften.

A particularly good class of viscosity reducers is the diacrylates. The crosslink density is low but responsivity is still good, although many have adverse dermatitic and toxicity properties. Examples are:

   1,4-Butane-diol diacrylate
   Neopentyl glycol diacrylate
   Di-ethylene glycol diacrylate
   1,6-Hexane diol diacrylate

The monomer component is often a blend, allowing a compromise between viscosity and photopolymerization characteristics. Methacrylates are slower in cure than acrylates because they do not have a labile hydrogen atom.

Electron-deficient unsaturated monomers such as the acrylate esters give good photoresponse. Vinyl monomers with electronic unsaturation are polymerized poorly by most photo-initiator systems, especially those operating by electron transfer. They can operate effectively if their reactivity ratios are such that they readily copolymerize through the propagating acrylic free radicals activated during exposure to actinic radiation.

Vinyl acetate is useful, co-reacting efficiently with acrylic free radicals. It has an extremely labile hydrogen, functional in chain transfer, and also high solvating capabilities. Adverse characteristics are its volatility and flammability hazards. The latter is significantly reduced when formulated into a coating system where the colligative properties of vapour pressure reduction are evidenced.

Rohm and Haas have published some data[95] concerning shrinkage and volatility characteristics of some common monomers. This is shown in Table 4.

4.3.3.3 Allylic Monomers

Examples of diluents possessing allylic unsaturation are:
1. *Triallyl cyanurate*

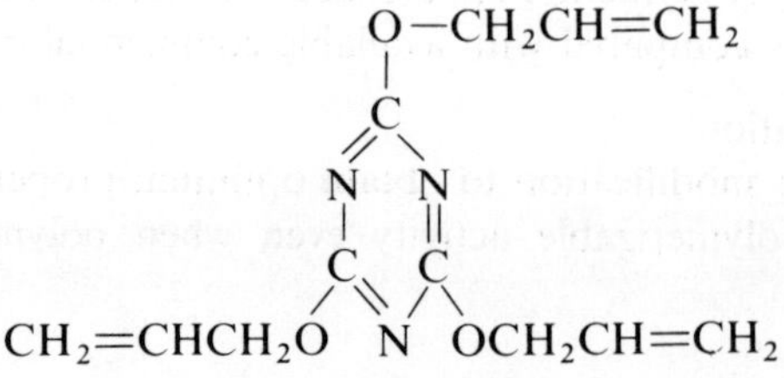

$$\begin{array}{c} O-CH_2CH=CH_2 \\ | \\ C \\ \diagup \diagdown \\ N \quad\quad N \\ | \quad\quad\; || \\ C \quad\; C \\ \diagup \diagdown \diagup \diagdown \\ CH_2=CHCH_2O \quad N \quad OCH_2CH=CH_2 \end{array}$$

2. *Trimethylol propane triallyl ether*

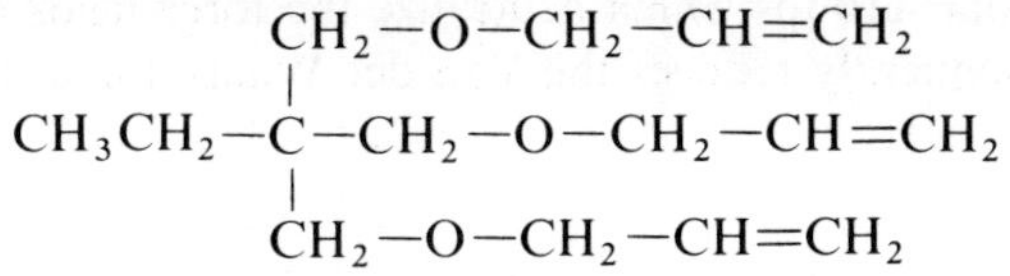

$$\text{CH}_3\text{CH}_2-\underset{\underset{\text{CH}_2-\text{O}-\text{CH}_2-\text{CH}=\text{CH}_2}{|}}{\overset{\overset{\text{CH}_2-\text{O}-\text{CH}_2-\text{CH}=\text{CH}_2}{|}}{\text{C}}}-\text{CH}_2-\text{O}-\text{CH}_2-\text{CH}=\text{CH}_2$$

Table 4 Shrinkage and volatility characteristics of monomers.[95] (Adapted with permission from Rohm and Haas Co.)

	Shrinkage (%)	Volatilization rate (mg/min)
Styrene	—	19
Vinyl acetate	21.7	—
Butyl acrylate	—	17
Methyl methacrylate	21.1	—
Cyclohexyl acrylate	12.5	1.9
2-Ethylhexyl acrylate	—	0.5
Isobornyl acrylate	8.2	0.2
Isodecyl acrylate	—	0.08
Neopentylglycol diacrylate	14.2	0.07
Hexanediol diacrylate	—	0.02

4.3.4 Plasticizing Diluents

There are numerous plasticizing diluents. Some examples of these that have been incorporated at low levels in some systems are:

1. Butyl acetate
2. Butyl cellosolve
3. Butyl carbitol acetate
4. Dipentene or pine oil
5. Tri-butyl phosphate
6. Hexadecanol
7. Diallyl-phthalate and similar plasticizers

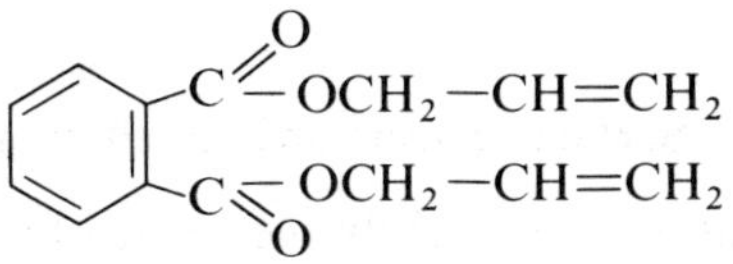

$$\underset{}{\text{C}_6\text{H}_4}\begin{Bmatrix}\text{C}(=\text{O})\text{OCH}_2-\text{CH}=\text{CH}_2\\ \text{C}(=\text{O})\text{OCH}_2-\text{CH}=\text{CH}_2\end{Bmatrix}$$

8. Sucrose acetate isobutyrate (viscous material that cuts viscosity)
9. Epoxy ester of iso-octyl tallate
10. Benzophenone (a solid photo-initiator that causes viscosity reduction).

Many of these are, however, very volatile.

Plasticizers are compounds introduced to a system to modify its flow properties and often to reduce the final film brittleness. There are two categories of plasticizers, primary and secondary.[96]

4.3.4.1 Primary Plasticizers

These contain polar groups which neutralize the force fields of polymer polar groups and consequently reduces the Van der Waals' forces between adjacent polymer chains.

4.3.4.2 Secondary Plasticizers

These are inert materials without polar groups that exist dispersed throughout the polymer providing mechanical separators between the polymer chains and thus reducing the Van der Waals' forces of attraction between them.

Plasticization may also be designated as internal or external.

Internal plastization occurs when the polymer chain is modified by incorporating small quantities of another monomer during photopolymerization. This second monomer needs bulky side groups. Vinyl acetate is a typical example, the sporadic occurrence of the bulky acetate groups in the polymer formed, providing separators and consequently reducing the intermolecular forces between adjacent chains which in turn increases the degree of viscous movement. Here an increase in plasticity requires the formation of a new copolymer rather than the use of a plasticizer as a chemically non-participating substance.

External plasticization occurs when the plasticizer is dispersed in molecular form throughout a polymer to form a viscous solution. A photopolymerizable coating must have good photocure response and application properties depending on the end use.

4.4 FORMULATION

Formulation governs the properties in the final cured film such as surface defects and levelling characteristics, mechanical properties, chemical resistance, gloss, adhesion to various substrates, and cure response (particularly if the system is pigmented or not).

Some of the effects of these variables are outlined below.[97]

4.4.1 Surface Defects and Levelling of Film

The method of application is frequently the cause of surface irregularities and these defects must be minimized to a degree not perceptible to the naked eye. The application method is often via roller coaters and viscosities of the order up to 200 centipoise at 25 °C are necessary for acceptable coatings.

Applied surface coatings attempt to form a uniform film under the influence of surface tension forces. A differential form of a levelling equation has been formulated by Orchard,[98] who postulated the use of a sine curve (for brushed paint films). For a Newtonian liquid this is:

$$\frac{da}{dt} = \frac{-h^3}{3} \cdot \frac{\sigma}{\eta} \cdot \left(\frac{2\pi}{\lambda}\right)^4 a$$

where a = the amplitude of the brushmark
σ = surface tension of system
h = film thickness (mean)
λ = wavelength (of brushmark in paint systems)
η = viscosity.

The process of levelling has an integrated form of the above equation, used by Dodge:[99]

$$a_t = a_0 \exp\left[-\frac{(2\pi)^4 \sigma h^3}{3\lambda^4} \int \frac{dt}{\eta}\right]$$

where a_0 = amplitude at time $t = 0$
a_t = amplitude at time $t = t$.

For photopolymerizable coatings, the time factor is small and the viscosity is assumed to be constant. The equation above under these conditions simplifies to

$$a_t = a_0 \exp\left[-\frac{(2\pi)^4 \sigma h^3 t}{3\lambda^4 \eta}\right]$$

The quantity enclosed between the brackets should be kept at a maximum to obtain a minimum amplitude. For adequate levelling, the surface tension is best made large and the viscosity small, the application waviness minimized, and a thicker film coating applied. Surface tension and viscosity are regulated by the coating formulator whereas the film coating thickness and application irregularities are dependent upon equipment. Camina and Howell[100] have performed a detailed study concerning the levelling of paint films. For non-Newtonian behaviour, Murphy[101] derived the equation for pseudo plastic systems to which the power law can be applied over the low shear rate range involved in the levelling process. The power-law equation is:

$$\tau = K\dot{\gamma}^N$$

where τ = shear stress
$\dot{\gamma}$ = shear rate
K and N are constants.

Murphy's equation[101] for a power-law fluid is:

$$\frac{da}{dt} = \frac{-N}{2N+1} \left(\frac{\sigma a}{K}\right)^{1/N} \left(\frac{2\pi}{\lambda}\right)^{(3+N)/N} h^{(2N+1)/N}$$

This equation is found to reduce to Orchard's equation when $N = 1$ and $K = \eta$. It can be seen that the levelling rate decreases as N decreases. Smith et al.[102] have derived an equation for the maximum shear stress τ_m existing in a film:

$$\dot{\tau}_m = \frac{8\pi^3 \sigma a h}{\lambda^3}$$

If the yield value is greater than the maximum, shear-stress levelling will not occur. Smaller yield values will stop levelling when the maximum shear stress in a film becomes equivalent to the yield value.

The monomer diluent level determines the viscosity of photopolymerizable coatings. Additives as used in conventional systems may be incorporated to regulate surface tension providing they have no detrimental effect on cure response of the system.

The coating viscosity may be a function of shear rate and for a thixotropic system, it will be time dependent. As the time interval between full cure and application is small, photopolymerizable coatings will require exceptional levelling properties, and a low viscosity at very low shear rates is necessary. A lengthy, restrictive viscosity build with time is also preferred for thixotropic systems.

Camina and Roffey[103] describe time-dependent viscosity build for thixotropic systems in terms of power-law behaviour for paint undercoats.

The thickness of the film and surface waviness are determined by the application equipment. As cure response is a function of film thickness, a situation may arise where the film levelling properties are not coincident with the film cure needs. Surface waviness may be optimized with tandem precision roll coaters, or reverse roll coaters, when the end use permits this equipment.

The latter methods determine the rheology in a manner such that viscosity and surface tension effects discussed above are lessened.

Inadequate wetting is a surface defect often observed in photopolymerizable coatings and shows itself in beading or crawling. Cratering and blistering also may occur, owing to the fact that the time gap between application and cure is small. Time-dependent defects are of lesser importance. Crawling and cratering may be overcome by the use of additives. Blistering may be the result of either air or solvent entrapped in the film. Blistering may be aggravated by the fast setting time of photopolymerizable coatings, in particular if trapped solvents are driven upwards from either the substrate or basecoat by the infra-red radiation from the light source. The higher molecular weight styrene/polyester systems seem inherently more sensitive to application troubles, probably owing to a lower order of polymer/solvent compatibility. A wider solvent selection is available to acrylics and they have greater latitude with respect to molecular weight, permitting solution to the rheological problems with skilled formulation.

4.4.2 Mechanical Properties

The mechanical properties required of photopolymerized films[104] include hardness, toughness, impact resistance, and flexibility. The end application determines those properties which are critical and a compromise is frequently necessary to satisfy any single performance needed. This equilibrium has to be related to the critical response and adhesion parameters.

4.4.2.1 Hardness

Generally this may be defined as the ability of a film to resist surface abrasion. Hardness is necessary for wood and metal coating and may be obtained by:

(a) increasing the aromatic, cycloaliphatic, etc., ring density;
(b) increasing the crosslink density;
(c) increasing the glass transition temperature (T_g) of the film;
(d) Stoving, subsequent to photo-cure for metal coatings.

Polyester systems often have a high ring density because the reactive diluent is frequently aromatic in nature (e.g. styrene). Extra ring content is preferable in the polymer and can be obtained by the incorporation of phthalic and similar ingredients.

Augmenting the crosslink density of the film confers greater hardness. Acrylic systems permit greater design latitude as multifunctional monomers are readily obtained. Also, reactive oligomers may be readily synthesized with variable unsaturation available.

As a surface consideration, abrasion resistance is a function of the coefficient of friction of the film surface. Related to the film bulk, abrasion resistance correlates to film toughness where hydrogen bonding can play an important part. The coefficient of friction for a photopolymerized film may be altered considerably depending on the cure conditions. A greater cure often lowers the coefficient of friction and frequently increases the abrasion resistance. Reductions of coefficient of friction are affected by both atmosphere and exposure, and additives such as waxes, etc.

Generally, polymers are either soft, flexible substances or hard, brittle, glassy materials. If the former are cooled sufficiently, a temperature is attained where they become hard and glassy. The temperature at which this change in properties takes place is known as the glass transition temperature (T_g.).

In the case where the polymer glass transition temperature is greater than room temperature, those materials which are normally hard will become soft and pliable when heated (thermoplastic).

The glass transition temperature is a property associated with the amorphous regions in a polymer and may be described as the temperature at which, on applying thermal energy, the polymer chains gain enough heat energy to vibrate in a co-ordinated fashion. Apart from limited vibrations about equilibrium positions, below this temperature the atoms are in a rigid frozen state and the polymer is consequently hard and glassy. At temperatures above the glass transition, the polymer becomes rubber-like in properties. Incorporation of plasticizers and other additives decrease the Van der Waals' forces magnitude to give a reduction in T_g.

If increased crosslinking increases T_g of the photopolymerized film, it can be expected that the T_g and molecular-weight relationships can relate to the expression developed by Fox and Flory:[104]

$$T_g = T_g^\infty - K_g M^{-1}$$

where T_g^∞ is the limiting T_g at high molecular weight

K_g is a constant (where molecular weight $M < 10^{+4}$) depending on the photopolymer system.

An increase in the ring density can increase T_g, and methacrylates can also harden the film by increasing T_g. The film can be softened by long hydrocarbon chains by lowering the T_g, which is determined by polymer polarity and chain stiffness.

4.4.2.2 Toughness

This may be defined as the surface coating's ability to absorb energy prior to or during fracture. Two of the chief properties governing toughness are tensile strength and elongation.

High crosslinking in a film causes brittleness. Elongation is at a minimum level owing to these restrictive crosslinks and a significant quantity of internal stress is propagated whilst these internal stresses are relieved, and consequently film rupture may occur. Stress cracking and functionality (i.e. the greater functionality the more rupture) appears to be correlated. There would seem to be an overall functionality limit that is also governed by film thickness. The molecular weight of the prepolymer unit may increase crack resistance by separating the distance between crosslinks. Toughness for acrylic systems should be balanced with cure rate required. Highly functional systems can then become hard but not necessarily tough. Polyester systems do not seem to be so functionality dependent. Adequate tensile and elongation properties should be incorporated into the backbone of a polyester resin to meet the necessary film toughness requirements.

4.4.2.3 Impact resistance and flexibility

These parameters are of particular importance to metal coatings and are determined by the visco-elastic character of the polymer. Factors such as time after impact and temperature are important. The fast cure response required for these coatings probably necessitates a high amount of functionality which will tend to increase hardness and reduce both impact resistance and flexibility.

Polyester systems are often inadequate for high-speed cure application when flexibility is wanted, and highly functional acrylic systems are favoured. Rapid-responding monofunctional monomers are necessary to meet the balance of cure/flexibility. 2-Hydroxyethyl acrylate and 2-hydroxypropyl acrylate are useful but have unfavourable skin irritancy and toxicity aspects. A flexibilizing monomer is 2-ethyl hexylacrylate but it tends to slow the cure rate and has the disadvantage of high volatility.

Adhesion to the substrate and film toughness can contribute to the flexibility and impact resistance of a specific film/substrate combination.

4.4.3 Chemical resistance

The chemical resistance of a photopolymerized coating is like that of other highly crosslinked systems. Coatings may withstand at least 50 rubs with methyl ethyl ketone in thin film when adequately cured. Polyester/styrene systems have solvent resistance to both polar and aqueous solutions. Acrylic systems have to be carefully chosen for resistance to aqueous media such as hydroxyls, carboxyls, and other hydrophilic groupings should be avoided.

An increase in crosslink density or exposure time can improve resistance to polar media but it is difficult to overcome the solvent sensitivity especially when hydrophilic species are at a high level inside the film.

4.4.4 Gloss

The extremes of matt and high gloss[105-106] are often wanted in photopolymerizable systems. High gloss is simpler to achieve than matt systems where viscosity problems can occur due to the incorporation of matting agents such as extender pigments, which in turn can affect application and general rheology. Some factors affecting gloss include:

(a) the nature of the matting agent and concentration;
(b) atmosphere;
(c) lamp focus and intensity;
(d) distance from source;
(e) coating cure rate.

In general, the greater the particle size of the matting agent or the broader the distribution range, the more efficient the matting obtained. A high loading of matting agent is necessary for photopolymerizable systems as a short time is permitted by the cure process for the matting agents to agglomerate and orientate themselves as surface perturbations.

Acrylic systems are often much harder to modify with matting agents than polyester/styrene types, probably because the absence of hydrogen bonding in the reactive solvents allows more freedom of movement of the pigment particles prior to curing and consequently augments matting efficiency. These systems tend to cure at a slower rate, allowing more time for surface disturbances before films set.

For aerobic curing of a matted photopolymerizable coating, the gloss is often noticeably lower than when the exact formulation is cured under an anaerobic inert atmosphere such as nitrogen.[107] A useful approach could be to cure the coating, first in air to establish gloss and then subsequently with a UV/inert atmosphere exposure to obtain desired film properties. As well as atmosphere considerations, lamp focal distance, intensity, and also source distance determine. gloss. Lamp decay or variations may cause a range of glosses for a specific formula and set of cure conditions.

Gloss[108] depends to a large extent on the rate of cure of the air/surface interface as opposed to that of the underlying portion of the film. It is debatable whether the time between film application and film set (open time) controls surface disturbances through volatility of monomer diluents, or if agglomerations are affected by longer dwell times for the cure process.

The optimum gloss properties are a result of a combination of the physical effects and the variations in cure rate related to the system's chemistry, and gloss is a complex combination of a multitude of variables, currently often approached by trial and error in formulation technique.

4.4.5 Adhesion to Various Substrates

Adhesion is governed by several parameters:[104]
(a) interfacial contact and surface tension;
(b) curing time/temperature relations;
(c) shrinkage forces;
(d) coating/substrate interactions.

Adequate wetting by close contact of the surface coating with the substrate is essential for the attainment of satisfactory adhesion. Interfacial contact is necessary as the forces of abstraction bridging the interface are effective for short distances, up to about 10 nm.

In order for the surface coating to wet the solid substrate, the critical surface energy of the solid (γ_S) has to be larger than the surface energy of the liquid (γ_L).

For thermally cured conventional systems, interfacial contact is relatively easily achieved owing to the facts that:

1. γ_L is inversely proportional to the temperature.
2. The surface coating is kept in the liquid phase whilst solvent removal occurs at raised temperatures.
3. The temperature is significantly higher than the glass transition temperature (T_g) of the film.
4. The film is annealed whilst cooling, therefore relaxing the residual cure stresses.

Photopolymerizable systems do not have an advantage over these thermal assists unless given a post-cure treatment at raised temperature. Infra-red radiation from the lamp source may be adequate if the photo-cure response is slow enough. Frequently the time/temperature relationship is insufficient to create adequate adhesion forces, in particular on metal substrates. This same time/temperature aspect may be significant to the wetting process before photo-cure as equilibrium may never be obtained prior to the film being set under the actinic radiation source.

In the cases when wetting is obtained, formulations for rapid photo-cure may develop less adhesion than expected. Fast photo-cure is frequently obtained by incorporation of a high level of either maleate/fumarate or acrylate functions in the surface coating. Great internal stresses are caused when the film is photo-cured as molecular distances are changed to atomic distances in a very short time interval. These shrinkage stresses in high-density crosslinked surface coatings are frequently strong enough to tear the coating off the substrate and free film character occurs.

If a photopolymerizable top coat is applied over a conventional basecoat, the shrinkage forces are likely to cause failure at the weakest interface. A basecoat might have excellent adhesion prior to coating but then become readily detached from the substrate after a topcoat is applied. An optimal formulation equilibrium is necessary to donate adequate cure rate with a minimum of residual stress forces to avoid adhesion failure. Care must be observed in the reverse case, that is when a conventional overprint varnish is applied over a UV ink, as solvent mobility from the varnish may occur into the latter, destroying adhesion.

Substrate/coating interactions are both chemical and physical in nature. Either ionic or covalent bonding between the coating and substrate can evolve powerful adhesion forces. When these bonds are made, the energy required to sever them may sometimes be greater than the cohesive strength of the material itself. Owing to the transient time gap and relatively low temperature, covalent bonding forces are not easy to create for photopolymerizable systems. They can, however, be a novel way of overcoming problematic adhesion properties. Ionic bonds are more readily formed but frequently are less chemically stable.

Some physical conditions of significance to adhesion include topography (the surface nature) of the substrate and physical absorption properties of the coating. Surface roughness promotes adhesion. If the monomer solvent is physically absorbed then promotion of weak interfacial forces may occur if the solvent substrate interactions are large and polymer/monomer solvent interactions are small, thus encouraging preferential absorption. It is of paramount importance that the polymeric constituents associate satisfactorily with the substrate for optimum adhesion.

To optimize adhesion in photopolymerizable systems the following may be considered:

1. Minimize coating surface tension and maximize substrate surface energy for adequate interfacial contact.
2. Minimize shrinkage forces (reduced functionality) to the extent necessary for required cure response.
3. In place of selected absorption of monomer solvent, advocate association of polymeric material with the substrate.
4. Employ extra thermal pre- and post-treatments if feasible.

Extra needs such as good adhesion on exposure to high humidity, boiling water, or extreme bending often poses extra problems in formulation technology.

Metal decorating is problematic for good adhesion as this is difficult to achieve very satisfactorily without a post-bake and a pre-treatment depending on the type of metal used. The usual pre-treatment is 'flaming' to remove the rolling oils such as palm oil, sebacates, etc., which destroy adhesion. For systems requiring no post-bake, photopolymerizable resins and monomers are best used that have low shrinkage characteristics after and during photo-cure as the large stresses thus induced into the film may render adhesion to be poor.

Adhesion to aluminium may be improved by pre-treatments with etch solutions such as orthophosphoric acid/alcohol and persulphate/acid combinations.

Lack of adhesion to plastics is similar to that found for conventional ink systems. For example, the problem with polythene is one of surface oxidation. This is found to be improved by the usual pre-treatments to make it polar, such as:

1. Gas flame.
2. Chlorination.
3. Electron (corona) discharge.

A small amount of solvent to give 'bite' into the plastic is sometimes tolerated in the photopolymerizable coating.

4.4.6 Pigmentation of Photopolymerizable Systems

This parameter affects photo-curing in many ways.[109–112] The presence of pigment, by virtue of its physical properties will affect the photo-initiator/sensitizer absorption characteristics by competing with the latter for the incident radiation. This can be understood in terms of scattering effects, film thickness, substrate type, amount of photo-initiator/sensitizer concentration, and the nature of its colour (i.e. yellow, magenta, cyan) or non-colour such as black, white, or whether it is an extender. The latter often possesses a refractive index close to that of the media present and is fairly transparent.

Competition for incident radiation into the film occurs because of additive absorption effects as depicted by the equation below at constant wavelength:[113]

$$I_A = I_0(1 - \exp[-(\varepsilon_1 c_1 + \varepsilon_2 c_2 + \varepsilon_3 c_3 + \cdots + \varepsilon_n c_n)l]$$

where n is any number of absorbing components, l is the film thickness, I_0 the incident radiation, and I_A is the total absorbed radiation. Consequently, the fraction of light absorbed by a component x is:

$$I_{A_x} = I_A \left(\frac{\varepsilon_x c_x}{\varepsilon_1 c_1 + \varepsilon_2 c_2 + \varepsilon_3 c_3 + \cdots + \varepsilon_x c_x} \right)$$

The significance of these formulae is that it may be seen that absorbing species will often cause poor absorption at the bottom of the film and subsequent poor cure and adhesion. Surface wrinkling may occur if too much radiation is absorbed at the film surface due to volume shrinkage during photopolymerization.

Complicated expressions may be developed based upon the Kubelka Munck equation.

Scatter of actinic radiation in a system is due to the difference in refractive indices of the pigment and media present, the size of the pigment particles, the pigment concentration, and the nature of the radiation wavelength, all of which are interrelated in their effect on scatter. Scatter increases the effective radiation path length throughout the film and also the internal reflectance and diffusion. The internal reflectance is very pronounced in pigmented systems on a reflective substrate (up to a certain pigment level), as it increases the path length of light more than in clear systems which themselves obtain more radiation passing through the film, in contrast to non-reflective substrates.[113]

Photoreactivity of a pigment will also cause an effect on the systems cure rate. Photo-sensitization quenching in this sense has already been discussed in Chapter 3. Pigmentation level will also affect the rheology, surface characteristics (wrinkling or rivelling), matting efficiency, and pot stability. For the latter, pigmentation often causes thermal instability in photopolymerizable coatings, due to after-treatments.

Application problems can obviously arise when a satisfactory clear coating is pigmented, mainly due to a rise in viscosity due to wetting and dispersion characteristics of the pigment/media interaction. Flow-out consequently suffers as there is inadequate flow time when the film is 'set' after being passed under a lamp. Heat curing systems are able to flow out as the heat will reduce the initial viscosity, making the fluid more mobile.

Matting pigment extenders such as silica and talc tend to give less gloss reduction in photopolymerized systems compared to conventional solvent-based types, as the solvent in the latter causes pigment to migrate to the surface, where the solvent evaporates. This causes stresses in the film surface and subsequent shrinkage and unevenness.

For ink films of around 3 μm in thickness applied to glass slides and cured at variable rates, the variation of log cure rate in m/min/lamp with log relative intensity as a function of the colourant used, i.e. a process set, magenta, yellow, cyan or black, the rate of cure was found not to vary with the 1.0 or 0.5 power of the intensity, but followed a higher power, dependent and differing according to the colourant. The cure rate was found to be in the order

$$\text{magenta} > \text{yellow} > \text{cyan} > \text{black}$$

This is depicted in Figure 57.[114]

For tack-free and thumb twist–free films, exactly the same cure order was found but the rate of cure for a given colourant was always faster for the tack-free state than the thumb-twist level.

The variance in cure rate seen for the different colourants have been attributed to different absorbances of these colourants over the wavelength range 220–350 nm.

Figure 58 depicts the percentage transmission wavelength curves of 10 per cent suspensions of these colourants in minerals oil. The curves vary somewhat with wavelength but magenta generally shows the highest percentage transmission at a given wavelength, followed by yellow, cyan, and black. This is also the decreasing cure rate colour order found experimentally.[115]

The presence of colourant also affects the behaviour of photo-sensitizer, presumably owing to competitor reactions and/or reducing transmission. The rate of cure decreases slightly with increasing film. The decrease of log rate of cure

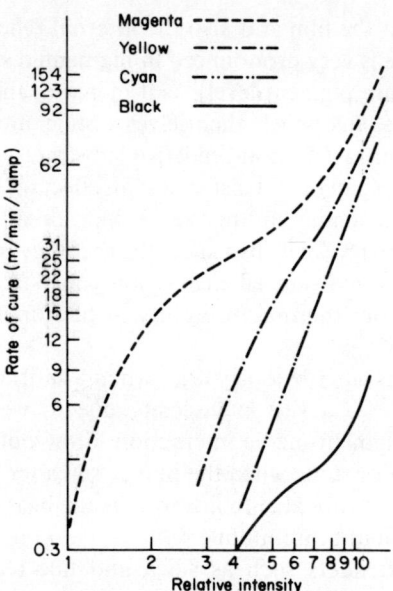

Figure 57   Variation of log rate of cure with log relative intensity of UV irradiation as a function of pigment colour. (Reproduced by permission of Technology Marketing Corporation.)

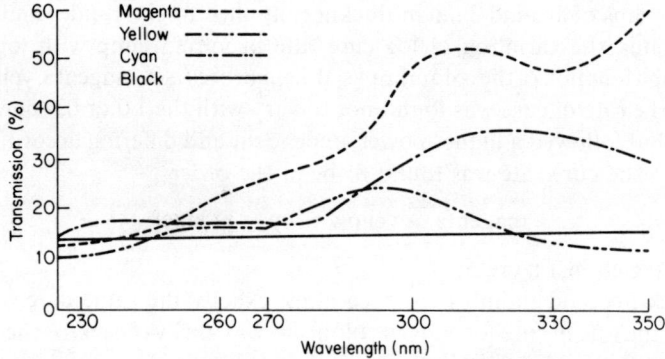

Figure 58   Some typical transmission characteristics of a process set of colours (10% pigmentation level). (Reproduced by permission of Technology Marketing Corporation.)

with film thickness was linear and approximately the same for the four colourants; in the range 2–12 μm thickness the rate of cure was about halved for every 5 μm film thickness increase.

4.4.7   Cure Response of a System

Photocure may be defined as the phase conversion by actinic radiation of a free-flowing liquid to a tack-free solid. More than a tack-free state on the film surface is often necessary, determined by the specification of the end user.

Three empirical practical levels of cure can be simply ascertained by using the finger.[115] These are:

(a) the surface, tack-free to the touch;
(b) thumb-twist, does not remove film (bulk cure);
(c) scratch resistant to the fingernail.

(a) *Tack free.* This can be defined as the absence of stickiness or ink transfer when the ink film is touched or comes into contact with another surface. For a very porous substrate, this type of 'cure' is often obtained with minimal UV irradiation aided by some absorption by the substrate. For a less porous or impervious substrate, a dried surface film may cover a semi-cured film, which can be smeared under pressure.
(b) *Thumb-twist free.* This requires the application of a downward pressure of the order of about 5 kg, simultaneously with a rotary twisting motion being applied to the film with the thumb. An unbroken film is then considered through cured.
(c) *Scratch free.* A fingernail edge is used under pressure to scratch the film. An unbroken film if it passes this severe test is regarded as fully cured and usually ready for immediate in-line processing (die-cutting, scoring, etc.).

The number of rubs with methyl ethyl ketone needed to move the film provides an indication of through cure or degree of crosslink density. Alternative methods such as dilatometry, spectroscopic analysis,[116] gas liquid chromatography, or estimation of monomer left by weight loss can be used. Cure assessment by the thumb test, scratch resistance, solvent resistance, hardness, etc., are all subjective tests in essence. Infra-red spectroscopy, however, can provide a convenient objective means of assessment.

Infra-red spectroscopy is objective as it provides a method of measuring the actual conversion of functional groups.[117-121] The acrylate functional group is the entity studied, as throughout the cure process this group:

$$CH_2=CH-\overset{\overset{O}{\|}}{C}-O-$$

changes to the grouping

$$-CH_2-\underset{|}{CH}-\overset{\overset{O}{\|}}{C}-O-$$

The infra-red spectrum subsequently modifies to accommodate this change. The vinyl group absorption peaks decrease, whereas an increase occurs with the $-CH_2-$ group. The acrylate $CH_2=$ twist in the wet film is generally a fairly large well resolved peak occurring at about 800–810 cm^{-1}. Changes in the absorbance can be used to monitor quantitatively the degree of cure via the percentage of original double bonds remaining after cure. A double bond peak at about 1635 cm^{-1} may also be used to monitor the reaction.

The cure response is determined by many factors, including the nature of the lamp, photo-initiator/sensitizer and film thickness interdependence, type of substrate, atmosphere, temperature, and the chemical formulation of the system.

4.4.7.1 Lamp

The reliance of through-cure on the distance of the lamp source to the surface coating, and consequently the irradiance, has been found for clear coatings to be basically constant. Surface cure, however, increases greatly on increasing the density of radiant power. This may be interpreted as below.

Intense irradiance can propagate a proportionally greater level of free radicals than diffuse irradiance is expected to. At the film surface this can lead to an enhancement of the photopolymerization reaction over the competing oxygen scavenging. In the bulk film where the oxygen gas level can be expected to be small, the photopolymerization rate is governed by the total available energy instead of its distribution. Photo-cure seems to be more greatly affected by the irradiance value than the exposure time and this implies that for a specific level of cure, throughput rate can be more efficiently increased by augmenting the lamp power rather than by increasing the number used.

The dependence of cure rate on intensity is found in some cases, at low conversion levels for non-pigmented surface coatings, to be:

Initial rate of cure $\propto$ (intensity of incident light)$^{1/2}$ $[S]^{1/2}$ where $[S]$ is the photo-sensitizer concentration.

4.4.7.2 Photo-initiator/Sensitizer and Film Thickness Relationship

For photo-initiators/sensitizers absorbing at short wavelengths such as 254 nm and longer wavelengths at 330 nm it appears that the latter determines through cure whereas the former is responsible for much of surface curing.[122]

The incorporation of larger amounts of a photo-initiator/sensitizer absorbing at 254 nm in a film tends to give more efficient surface cure. This means a faster throughput rate for an equivalent energy intake, but care must be taken or else surface rivelling or wrinkling may occur.[123]

Film thickness is an important factor, significant effects being seen in extremely thick or thin films. In the case of thin films ($<5\ \mu\mathrm{m}$) reduced scratch and solvent resistance can frequently be seen. In thick films, undercure can be expected at the coating/substrate interface, dependent to a degree on the photo-initiator/sensitizer's extinction coefficient. A large extinction coefficient often results in surface cure only.

For thick surface coatings, excessive amounts of photo-initiator/sensitizer, while allowing efficient curing of the film to the tack-free state, may be counterproductive for the overall film curing.

For a low-level photo-initiator/sensitizer concentration, radical fragments can be generated somewhat uniformly in the reaction system. As the concentration is increased, more fragments can be formed and an increase in photopolymerization would be expected. This is not borne out in practice as at proportionally higher photo-initiator/sensitizer concentrations, more and more radicals appear to be generated in the areas nearer to the UV lamp source and this uneven distribution can result in an overall lowering of the polymerization rate.

For a film coating this would correspond to less photochemical activity in the critical bottom layer and consequently a slower rate of through-cure.

With the exception of significantly thin films, surface cure is independent of the substrate reflectance, but for through-cure, substrate reflectance is an important factor in determining throughput rates.

The relative energy requirements for top and bottom cure using a benzoin butyl ether photoinitiator for a 12 μm coating film has been found to be for the surface 1 μm layer and the bottom 1 μm layer to be of the order 20:1. The additional energy for surface cure is necessary to form an adequate excess of radicals to supercede the oxygen inhibition.

4.4.7.3 Atmosphere

An inert atmosphere reduces the energy necessary to obtain a specific level of cure.[107,124,125] Removal of oxygen permits the improvement of surface scratch resistance, probably by reducing the chain terminations at the atmosphere/coating interface. Surface cure may also affect slip and block resistance.

The effect of oxygen is discussed elsewhere (see Chapter 3).

Plews and Phillips[117] have investigated the effect of air inhibition on photopolymerizable systems. Poor cure due to oxygen inhibition is shown in thin films and passes through an optimum value as the film thickness increases. Rubin[126] found that increasing the intensity of irradiation reduces the effect of oxygen inhibition. Also, decreasing the intensity of irradiation has no effect on the cure of films in nitrogen.

A combination of a photo-initiator such as 2-chloro-thioxanthone (2CTX) and ethyl *para*-dimethyl aminobenzoate (EPDMAB) is often less affected than an initiator such as diethoxyacetophenone (DEAP) by oxygen inhibition.

Oxygen is a free radical in its ground state and is a very effective quencher of excited states. 2CTX alone is a poor acrylate photo-initiator as the rate at which it can create monomer radicals by hydrogen abstraction from the acrylate group is slow in comparison to the rate at which its excited state is quenched by oxygen. Tertiary amines such as EPDMAB are very efficient hydrogen donors, minimizing oxygen quenching.[127]

Also, as oxygen is a ground-state free radical, it may react with other intermediate free radicals in the photopolymerizing process, the peroxy radical formed being a non-effective chain initiator.

$$R\cdot + O_2 \rightarrow RO_2\cdot$$

If an effective hydrogen-atom donor such as an amine is present, an efficient radical can be liberated by a hydrogen transfer reaction.[128]

$$RO_2\cdot + -CH_2-NRR' \rightarrow RO_2H + -\dot{C}H-NRR'$$

For the 2CTX/EPDMAB combination, the α-amino radical derived from the excited charge complex will act as an effective scavenger of oxygen by a chain process throughout which amino radicals are constantly regenerated, maintaining their concentration for the chain initiation process.

The initial hydrogen abstraction process for the above combination may be written:

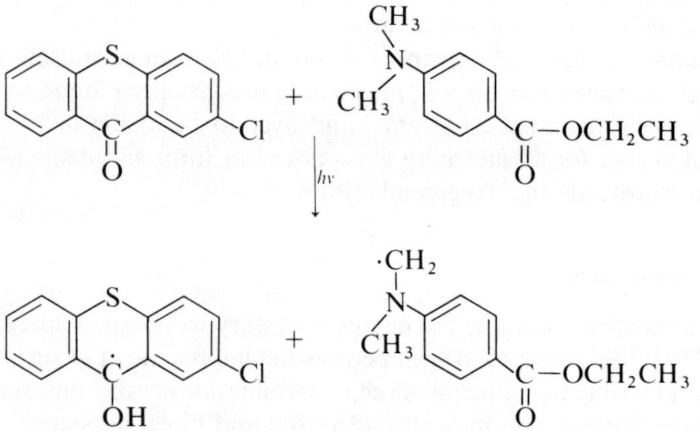

[Structural scheme: 2-chlorothioxanthone + ethyl 4-(dimethylamino)benzoate $\xrightarrow{h\nu}$ corresponding ketyl radical (C–OH) + α-amino radical (·CH$_2$–N(CH$_3$)–Ar–C(=O)OCH$_2$CH$_3$)]

but it is likely to be more complicated, involving a charge transfer complex.[129]

Amines are also known to enhance the surface cure of formulations containing DEAP and 2,2-dimethoxy 2-phenylacetophenone.[122]

Collins and Costanza[130] have found that the photopolymerization of 1,6-hexanediol diacrylate (HDDA) with benzoin isobutyl ether (BIBE) is inhibited by dissolved oxygen. This could be eliminated by the addition of N,N'-dimethylaminobenzaldehyde and eosin Y.

The removal of oxygen interference to initiation may be achieved by forming a strong oxyplex that irreversibly fixes the dissolved oxygen. The aromatic portion of the p-dimethylaminobenzaldehyde (DMABA) molecule assumes a radical anion-like character on photoexcitation which strongly favours oxyplex formation and consequently leads to irreversible oxidation:

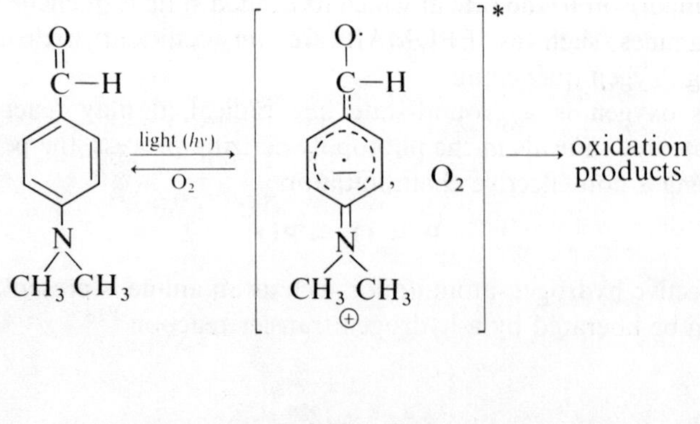

[Scheme: p-(dimethylamino)benzaldehyde $\underset{O_2}{\overset{\text{light }(h\nu)}{\rightleftarrows}}$ [radical anion-like excited complex with O_2]* $\longrightarrow$ oxidation products]

An enhancement of the benzoin isobutyl ether photo-initiator efficiency can be achieved by using a triplet sensitizer, (unpaired spins of fragment radicals provide an energy barrier to recombination), eosin Y, to cause the benzoin ether to fragment from the triplet state rather than the singlet state (paired spins and therefore favourable energy requirements for recombination) which can curtail primary radical 'solvent cage' recombination.[131,132]

To summarize, then, for this system with no additives, radical fragments are lost by oxygen deactivation and cage recombinations. With DMABA present, oxygen deactivation is eliminated and radical fragments are lost by cage recombination. When both DMABA and eosin-Y are present, oxygen deactivation is eliminated and cage recombination is suppressed causing an increase in initiation efficiency.

Collins and Costanza also found that during the course of the polymerization, the reaction became retarded and ultimately lead to unreacted acrylic unsaturated groups or residual unsaturation. This possibly occurs as the polymer network formed is highly crosslinked and contains occluded acrylic groups. It was thought that after the network reaches the ultimate residual unsaturation, triple bond species arise from the intramolecular photochemical cycloaddition of the unsaturated portion to the keto portion of the acrylic groups that are isolated within the network as shown below:

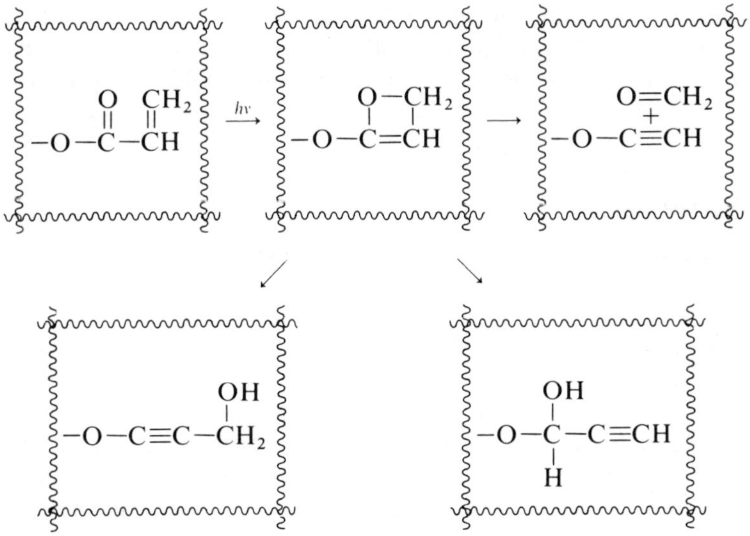

$$\begin{bmatrix} -O-\underset{\parallel}{C}-\underset{\parallel}{CH} \\ O \quad CH_2 \end{bmatrix} \xrightarrow{h\nu} \begin{bmatrix} -O-C=CH \\ | \quad | \\ O-CH_2 \end{bmatrix} \longrightarrow \begin{bmatrix} -O-C\equiv CH \\ + \\ O=CH_2 \end{bmatrix}$$

$$\begin{bmatrix} -O-C\equiv C-CH_2 \\ | \\ OH \end{bmatrix} \qquad \begin{bmatrix} -O-\underset{|}{\underset{H}{C}}-C\equiv CH \\ | \\ OH \end{bmatrix}$$

These triple bond species have been found by infra-red studies, as during cure, the growth of new absorptions have been observed at 2340 cm^{-1} and 2120 cm^{-1} which are typical of alkyl substituted acetylenic groups.

Oxetane intermolecular species as intermediates are known in the literature.[133]

The overall residual unsaturation of hexanediol diacrylate was successfully lowered from 13 to 3 per cent by use of a diluting comonomer such as butoxyethylacrylate. Vicinal hydrogens in ethers are susceptible to abstraction by radicals.[134,75] It is postulated that chain transfer to molecules possessing the ether groups permits the radical chain process to perpetuate without forming a crosslink site which would rapidly increase the network structure of the reaction medium and may ultimately cause occluded unsaturated groups.

4.4.7.4 Substrate

Factors such as conduction of the substrate and any convection cooling for the lamp source complicate the situation. This integration of the radiation, conduction, and convection thermal effects can affect scratch, block resistance, and in some instances residual film odour.

The substrate frequently affects both surface and the bulk and through-cure.[113] Metal substrates by reflection tend to cause bulk through-cure. Cold substrates or moist ones frequently have a retarding influence. For multi-coat layers, solvent release from an underlying coating film may result in a total undercure of the topcoat. Likewise, attack of the basecoat by the monomer solvent of the topcoat can inhibit its cure to the extent that film properties may become unacceptable.

A reflecting substrate can influence cure depending on the kind of work printed, such as solid or half-tone. A half-tone dot, being small, will receive more reflected radiation underneath it than solids, and would be expected to cure faster.

4.4.7.5 Temperature

The temperature affects the cure rate and consequently the final level of film conversion. The high-intensity, medium-pressure mercury arc lamp ubiquitously employed for photo-curing produces a mixture of ultraviolet and infra-red radiation, and both of these determine the coating's response. Triplet excited states resulting from photo-initiation are of the order 10^{-3} to 10^{-6} s in lifetime and the initiation rate is fast. The ensuing propagation reaction is partnered by a thermal release from polymerization and to this heat energy is also added the infra-red heat from the lamp source, and the total is absorbed by the coating. Elevated temperatures speed up radical reactions and then temperature effects are manifested both in the acceleration and degree of cure.

The final cure stage, that is, conversion to solid film, is probably dependent upon the ability of the monomer to diffuse through the polymer to active sites, and the glass transition of the polymer network may be an important temperature-dependent factor. This is partly because during the formation of the polymer network, less monomer is left to function as a plasticizer and a point is attained where the system becomes glassy. The higher the reaction temperature the less the quantity of monomer needed to keep the non-glassy state, and consequently the higher the limited percentage reaction.

It has been found that in order for small amounts of residual styrene monomer ($<15\%$) to remain in some systems, photo-curing should be carried out at a temperature of at least 50 °C.

4.4.7.6 Theoretical treatise

Many of the parameters previously outlined simultaneously affecting the cure of a coating have been investigated by Herbert Rubin of Inmont, and his theory and paper are reproduced below by kind permission from TAGA for the role of radiant absorption by photo-initiators in UV curing.[135]

This theory takes account of the UV source spectrum, the spectral absorptions of photo-initiator and oligomer, coating thickness, and substrate reflectance, and yields the spectral radiant power absorbed in the layers of a clear coating. Distinctions between surface cure and bottom cure and the role of the absorption peaks of the photo-initiator are presented with experimental verification.

Kinetic evaluations have pointed out significant differences between thin layers and massive systems. These are concerned with the existence of gradients of light absorption across a film. This means that the concentrations of reactants and products in polymerization also depend upon time and depth. For this reason, the solution of the kinetic expression becomes very complex. In the overall photopolymerization process the absorption of ultraviolet radiation by the photo-initiator is a critical step. This takes place in the presence of oligomers and additives which may interfere by optically masking some of the source radiation in one or another of its wavelength bands. A first-order means of treating this problem which focuses on this critical step is given. It takes into account the concentration and spectral absorption of the initial components, the spectral radiance of the UV source, the film thickness, and the substrate reflectance. The basic premise is that the properties of a cured coating are strongly dependent upon the UV light absorbed by the photo-initiator in the system at zero time, that is, on the first instant of exposure to the source.

A simple model for light interaction with a film is given in Figure 59. Light rays undergo reflection at every interface between media whose refractive indices differ; therefore, light is reflected and lost for photopolymerization at the air–coating interface. Further, part of the ray which has passed through the coating is absorbed by the substrate and is lost and the remaining part is reflected so that it can again traverse and be absorbed by the coating. This ray will in turn be partly reflected and partly transmitted at the coating–air interface. This process, in theory, occurs indefinitely. A complete expression for the case of an infinite number of passages will be given. In most cases only a few passages need be considered because the light ray is rapidly attentuated. For a single component in a coating of thickness l cm, concentration C moles/litre, and decadic light absorption coefficient ε litre/mole-cm, the transmission for one passage is given by the Beer–Lambert law:

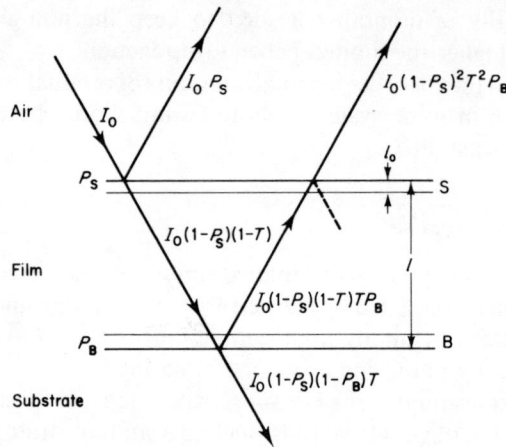

Figure 59 Light absorption of coating.[135] (Reproduced by permission of the Technical Association of the Graphic Arts.)

$$T = 10^{-\varepsilon Cl}$$

T is zero for a totally opaque coating and is unity for a perfectly clear coating. The reflectances of the air–coating and coating–substrate interfaces, are respectively, P_S and P_B. I_0 is the relative spectral power output (radiance) of the lamp. Interest is in the irradiance, that is, the radiant power incident on a unit area of coating. The approximation is made that the irradiance is directly proportional to I_0. In this work attention is focused on absorption in two very thin-layered portions of the coating; namely, S, the surface layer, just under the air–coating interface, and B, the bottom layer, just above the substrate–coating interface. For surface cure, it is essential that the S layer absorb at least a minimum energy for cure to occur. Likewise, for through-cure attention need only be given to the energy required to just cure the B layer. To select the layers we introduce the fractions $f_s = l_0/l$, where l_0 is the thickness of the S layer, and $f_b = 1 - l_0/l$, where l_0 is the same thickness for the B layer. By summing the absorptions which occur in each passage, infinite geometric series are obtained, as shown below.

(i) Light absorption in a partially transparent film

Assume that a parallel beam of light of intensity I_0 is normally incident upon an air–coating interface whose reflection coefficient is P_S. Then the amount of light reflected externally is $I_0 P_S$, and $I_0(1 - P_S)$ is the remaining amount which enters the coating. If the Beer–Lambert law is obeyed, the amount of light which passes through the coating to the substrate interface is given by $I_0(1 - P_S)T$ and the amount absorbed by the coating in this first forward pass is $I_0(1 - P_S)(1 - T)$, where $T = 10^{-\varepsilon Cl}$ is the optical transmission of the coating.

On incidence at the coating–substrate interface, whose reflection coefficient is P_B, $I_0(1 - P_S)TP_B$ is reflected back into the coating while $I_0(1 - P_S)(1 - P_B)T$

enters the substrate and is lost. At each air–coating interface, ever-diminishing fractions of light pass through to the air and are lost while some are reflected back into the coating to undergo partial absorption. Further, at each coating–substrate interface a similar process occurs.[136,137] These steps may be summed and yield the following series. For light absorbed by the film:

$$I_{film} = I_0(1 - P_S)(1 - T) + I_0(1 - P_S)(1 - T)TP_B + \cdots$$
$$+ I_0(1 - P_S)(1 - T)[T^2P_SP_B] + I_0(1 - P_S)(1 - T)TP_B[T^2P_SP_B] + \cdots \quad (1)$$

$$I_{film} = I_0(1 - P_S)(1 - T)\sum_{n=0}^{\infty}[T^2P_SP_B]^n$$
$$+ I_0(1 - P_S)(1 - T)TP_B\sum_{n=0}^{\infty}[T^2P_SP_B]^n \quad (2)$$

where n is the order of the passage; $n = 0$ is the first passage. The first term on the right of Eqn. (2) is the sum of all the absorptions which occur on the forward passage as the light beam travels from air-to-substrate interface. The second term corresponds to the absorptions in the reverse direction. The light absorbed by the substrate is given by

$$I_{sub} = I_0(1 - P_S)(1 - P_B)T\sum_{n=0}^{\infty}[T^2P_SP_B]^n \quad (3)$$

and the light lost on passage from coating to air is

$$I_{air} = I_0P_S + I_0(1 - P_S)^2T^2P_B\sum_{n=0}^{\infty}[T^2P_SP_B]^n \quad (4)$$

Since the summation terms are an infinite geometric series, we have

$$\sum_{n=0}^{\infty}[T^2P_SP_B]^n = \{1 - T^2P_SP_B\}^{-1} \quad (5)$$

Therefore,

$$I_{film} = I_0(1 - P_S)(1 - T)(1 + TP_B)\{1 - T^2P_SP_B\}^{-1} \quad (6)$$
$$I_{sub} = I_0(1 - P_S)(1 - P_B)T\{1 - T^2P_SP_B\}^{-1} \quad (7)$$
$$I_{air} = I_0P_S + I_0(1 - P_S)^2T^2P_B\{1 - T^2P_SP_B\}^{-1} \quad (8)$$

As a check, it can be readily shown that the initial irradiance is equal to the sum of the various losses:

$$I_0 = I_{film} + I_{sub} + I_{air}$$

To determine the radiant power absorption in any partial layer of thickness l_0 a simple process of subtraction is employed. It is, however, essential to keep the direction of light travel (forward or reverse) and the position of the partial layer (surface or bottom) in mind. For example, consider the surface layer which extends to a distance l_0 beneath the air–coating interface. The transmission

(Beer–Lambert) of this S layer is simply $T_0 = 10^{-\varepsilon C l_0}$ and the amount of light absorbed is the product of incident light and $1 - T_0$. The incident light is generally a function of the number of passages and the direction of the passage. On the first passage

$$I_0(1 - P_S)[1 - 10^{-\varepsilon C l_0}]$$

is absorbed in l_0 and on the return passage it is

$$I_0(1 - P_S)TP_B[\{1 - 10^{-\varepsilon C l}\} - \{1 - 10^{-\varepsilon C(l - l_0)}\}].$$

By using the fraction $f_S = l_0/l$, these expressions become

$$I_0(1 - P_S)[1 - T^{f_S}]$$

for forward passage and

$$I_0(1 - P_S)TP_B[T^{1 - f_S} - T]$$

for the return.

A similar treatment of the bottom layer B results in $I_0(1 - P_S)(T^{f_B} - T)$ for the first forward passage and $I_0(1 - P_S)(TP_B)[1 - T^{1 - f_B}]$ for the first return passage. Here we use $f_B = 1 - l_0/l$ or $f_B = 1 - f_S$. The expressions for the radiant power absorbed by the surface layer S are, from Eqns. (2) and (6),

$$S^1 = I_0(1 - P_S)(1 - T^{f_S}) \sum_{n=0}^{\infty} [T^2 P_S P_B]^n$$

$$+ I_0(1 - P_S)(T^{1 - f_S} - T)TP_B \sum_{n=0}^{\infty} [T^2 P_S P_B]^n \quad (9)$$

or

$$S^1 = \frac{I_0(1 - P_S)[1 - T^{f_S} + TP_B\{T^{1 - f_S} - T\}]}{1 - T^2 P_S P_B} \quad (10)$$

Similarly, we obtain for the radiant power absorbed in the partial layer B at the bottom of the coating

$$B^1 = I_0(1 - P_S)(T^{f_B} - T) \sum_{n=0}^{\infty} [T^2 P_S P_B]^n \quad (11)$$

$$+ I_0(1 - P_S)(1 - T^{1 - f_B})TP_B \sum_{n=0}^{\infty} [T^2 P_S P_B]^n$$

or

$$B^1 = \frac{I_0(1 - P_S)[T^{f_B} - T + TP_B\{1 - T^{1 - f_B}\}]}{1 - T^2 P_S P_B} \quad (12)$$

Equations (10) and (12) can be extended to a multicomponent system by substituting $T = T_1 T_2 \ldots T_n$ for n components. In Rubin's paper[135] our interest is in the case of two components, photo-initiator 1 and oligomer 2, and $T = T_1 T_2$.

Since absorption by an effective photo-initiator will produce much more photopolymerization than the oligomer by itself, we need to determine only that fraction of the total light absorbed which is absorbed by the photo-initiator (component 1).

This fraction is given by the weighting factor, the so-called inner filter factor[138,139]

$$\frac{\varepsilon_1 c_1}{\varepsilon_1 c_1 + \varepsilon_2 c_2} \tag{13}$$

The expressions (which are wavelength dependent since, I_0, ε_1, P_S, P_B, and T are all wavelength functions) become

$$S = \frac{\varepsilon_1 c_1}{\varepsilon_1 c_1 + \varepsilon_2 c_2}\left(\frac{I_0(1 - P_S)}{1 - (T_1 T_2)^2 P_S P_B}\right)[1 - (T_1 T_2)^{f_S} + T_1 T_2 P_B\{(T_1 T_2)^{1-f_S} - T_1 T_2\}] \tag{14}$$

$$B = \frac{\varepsilon_1 c_1}{\varepsilon_1 c_1 + \varepsilon_2 c_2}\left(\frac{I_0(1 - P_S)}{1 - (T_1 T_2)^2 P_S P_B}\right)[(T_1 T_2)^{f_B} - T_1 T_2 + T_1 T_2 P_B\{1 - (T_1 T_2)^{1-f_B}\}] \tag{15}$$

These equations take the basic variables into account and are generally applicable. Considerable simplifications are possible for various conditions. For example, if the substrate is totally absorbing, $P_B = 0$, and we obtain

$$S = \frac{\varepsilon_1 c_1}{\varepsilon_1 c_1 + \varepsilon_2 c_2}(I_0(1 - P_S)[1 - (T_1 T_2)^{f_S}]) \tag{16}$$

$$B = \frac{\varepsilon_1 c_1}{\varepsilon_1 c_1 + \varepsilon_2 c_2}(I_0(1 - P_S)[(T_1 T_2)^{f_B} - T_1 T_2]) \tag{17}$$

These equations are the first terms in the respective geometric series which lead to Eqns. (14) and (15). The most useful approximations are for a single passage of light for S, and a single passage to the substrate and return for B. First-order approximations are then obtained which generally agree by computation involving practical values of the variables to within a few per cent of the values found from Eqns. (14) and (15). These expressions are:

$$S = \frac{\varepsilon_1 c_1}{\varepsilon_1 c_1 + \varepsilon_2 c_2}(I_0\{1 - (T_1 T_2)^{f_S}\}) \tag{18}$$

$$B = \frac{\varepsilon_1 c_1}{\varepsilon_1 c_1 + \varepsilon_2 c_2}(I_0[(T_1 T_2)^{f_B} - T_1 T_2 + T_1 T_2 P_B\{1 - (T_1 T_2)^{1-f_B}\}]) \tag{19}$$

The $(1 - P_S)$ term representing first surface reflection for normal incidence is a constant factor and may be omitted.

On comparison we see that Eqns. (16) and (18) are identical. In Eqn. (16), the substrate is totally absorbing, $P_B = 0$, and in the other, the approximation is equivalent to a strong absorption over the light path totally attenuating the returning light beam. Equations (18) and (19) suffice for most calculations in practical systems where photo-initiator concentration and coating thickness are not vanishingly small.

(ii) Spectral data and their significance

In order to utilize these equations attention must be given to providing input information in a tractable form. The lamp spectral distribution and the absorption coefficients of components in the films are usually obtained in continuous form as functions of wavelength and therefore a problem arises because the expressions have to be integrated or summed over the wavelength bands of interest. Absorption spectra pose no difficulty since they are usually smooth and relatively featureless in the UV. Mercury arc lamp spectra are, however, composed of spikey lines superimposed upon a continuum and cannot be used in that form. To avoid this, the lamp data are converted to 5 nm wavelength increments and the equations for the radiant power absorption are computed for these bandwidths.

Spectral data on the UV normal lamp was taken from the paper by Rössler.[140] His measurements were made on a mercury lamp whose spectral distribution is like that of the medium-pressure mercury discharge lamp in current use. However, the lamp operates under less power and is shorter; therefore, Rössler's data give the relative spectral distribution of I_0 appropriate for this study. Rössler's measurements are given in tabular form in terms of spectral power concentrations (mW/nm), that is, the power emitted over a small bandwidth. Their conversion to 5 nm increments is performed by averaging the continuum values at the ends of the bandwidths, multiplying by the bandwidth, and adding the line components. These lamp data are shown in Table 5.

The photo-initiator in this investigation was benzoin butyl ether (BEB) (component 1) and the oligomer (2) was a commercial acrylate resin. Ultraviolet absorption spectra were measured by a Cary spectrophotometer, Model 14. Figure 60 is taken from an isopropanol solution of BEB and serves to highlight details obscured in coatings. The absorption coefficients, ε_1 and ε_2, in Table 6 were obtained from coatings of a binary solution of BEB and oligomer and from neat oligomer by algebraic manipulation with corrections for coating thicknesses and concentrations. The data for each 5 nm band were calculated by averaging the values at the end points of each interval.

From Figure 60, which is a semi-log plot of the absorption coefficient ε_1, of BEB vs. the wavelength, we find two spectral regions of interest. There is a long wavelength absorption maximum centred around 330 nm with a value of $\varepsilon_1 \approx 250$. It is abbreviated as ε_1 (330 nm)$_{max} \approx 250$. This peak is believed to arise from the n–π* spectral transition involving the electrons in the carbonyl group in BEB and is also characteristic of other benzoin ethers. In this transition, one of the non-bonding electrons (n) on the oxygen atom undergoes a transition to an anti-bonding orbital of the π type (π*) in the C–O bond. This is the lowest energy

Table 5 Sum of continuum and line spectral radiant power in 5 nm bands for the UV-normal lamp (Mercury)

Wavelength band	Watts	Wavelength band	Watts
200–205	0	315–320	0.269
205–210	0.008	320–325	0.250
210–215	0.060	325–330	0.230
215–220	0.165	330–335	0.756
220–225	0.263	335–340	0.184
225–230	0.310	340–345	0.165
230–235	0.465	345–350	0.145
235–240	0.948	350–355	0.125
240–245	0.303	355–360	0.105
245–250	1.153	360–365	0.088
250–255	4.817	365–370	7.60
255–260	2.660	370–375	0.061
260–265	0.790	375–380	0.054
265–270	2.47	380–385	0.005
270–275	0.255	385–390	0.004
275–280	0.523	390–395	0.064
280–285	0.980	395–400	0.004
285–290	0.645	400–405	3.12
290–295	0.323	405–410	0.47
295–300	1.408	410–415	0.004
300–305	2.445	415–420	0.004
305–310	0.295	420–425	0.003
310–315	5.336	425–430	0.003

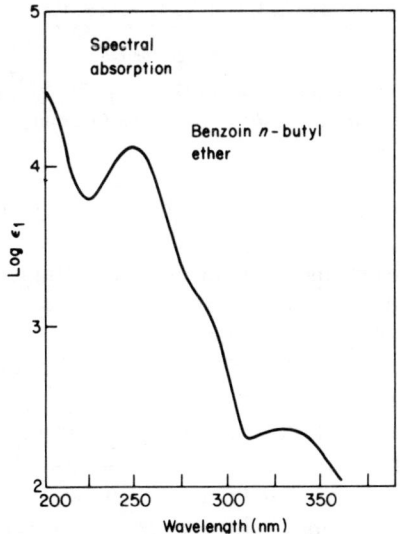

Figure 60 Absorption spectrum of benzoin n-butyl ether.[135] (Reproduced by permission of the Technical Association of the Graphic Arts.)

transition in the UV region and produces a photochemically active state of BEB.

There is also a short wavelength peak $\varepsilon_1 (250\,\text{nm})_{max} = 1.3 \times 10^4$. It has been identified as a $\pi-\pi^*$ transition since the electron excited originates in the π orbital of the C–O bond. This peak is about 50 times more intense at its maximum than the n–π^* peak and is accordingly more absorbing. Even though the 250 nm peak does not represent a photochemically active state because of its short lifetime, it functions by very rapidly transferring its excitation energy to the n–π^* state and is in this way quite significant.

The spectral characteristics of the acrylate oligomer are shown in Figure 61.

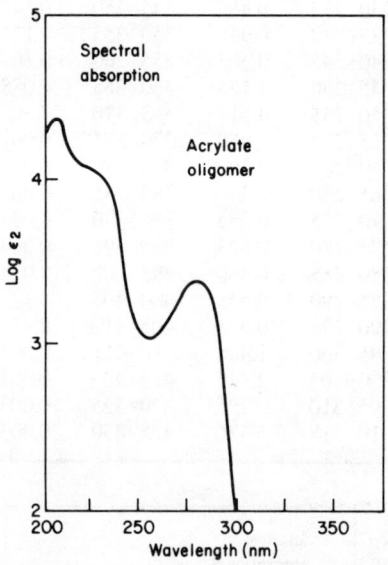

Figure 61 Absorption spectrum of acrylate oligomer.[135] (Reproduced by permission of the Technical Association of the Graphic Arts.)

Here we find several prominent peaks which have been identified by comparison with other compounds containing similar chemical groups. There is one maximum around 280 nm and another around 230 nm which shows up as a shoulder. These are identified as secondary and primary bands arising from substituted benzene rings. They serve to filter and mask some of the UV irradiance and are not photochemically active. The highest energy band with a peak around 205 nm arises from the acrylate group. It is quite intense and has been shown to have some photochemical activity in polymerization. The UV spectrum of this oligomer shows more detail in solution than in a neat film and the peak heights differ somewhat in relative proportion. Conspicuous by its absence is significant oligomer absorbance at wavelengths above 300 nm. Essentially this means that it is fairly clear and offers little interference to absorption by the 330 nm photoinitiator band.

Reference to Table 5 shows the uneven nature of the mercury spectrum. The most important bands are in the vicinity of 254 nm, 313 nm, and 365 nm. The interaction of the light from the source and its absorption by the coating has been described by Eqns. (14) and (15) for S and B. It should be recalled that these equations are wavelength dependent and if either I_0 or $\varepsilon_1 C_1$ is sufficiently small at any wavelength, little radiant power will be absorbed. Further, if ε_2, or more generally $\varepsilon_2 C_2$, is large at a particular wavelength, there will be less light available for absorption by the photo-initiator at that wavelength.

(iii) Computation of S and B

The determination of the radiant power absorbed by the partial layers requires various data inputs in addition to the spectral information described. It is convenient in dealing with coatings to use weight fractions w_1 and w_2 for components 1 (BEB) and 2 (acrylate oligomer), while molar concentrations C_i are required for transmission terms,

$$C_1 = \frac{1000 w_1 d}{MW_1}$$

$$C_2 = \frac{1000 w_2 d}{MW_2}$$

$$d = \frac{d_1 d_2}{w_1 d_2 + w_2 d_1}$$

where d is the solution density (g/ml), MW are the molecular weights (g mole), and $d_1 = 1.06$ g/ml and $d_2 = 1.191$ g/ml are the densities of the pure components. Calculations have been made in which the weight fraction of BEB varies from 0.01 (1%) to 0.32 (32%). The reflection coefficients at interfaces are given by the Fresnel formulae and are dependent only on the indices of refraction for normal incidence at boundaries between transparent media. Surface reflectance P_S may be taken as 0.04, which is typical for air–glossy polymer interfaces in the visible spectrum. The reflectance at the oligomer–substrate interface has not been determined and is taken as $P_B = 0.5$. The influence of substrate reflectance is taken up later.

The thickness of the partial layers at the top and bottom of the coating is arbitrarily taken as $l_0 = 1$ μm. The choice requires only that l_0 be small enough to represent regions which are responsible for the local physical properties and to significantly differentiate between top and bottom layers. Total coating thicknesses ranging from 1 μm to about 1000 μm have been used in the computations.

(iv) Results and discussion

A selection of computed values based on Eqns. (14) and (15) have been made to illustrate the functional dependences of S and B upon coating thickness l_1, and weight fraction of photo-initiator w_1 in specific wavelength regions. These are illustrated in Figures 62–65.

From Figure 62, in the wavelength band around 250 nm it is observed that S, the radiant power absorbed in the top 1 μm of the coating by BEB, is independent of the total coating thicknesses above 2 μm. This means that at thicknesses above that value there is negligible light energy returning from the substrate after the first forward passage. As the weight fraction of BEB increases from $w_1 = 0.01$ to 0.32 there is an eighteen-fold increase in the value of S.

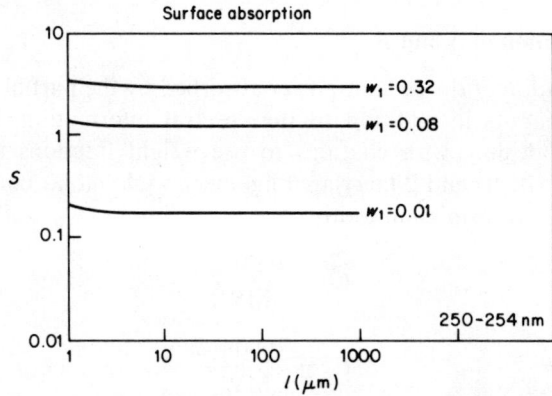

Figure 62 Radiant power absorption in surface layer vs. coating thickness (250–254 nm).[135] (Reproduced by permission of the Technical Association of the Graphic Arts.)

In Figure 63, illustrating the S vs. l relation at 365 nm, qualitatively the same behaviour is shown. However, there is significant light return from the substrate for thicknesses up to 64 μm. The radiant power absorbed in the 365 nm band at the surface is less than one-tenth that found for the 250 nm band at equivalent

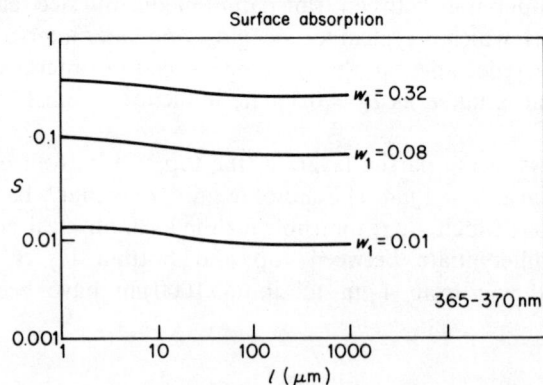

Figure 63 Radiant power absorption is in upper micron layer vs. coating thickness (365–370 nm).[135] (Reproduced by permission of the Technical Association of the Graphic Arts.)

concentrations of photo-initiator. This is primarily because $\varepsilon_1(250) > \varepsilon_1(365)$, as shown in Table 6 and Figure 60, while $I_0(250)$ is roughly the same as $I_0(365)$.

Table 6 Spectral absorption coefficients, ε_1, benzoin butyl ether, and ε_2 acrylate oligomer

$\Delta\lambda$	ε_1	ε_2	$\Delta\lambda$	ε_1	ε_2
215–220	—	13920	305–310	390	79
220–225	1930	12575	310–315	326	71
225–230	1498	12021	315–320	295	71
230–235	1677	10914	320–325	274	79
235–240	2573	7592	325–330	274	79
240–245	3923	3693	330–335	253	79
245–250	4894	1827	335–340	243	79
250–255	5284	1273	340–345	221	79
255–260	4820	1210	345–350	190	79
260–265	3744	1400	350–355	179	63
265–270	2552	1764	355–360	148	63
270–275	2099	1795	360–365	137	55
275–280	939	2539	365–370	116	55
280–285	548	2547	370–375	95	55
285–290	569	1859	375–380	74	63
290–295	738	791	380–385	63	63
295–300	654	198	385–390	42	71
300–305	506	95	390–395	42	71
			395–400	32	71

Accordingly, one might expect the 250 nm band to be more important in surface cure than the 365 nm band.

The uppermost curves in Figures 64 and 65 demonstrate that S and B merge as the coating thickness becomes very small. It can be shown that as l approaches l_0 and therefore $f_S = 1$ and $f_B = 0$, the S and B expressions in Eqns. (14) and (15) become identical. In this case, of course, the S and B layers are both equal to each other and comprise the total coating.

In preparation of the graphs in Figures 62–65, only 5 nm band widths were used because each band encloses a very intense line from the mercury arc, and each falls under or near a UV absorption maximum in the photo-initiator spectrum. A more correct method for obtaining the S and B values is to sum the contributions under the whole absorption band, and this is illustrated as a function of coating thickness for 2 per cent BEB in Figure 66. The points in circles represent the computation of B using only a 5 nm bandwidth at an intense mercury line, namely 365–370 nm. They give a good approximation to the more correct computation shown by the solid line and accordingly the use of $\Delta\lambda = 5$ nm appears to be justified as a simplification. This method may be used in other oligomer–photo-initiator systems when the bandwidth chosen is near a photo-initiator peak and encompasses a prominent mercury arc line. The other curve is a plot of the radiant power in the bottom 1 μm obtained by summing B over the short wavelength (225–280 nm) absorption peak of BEB.

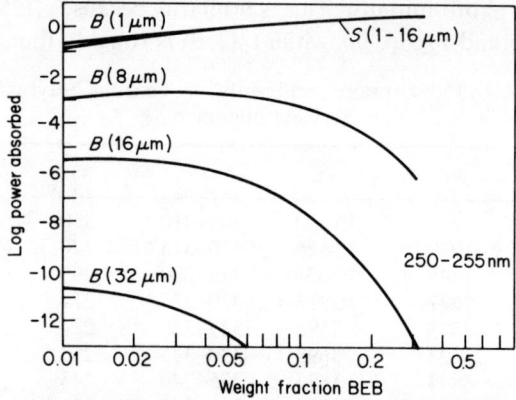

Figure 64 Radiant power absorption vs. weight fraction BEB (250–255 nm).[135] (Reproduced by permission of the Technical Association of the Graphic Arts.)

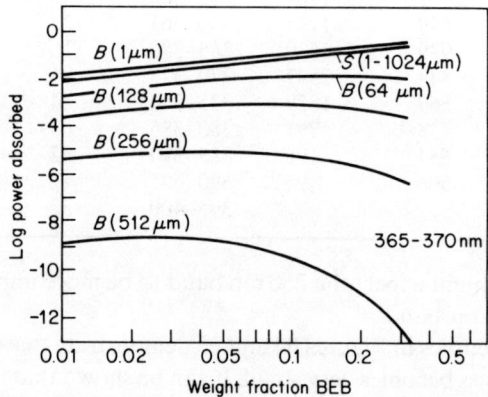

Figure 65 Radiant power absorption vs. weight fraction BEB (365–370 nm).[135] (Reproduced by permission of the Technical Association of the Graphic Arts.)

(v) Dependence of cure on wavelength

The graphs in Figure 66 were heuristic and led to a determination of the wavelength dependence of cure. In this figure ($w_1 = 2\%$), B, the radiant power absorbed in the bottom 1 μm, has a strikingly different dependence upon coating thickness in the 4–16 μm range according to whether the 225–280 nm band or the 315–380 nm band is selected. Now it is clear that the energy, $E(\Delta\lambda)$, required just to through-cure a coating should have the same magnitude, independent of total coating thickness, in that wavelength band which governs the cure process. This is so because through-cure does not occur until the bottom layer has absorbed sufficient radiant energy. $E(\Delta\lambda)$ is found from

$$E(\Delta\lambda) = \frac{B(\Delta\lambda)}{V} \qquad (20)$$

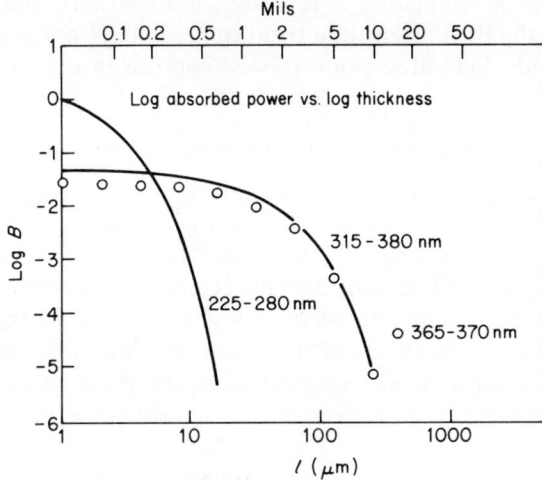

Figure 66 Radiant power absorption in bottom μm of coating vs. thickness.[135] (Reproduced by permission of the Technical Association of the Graphic Arts.)

where $\Delta\lambda$ refers to the wavelength region, and V is the effective conveyor speed required for cure. Since the reciprocal of the conveyor speed is directly proportional to the exposure time, and since $B(\Delta\lambda)$ is the radiant power in watts absorbed by the photo-initiator (in the bottom 1 μm) for a given value of l, $E(\Delta\lambda)$ is directly proportional to the energy absorbed in joules. A similar expression serves to calculate the energy required for surface cure.

A series of coatings of different film thicknesses composed of 1 part of a 9.98 per cent solution of BEB in acrylate oligomer mixed with 2 parts of dichloromethane (methylene dichloride), were deposited on polyethylene-coated paper board with wire-wound rods. The films were cured under one standard 200 watts/inch 80W/cm Hanovia lamp on CUVEX (Controlled Ultra-violet Exposure device) (Rubin)[126] and the cure speeds just required to cause through-cure were determined. Table 7 summarizes the results.

Table 7 Dependence of through-cure energy on wavelength

l			V			
μm	(mils)	WW Rod	(feet/min)	(metre/min)	E(250–255 nm)	E(365–370 nm)
5.6	0.22	#10	1100	335	5.3×10^{-6}	6.4×10^{-5}
11.2	0.44	#20	850	259	2.4×10^{-8}	7.4×10^{-5}
20.3	0.80	#46	596	182	6.7×10^{-15}	5.8×10^{-5}

The B(250 nm) and B(365 nm) values needed for this table were derived from graphs such as Figures 64 and 65 by picking off the B values at $w_1 = 0.0998$ and preparing the required B vs. l plots. This table clearly shows the constancy of the

$E(365\,nm)$ values as compared with those of $E(250\,nm)$, and supports the conclusion that the BEB absorption band centred at 330 nm is the wavelength region in which the light absorption process controls through-cure.

(vi) Dependence of cure on photo-initiator concentration

The influence of photo-initiator concentration upon cure speed is also demonstrated by the theory. The behaviour of the light energy in the 250–255 nm band which is highly absorbed by the surface layer is shown in Figure 64. S increases with w_1 up to 32 per cent BEB but is fairly independent of l. Therefore, as far as surface cure alone is concerned, higher cure speeds (less energy) are predicted when larger amounts of BEB are used, no matter what the total coating thickness. On the other hand, examination of the B curves in Figure 65 for 365–370 nm shows that the bottom layer absorption depends both upon w_1 and l. For thickness up to about 25 or 50 μm, the bottom layer absorption increases with increasing w_1 and cure speeds may be expected to follow suit. At 63.5 μm, the value of B reaches a maximum at 16 per cent BEB. With even higher values of l the position of the maximum occurs at lower w_1 values, namely 4 per cent at 254 μm and 2 per cent at 508 μm. We see then that excessive use of photo-initiator in thick coatings may be counterproductive and compositions should be designed with thickness and nature of cure in mind. Pigments and other absorbing components effectively increase the light path and coatings containing these may be expected to behave approximately like clears at considerably larger thicknesses.

(vii) Relative energy required for top and bottom cure

Inhibitive processes which slow down the rate of surface cure show up in some formulations in which a through-cure occurs while the surface retains a tacky or greasy character. Oxygen inhibition, in which the oxygen molecule (excited singlet state) competes with photo-initiator for the reactive sites, has been suggested as a primary cause. The relative through-cure and surface cure speeds give some insight into this problem. For example, an approximately 15 μm coating of 9.98 per cent BEB in acrylate oligomer required conveyor exposure speeds of 1420 feet/min (433 metre/min) for surface cure and 989 feet/min (301 metre/min) for through-cure. When the energy absorbed in both cases is calculated by means of expressions like Eqn. (21) it is found that the surface cure (top 1 μm) requires about twenty times the energy as does the through-cure (bottom 1 μm). The extra energy apparently is required to produce an excess of photo-initiator fragments to overcome the inhibition at the surface.

(viii) Influence of substrate reflectance

The nature of the substrate may influence the cure rate by several means. If the substrate is porous, absorption of the coating can occur and effectively lessen the coating thickness. This will result in a somewhat faster cure. Another factor of interest here is the effect of coating reflectance on cure. Reflectance coefficients of commercial coating–substrate interfaces in the UV are not readily available in the literature. However, a number of values of reflectances from zero to one may be chosen and it is an easy matter to calculate their effect on cure. Surface cure is relatively unaffected by substrate reflectance since very little light returns to the surface from the substrate for films of reasonable thickness after the first passage. Therefore the only meaningful calculations involve through-cure.

As discussed, the condition for through-cure of a coating is that a certain minimum energy be absorbed in the bottom $1\,\mu m$. This energy should be independent of both coating thickness and substrate reflectance. Since B is a relative quantity only relative values of cure speeds are predictable. Using the condition of equal energy absorbed at cure for coatings on substrates of reflectance P_1 and P_2, the relative cure speeds (from Eqn. 20) are given by

$$\frac{V_1}{V_2} = \frac{B_1}{B_2} \tag{21}$$

where the B's are determined by calculation from Eqn. (15). The wavelength band used is 365–370 nm which is of predominant importance in through-cure. The following values are arbitrarily selected: $l = 1.6 \times 10^{-3}$ cm, $w_1 = 0.01$, and $f_B = 0.9375$. The results of these calculations are given in Table 8.

Table 8 Influence of substrate reflectance on through-cure

Substrate reflectance	Relative B(365–370 nm)(W)	Relative cure speed (feet/min)	(metre/min)
0	0.0053	67	20
0.1	0.0058	73	22
0.5	0.0079	100*	30
0.9	0.0115	145	44
1.0	0.0129	165	50

* arbitrary reference value

A range of 2.5 to 1 in through-cure speed is predicted for the variation in substrate reflectance from the condition of no reflectance to total reflectance, and it is clearly a significant factor at intermediate reflectances.

4.4.8 Typical Formulations

A typical conventional ink could be based upon a formulation such as:

Material	Function
Pigment or dyestuff	Colour
Heat-sensitive resin	Film-forming material

Driers	Catalyst
Solvents	Thinner (evaporates)
Additives	Slip, mist, wetting, anti-oxidant, dispersion control

In contrast, the composition of an ultraviolet curing printing ink is basically as follows:

Material	Function
Pigment or dyestuff	Colour
Pre-polymer Monomer	Film-forming materials
Photo-initiator	Light-sensitive chemical
Inhibitor	Pot-life stabilizer (anti-oxidant)
Reactive diluent	Rheology control (tack and viscosity)
Additives	Slip, mist, wetting, dispersion control, etc.

A coating can be similar without the presence of a pigment. The above comparison shows that for conventional inks, a great deal of volatile solvent may be exhausted to the atmosphere whereas with UV inks and coatings all the components react to form the film. Consequently, there is little or no atmospheric pollution with these inks as no volatiles are lost.

Most ingredients, such as specific resins, are closely guarded proprietary secrets. Some general types of formulation could be based on the principles outlined below for a varnish and a screen ink representing liquid-type coating systems and also for some lithographic inks representing paste inks. Gravure and flexographic systems are not shown as these markets do not as yet seem to have absorbed UV technology to any significant extent. Flexography is possibly not commercially viable owing to the high cost involved compared to conventional systems. Gravure represents technological difficulties as it requires high pigmentation, thick film and low viscosity characteristics and fast cure. Perhaps the latter is one of the most difficult tasks for successful UV curing.

Commercially, the formulae below would need modification with individual company expertise to obtain the correct application rheology in terms of tack and viscosity for a successful run on a printing machine. Consequently, no figures are given below for these. Only broad classes of materials are indicated, not specific resins, etc., from certain manufacturers, as this is where the skill and knowledge of the formulator are required.

4.4.8.1 UV Clear Coating Varnish

	Composition (%)
Acrylated polyurethan resin	35.0
Low-viscosity acrylated polyester resin	38.0

	Composition (%)
N,N'-Dimethylethanolamine	5.0
Neopentyl glycol diacrylate	15.0
Dimethoxy 2-phenyl acetophenone	2.0
Flow agent	1.0
Benzophenone	4.0
	100.0

Typical cure rate: 2 lamps/200 ft (61 metre) per minute.

4.4.8.2 UV Silk Screen Ink

	Composition (%)
Organic pigment	10.0
Extender	15.0
Silica	1.0
Trimethylol propane triacrylate	10.0
Acrylated polyurethan resin	30.0
Diethylene glycol diacrylate	27.7
Wetting/flow agent	1.0
Benzophenone	4.0
Michler's ketone	0.3
N,N'-Dimethylethanolamine	1.0
	100.0

Typical cure rate: 2 lamps/100 ft. (30 metre) per minute.

4.4.8.3 UV Paste Inks

(i) Offset litho white metal decorating ink

	Composition (%)
Titanium dioxide	50.0
Acrylated epoxy resin	30.0
Trimethylol propane triacrylate	10.0
Plasticizer	1.0
Michler's ketone initiator	0.3
Silica	2.0
Diethoxyacetophenone initiator	1.5
Benzophenone initiator	5.0
p-Methoxyphenol inhibitor	0.2
	100.0

Typical cure rate: 2 lamps/300 ft. per minute. (91 metre/min)
Thermal assist: 10–11 min. at 150 °C

Metal decorating inks can be run at viscosity and tack values higher than those for sheet fed carton and web offset work as no 'picking' of the substrate occurs.

(ii) Offset litho sheet-fed carton inks

	Composition (%)
Calcium 4B magenta pigment	16.0
Trimethylolpropane triacrylate	22.0
Acrylated epoxy resin	35.5
Inhibitor compound (5% p-benzoquinone/monomer) solution	2.0
Wax compound (30% wax/monomer)	6.0
Silica	1.0
Diluent	7.5
Benzophenone photo-initiator	5.0
Benzil photo-initiator	3.0
Michler's ketone	2.0
	100.0

Typical cure rate: 2 lamps/300 ft. per minute. (91 metre/min.)

Often initiators can be premilled into bases. In some cases enhancement of cure is reported by this technique as in the case of benzil/Michler's ketone.

4.4.8.4 Photoemulsion Coating for Screen Application

	Composition (%)
Polyvinyl acetate	36.0
Polyvinyl alcohol	10.0
Water	51.7
Cellosolve	2.0
Dye	0.3
	100.0

These photoemulsion coatings after application are dried (removal of water) with a warm air fan and then exposed to fairly long-wavelength radiation. Radiation exposure may be of the order of several minutes.

4.5 REFERENCES

1. Kinstle, J. F., *Polymerization by UV Radiation. Part II. Free Radical Homopolymerization in Liquid Systems*, J. Radiation Curing, **1**(2), 2–17, (April 1974).
2. Faulkner, R. N., *Light Induced Polymerization Processes & Newer Catalysts for Film Curing at Ambient Temperatures*, The Paint Research Station, Teddington, Middlesex, (August 1967).
3. McCloskey, C. M., and Bond, J., *Ind. Eng. Chem.*, **47**, 2,2125, (1955).
4. Du Pont de Nemours & Co., *U.S. Pat 2,892,716.*
5. Du Pont de Nemours & Co., *U.S. Pat. 2,929,710.*
6. Du Pont de Nemours & Co., *U.S. Pat 3,043,805.*
7. Du Pont de Nemours & Co., *Brit. Pat. 1,058,798.*
8. Du Pont de Nemours & Co., *U.S. Pat 2,893,868.*
9. A. E. Stanley Manufacturing Co., *Brit. Pat 944,322.*
10. General Motors Corporation, *Brit. Pat 1,042,908.*
11. Allen, P. W., et al., *J. Polymer Science*, **36**, 55, (1959).
12. Melville, H. W. et al., *Trans. Far. Soc.* **50**, 279, (1954).
13. Bamford, C. H. et al., *Trans. Far. Soc.* **61**, 267, (1965).
14. Bamford, C. H., and Paprotny, J., *Polymer*, **13**, 208, and references therein, (1972).
15. Bamford, C. H. et al., *Proc. Roy. Soc.*, **A284**(1399), 455–68, (1965).
16. Bamford, C. H., *Brit. Pat. 1,027,148.*
17. Bamford, C. H., et al., *Pure & Allied Chem.*, **12**, 1–4, 183, (1966).
18. Imoto, M., Otsu, T., and Yonezawa, J., *Makromol. Chem.*, **36**, 93, (1960).
19. Otsu, T., *J. Polymer Sci.*, **26**, 236, (1957).
20. Okawara, M., and Nakai, T., *Internat. Symposium on Macromol Chem.*, Tokyo, Preprint IV-16, (1966).
21. Oster, G. K., and Oster, G., *J. Polymer Sci.*, **48**, 323, (1960).
22. Guillet, J. E., and Norrish, R. G. W., *Proc. Roy. Soc. A*, **233**, 153–183, (1955).
23. Dhanraj, J., and Guillet, J. E., *Internat. Symposium on Macromol. Chem.*, Tokyo, Preprint III–11, (1966).
24. Feldmuhle Papier–Und. Zellstoffewerke, A. G., *Fr. Pat. 1,091,323.*
25. Smets, G., *Pure & Applied Chem.*, **4**, (2–4), 294, (1962).
26. Smets, G., *J. Polymer Sci.*, **55**, 767, (1961).
27. Gavaert Photoproducten N. V., *U.S. Pat. 3,252,966.*
28. D'Alelio, G. E., and Caiola, R. J., *J. Polymer Sci.*, **A1**(5), 287–306, (1967).
29. Scott Paper Co., *Brit. Pat. 1,037,372.*
30. D'Alelio, G. F., and Huemmer, T., *J. Polymer Sci.*, **A1**(5), 307–321, (1967).
31. Gevaert Photoproducten, *Brit. Pat., 1,074,234.*
32. Gevaert Photoproducten, *U.S. Pat. 3,278,305.*
33. Eastman Kodak Co., *Brit. Pat. 843,541.*
34. Eastman Kodak Co., *Brit. Pat. 843,542.*
35. Eastman Kodak Co., *Brit. Pat. 995,862.*
36. Delzenne, G. A., and Laridon, U., *IUPAC Internat. Symposium on Macromolecular Chem.* Brussels–Louvain Preprint 2/109, (1967).
37. Cohen, M. D., Schmidt, G. M. J., and Somtag, F. I., *J. Chem. Soc.*, 2000, (1964).
38. Stobbe, H., *Ber.* **586**, 2859, (1925).
39. Stobbe, H., *C.A.* **20**, 1612, (1926).
40. Griffin, G. W., Vellturo, A. F., and Furukawa, K., *J. Am. Chem. Soc.* **83**, 2725, (1961).
41. Minsk, L. M. et al., *J. Appl. Polymer Sci.*, **11**(6), 302–11, (1959).
42. Kosar, J., *Light Sensitive Systems*, John Wiley & Sons, p. 140, (1965).
43. Hepher, M., 'The Photoresist Story', *Journal of Photographic Science*, **12**, (1964).
44. Eastman Kodak Co., *U.S. Pat. 2, 751,373.*
45. Farbenfabriken Bayer, *U.S. 3,066,117.*
46. Gevaert Agfa, *Fr. 1,446,213.*

47. Unruh, C. C., *J. Appl. Polymer Sci.*, **11**(6), 358–62, (1959).
48. Eastman Kodak Co., *Brit. Pat. 1,056,786*.
49. Eastman Kodak Co., *Ger. Pat. 1,229,388*.
50. Kirsh, Yu. E., *Russ. Pat. 178,983-4* (Soviet Inventions Illustrated Gp III, p. 3.) (Oct. 1966).
51. Laws, A., Lynn, S., and Hall, R., *J.O.C.C.A.*, **59**, 193–196, (1976).
52. Boenig, H. V., *Unsaturated Polyesters: Structure & Properties*, Elsevier, N.Y., (1964).
53. Turner, G. P. A., *Introduction to Paint Chemistry*, Chapman & Hall Ltd., (1967).
54. Hulme, B. E., 'The Curing of Coatings with UV Radiation, Part II', *Paint Manufacture*, 12–16, (May 1975).
55. Hickner, R. A., 'Dow Chemical's Radiation Curable Materials', *Radiation Curing*, **2**(3), 9–13, (August 1975).
56. Fekete, F., and Keenan, P. J., *U.S. Pat.*, *3,373,075* (to H. H. Robertson Co.), (12 March 1968).
57. Fekete, F., Keenan, P. J., and Patnet, W. J., *U.S. Pat. 3,301,743* (to H. H. Robertson Co.), (31 Jan. 1967).
58. Fekete, F., Keenan, P. J., and Plant, W. J., *U.S. Pat. 3,256,222* (to H. H. Robertson Co.), (14 June 1966).
59. Miller, L. S., *U.S. Pat.*, *3,560,237*, (to Weyerhauser Co.), (14 July 1970).
60. Chandler, R. H., 'The UV Curing of Unsaturated Polyester Lacquers', *Paint Technology*, 21–22, (February 1970).
61. Laws, A., Lynn, S., and Hall, R., *J.O.C.C.A.*, **59**, 193–196, (1976).
62. Laus, R., *Rad. Curing*, **3**(A), 15, (1976).
63. De Lange, P. G., *Curing by Irradiation*, Verfroniek, p. 105–116, (March 1970).
64. Rybny, C. Y., and Vona, J. A., *J.O.C.C.A.*, **61**, 179–188, (1978).
65. Brown, R. A., *Paint & Varnish Production*, **64**(2), 17, (1974).
66. Guthrie, J. L., and Rendulic, F. J., *U.S. Pat. No. 3*, **787**, 303. Assigned to W. R. Grace & Co., (22 Jan. 1974).
67. Watt, R. W., *U.S. Pat. No. 3*, **794**, 576. Assigned to American Can Co. (26 Feb. 1974).
68. Kirk-Othimer, *Encyclopaedia of Chemical Technology* (2nd edn), Vol. 21, p. 57.
69. Celanese Chem. Corporation, *U.V. Applications Bulletins*.
70. Burlant, W., and Taylor, C., *U.S. Pat. 3,509,234*, (20 April 1970).
71. Smith, O. W., Weigle, J., and Trecker, D., *U.S. Pat. 3,700,643*, (to Union Carbide Corp.) (24 Oct. 1972).
72. Fekete, F., Keenan, P. J., and Plant, W. J., *U.S. Pat. 3,297,745*, (to H. Y. Robertson), (19 Jan. 1967).
73. Huemmer, T., and Miranda, T., *U.S. Pat. 3,719,638*, (to O'Brien Corp.), (6 Mar. 1973).
74. Gorman, J., and Toback, A., *U.S. Pat. 3,425,988*, (to Lectite), (4 Feb. 1969).
75. Rowe, W., *S.M.E. Technical Paper FC* 76–495, (1976).
76. McKillip, W., and Impala, C., *U.S. Pat. 3,396,210*, (to Ashland Oil), (6 Aug. 1968).
77. Boranian, A. G., and Terwilliger, B., *U.S. Pat. 3,924,023* (to GAF Corporation), (2 Dec. 1976).
78. McKillip, W., and Impala, C., *U.S. Pat. 3,396,210* (to Ashland Oil), (6 Aug. 1968).
79. Riddle, E. H., *Monomeric Acrylic Esters*, Reinhold Publishing Corporation, New York, (1954).
80. Morgan, C. R., Magnotta, F., and Ketley, A. D., *J. Polym. Sci.* (*Polym. Chem. Ed.*) **15**, 627, (1977).
81. Morgan, C. R., and Ketley, A. D., *ACS Div. of Org. Coat. & Plast.* Preprints, 165th meeting, **33**, 281, (1973).
82. Posner, T., *Chem. Ber.*, **38**, 646, (1905).
83. Coffman, D. D., *U.S. Pat. 2,508,005* (to Shell Development Co.), (16 May 1950).
84. Rust, F. F. and Vaughan, W. E., *U.S. Pat. 2,392,294* (to Shell Development Corp.), (1 Jan. 1946).

85. Morgan, R. C., 'Radiation – Curable Polymers for Closures', *Radiation Curing*, **1**(3), 11–14, (Aug. 1974).
86. Kehr, C. L., and Wszolck, W. R., *Preprints, Div. of Organic Coatings & Plastics Chem.*, American Chemical Society, **33**, 295, No. 1, (1973).
87. Gruber, G. W., *UV Curing: Science & Technology*. (Ed. S. Peter Pappas), Technology Marketing Corporation, (1978).
88. Gush, D. P., Ketley, A. D., 'Thiol/Acrylate hybrid Systems in Radiation-Curable Coatings – The best of both worlds', NPCA Chem. Coatings Conference II, Cincinnati, Ohio, (10 May 1978).
89. Rybny, B. C., and Vona, J. A., 'New Developments in UV Curable Coatings Technology', paper presented by Celanese Chemical Co., 2nd Newcastle UV Symposium, Durham University, (14–15 Sept. 1977).
90. Huemmer, T. F., *Industrial Finishing*, **46**(5), 34 (1970).
91. Pelgrims, J., 'Present Status of UV Curable Coatings Technology in the U.S.', paper presented by Union Carbide Corporation, 2nd Newcastle UV Symposium, Durham University, (14–15 Sept. 1977).
92. Younger, J. R., *J.O.C.C.A.*, **52**, 197–201, (1976).
93. Himics, R. J., *A.C.S. Polymer Preprints*, **33**(1), 274, (1973).
94. Himics, R. J., 'Novel Photosensitive Monomers & Polymers', *J. Radiation Curing*, **2**(3), 7–14, (July 1975).
95. Rohm & Hass Company, March 1976, adapted from tables 1 & 2 of Technical Release for Isobornyl Acrylate, CM-38.
96. Higgins, R. A., *Properties of Engineering Materials*, Hodder & Stoughton, London, p. 270, (1977).
97. De Poortere, M., Duncarme, A., Dufour, P., and Merck, Y., *J.O.C.C.A.*, **61**, 195–203, (1978).
98. Orchard, S. E., *Appl. Sci. Res. A.*, **11**, 451, (1962).
99. Dodge, J. S., *J. Paint Technol.*, **44**(564), 72, (1972).
100. Camina, M., and Howell, D. M., *J.O.C.C.A.*, **55**, 929–939, (1972).
101. Murphy, J., *P.R.A. Internal. Report*, RS/T/73/69.
102. Smith, D. N. P., Orchard, S. E., and Rhind-Tutt, A. J., *J.O.C.C.A.*, **44**, 618, (1961).
103. Camina, M., and Roffey, C. G., *Rheol. Acta*, **10**, 606–607, (1971).
104. Huemmer, T. F., *J. Radiation Curing*, **1**(3), (July 1974).
105. Huemmer, T. F., Wasowski, L. A., and Plooy, R. J., *J. Paint Technol.* **44**(572), 61, (1972).
106. Hahn, A. E., 'Dual Cure Gloss Control of Radiation Polymerized Coatings', *Radiation Curing*, **1**(2), 13–17, (May 1974).
107. Anonymous, 'UV Curing and Inert Atmospheres', *Radiation Curing*, 4, (Nov. 1976).
108. Bassemir, R. W., 'The Gloss of UV Cured Press Applied Films', *Radiation Curing*, 10, (May 1976).
109. Hahn, A. E., *Radiation Curing*, **1**(2), 13–17, (May 1974).
110. Hulme, B. E., *J.O.C.C.A.*, **59**, 245–252, (1976).
111. Hencken, G., *Farbe und Lack*, **81**, 916, (1975).
112. Marvuglio, P., Sharrock, R. F., and Kennedy, R. J., *J.O.C.C.A.*, **61**, 79–85, (1978).
113. Vincent, K. D., *Radiation Curing*, **1**(4), 11–12, (Nov. 1974).
114. Bassemir, R. W., and Bean, A. J., 'Parameters of UV Printing Inks', *Taga 1974 Proceedings*, Tech. Assn. of the Graphic Arts (Rochester, N.Y.), p. 133, (1974).
115. Bassemir, R. W., and Bean, A. J., paper presented at 26th Annual Meeting of TAGA, St. Paul, Minn., (13–15 May 1974).
116. Van Neerbos, A., *J.O.C.C.A.*, **61**(7), 247–248, (July 1978).
117. Plews, G., and Phillips, R., *J. Coatings Technol.* **51**(648), 69–77, (Jan. 1979).
118. Nishikubo, T., Imaura, M., Mizuko, T., and Takaoka, T., *Appl. Polymer Sci.*, **18**, 3445, (1974).
119. Phillips, R., (1978), *J. Oil and Colour Chemists Assoc.*, **61**, 233.

120. Collins, G. L., Young, D. A., and Costanza, J. R., *Journal of Coatings Technology*, **48**(618), 48, (1976). AFP-SME Technical Paper, 1976, FC 76-484.
121. Chang, Y. C., *Phot. Sci. Eng.*, **21**(6), 348, (1977).
122. Berner, G., Kirchmayr, R., and Rist, G., *J.O.C.C.A.*, **61**, 105–113, (1978).
123. Wicks, Z. W., Jnr., and Kuhhirt, W., *J. Paint Technology*, **47**(610), 49, (1975).
124. Osborn, C. L., Sander, M. R., and Tracker, D. J., *J. Polymer Science*, **10**, 3173, (1972).
125. Shulman, J. J., *Inert Atmosphere Proves Cost Effective for High Speed UV Curing*, Paperboard Packaging, (June 1976).
126. Rubin, H., *J. Paint Technology*, **46**(588), 74, (1974).
127. Pappas, S. P., *AFP–SME Technical Paper* FC 76-490, (1976).
128. Osborn, C. L., *J. Radiation Curing*, 2, (July 1976).
129. McGinniss, V. D., *A.C.S. Org. Coat. Plast. Prepr.*, **35**(1), 118, (1975).
130. Collins, G. L., and Costanza, J. R., *J. Coatings Technol.* **51**(648) 57–63, (Jan. 1979).
131. Tokumaru, K., Ohshima, A., Nakata, T., Sakuragi, H., and Mishima, T., *Chem. Letters*, 571, (1974).
132. Bartlett, P. D., and Porter, N. A., *J.A.C.S.*, **90**, 5317, (1968).
133. Mousseron-Canet, M., and Mani, J. C., *Photochemistry & Molecular Reactions*, p. 146, Israel Program for Scientific Translations, Jerusalem, (1972).
134. Walling, C., *Free Radicals in Solution*, p. 478, John Wiley, New York, (1957).
135. Rubin, H., 'U.V. Curing: Role of Radiant Absorption by Photo-initiators,' *Taga Proceedings*, 279–301, (1976).
136. Heavens, O. S., *Optical Properties of Thin Solid Films*, Academic Press Inc., (1955).
137. Walsh, J. W. T., *Photometry*, Constable & Co., Ltd., London, (1953).
138. Noyes, W. A., and Leighton, P. A., *The Photochemistry of Gases*, Dover, New York, (1966).
139. Michaelson, R. C., and Loucks, L. F., *J. Chem. Ed.*, **52**, 652, (1975).
140. Rössler, F., *Ann. Phys.*, Leipzig **34**, 1, (1939).

Note. Parts of the articles in question are reprinted from *Journal of Radiation Curing*®, Vol. 1, Nos. 2, 3, & 4, 1974, and *Radiation Curing*®, Vol. 1, No. 3, 1974, and Vol. 2, No. 3, 1975, published by Technology Marketing Corporation, 17 Park Street, Norwalk, CT 06851, USA. Copyright © 1974, 1975.

5
Ink technology and the application of radiation curing

5.1 THE PRINTING PROCESSES

5.1.1 Roller Coating

This is the chief means of applying coatings. A roller coater[1-3] is basically a series of rollers which in turn pick up a set amount of surface coating from a tray and apply it uniformly at a controlled viscosity across the surface of the substrate as it passes through the machine.

This is depicted schematically in Figure 67.

The coating material is pumped from a container reservoir into the feed tray. The feed roller rotates in this tray, picks up the material, and deposits it on to the transfer roller. From this it passes to the application roller and on to the sheet.

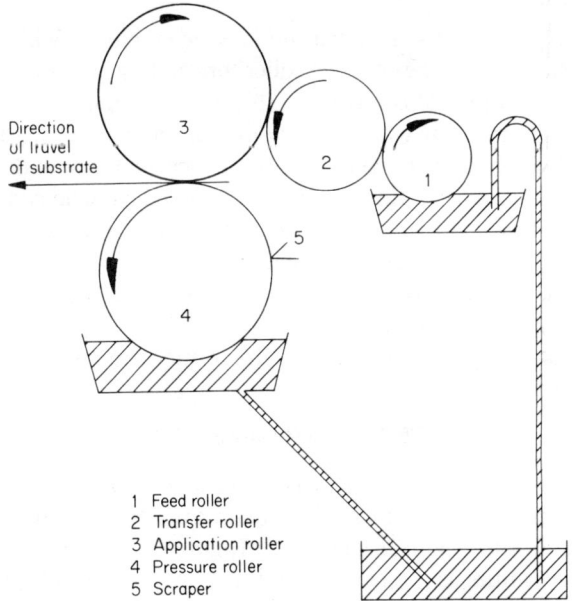

1 Feed roller
2 Transfer roller
3 Application roller
4 Pressure roller
5 Scraper

Figure 67 A roller coater. (Reproduced by permission of Arthur Holden & Sons Ltd.)

The quantity of coating material applied to the sheet is determined by the pressure between the feed and transfer roller and the transfer roller and the application roller. It may be regulated to produce an exact control of the film weight needed.

The sheet passes between the application roller and the pressure roller. Both of these are set in contact and the sheet is gripped firmly on the application roller during transit. As one sheet passes and before the next comes into contact with the application roller, coating material transfers to the pressure roller and a scraper blade is applied to this roller in order to remove the material and maintain the backs of the sheets free from material. This is very important as unwanted coating on the wrong side of a sheet can cause problems at a subsequent step of manufacture. The steel scraper blade has to be kept in perfect condition.

The application roller is generally a steel cylinder coated with a rubber compound, a gelatine composition or polyurethan. In this mode the roller will apply the coating material across the total sheet. In some instances it is required to keep areas free from coating material, for example, where soldering is needed or as part of a design. To achieve this the roller must be cut in order for these stencil areas to remain. A different roller is needed for each design. Some machines have a specially designed cylinder covered with a rubber or gelatine blanket with the required stencil area extracted. This blanket type of stencil or 'spot' coating has advantages in the sense that when there are frequent changes in design, a blanket can be changed more readily than a roller. Storage of these blankets is also much easier.

An alternative means of stencil coating may be employed which comprises a metal scraper running on the transfer roller, preventing the coating from being transferred to the application roller. This method may only be used for elementary designs and only where the reserve runs along the length of the sheet. It cannot be used where transverse reserves are needed across the sheet.

Solvent systems used in coating materials can often have a detrimental effect on the roller coverings. Care must be taken to ensure that compatible roller coverings and coatings are used together.

There are four main processes for printing inks, each having varying requirements of printing application.[1-3]

These four processes are: letterpress, lithography, gravure, and screen.

5.1.2 Letterpress[1-3]

In this relief printing the image area is raised above the plate surface. Ink is applied only to the type face which is the image area and then transferred to the paper substrate in the type face plane (see Figure 68).[6]

Letterpress printing is carried out on three common types of presses: platen, flat-bed cylinder, and rotary.

Lettepress printing can be recognized by a heavier ring of ink bordering around each letter. The ink is inclined to spread slightly from the pressure of the

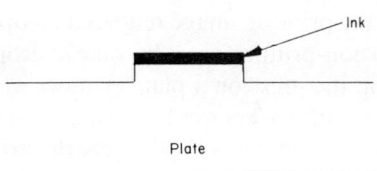

Figure 68   Letterpress (relief printing) process. (Reproduced by permission of Technology Marketing Corporation.)

plate upon the printed surface. A slight embossing often occurs on the reverse side of the paper.

Another form of letterpress printing known as flexography exists, where the plate is rubber instead of metal. Flexography is a form of rotary letterpress using flexible rubber plates and fast-drying fluid inks. The process is used for printing large areas of solid colour. Flexography can be uneconomical with high absorptive papers owing to the ink fluidity and subsequent penetration. Flexographic printing does have limitations. Hairline register, closed typefaces, and small type are best avoided. It is not adaptable to the production of fine screen process work or tints. The liquid ink tends to overflow the dots, making it difficult to obtain a clean halftone reproduction, but many jobs are appearing where screens up to 100 lines per inch (39 per cm) are used.

5.1.3 Lithography[1-3]

This is planographic printing where the image is essentially coplanar with the non-image area. The image area is hydrophobic (water repelling) or oleophilic (ink or oil receptive) whilst the non-image area is hydrophilic (water receptive) or oleophobic (ink or oil repelling). The image area is coated by the hydrophobic ink film and aqueous fountain solution is applied to the non-image area; the ink film is transferred to the blanket roll and then to the paper substrate.[3] This process is depicted in Figure 69.[6]

Offset lithography has therefore two basic differences with other processes:

1. It is based on the principle that grease (oil) and water are immiscible.
2. Ink is offset first from plate to rubber blanket and then from blanket to paper or other substrate.

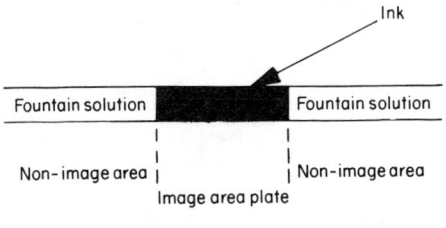

Figure 69   The lithographic process. (Reproduced by permission of Technology Marketing Corporation.)

The printing plate has the printing image rendered oleophilic (grease receptive) and water repellent; the non-printing areas become hydrophilic and ink repellent. The plate is mounted on the press on a plate cylinder which on rotation comes into contact successively with rollers wet by a water or dampening solution and rollers wet by ink. This dampening solution wets the non-printing areas of the plate and prevents the ink from wetting these areas. The ink wets the image areas which are transferred to an intermediate (blanket) cylinder covered by a rubber blanket. The image is picked up by the paper as it passes between the blanket cylinder and the impression cylinder. The transferring of image from the plate to a rubber blanket before transfer to the paper is known as the offset principle. Letterpress and gravure can also be printed by this method.

A major advantage of lithography is that the soft rubber surface (compared to metallic letterpress plates) allows a clearer impression on a wide variety of papers and other materials. Offset lithography permits extensive economic use of illustrations and with a minimum of press make-ready.

Offset lithographic printing can be recognized by a smooth print and also by the lack of any impression or ring of ink which is characteristic of letterpress.

As well as the letterpress process, lithography also has equipment for short, medium, and extremely long runs. Both sheet-fed and web-fed presses are used. Web offset is printed on presses possessing two printing units opposing each other so that both sides of the web are printed at the same time. One printing-unit blanket serves as the impression cyclinder for the opposing unit, and vice versa, and this process is called perfecting printing. Perfecting printing can also be achieved by turning the sheets or web over or by reversing the direction of rotation of successive printing units between printings.

Sheet-fed lithography includes general commercial printing, letter heads, advertising, books, poster, labels, folding boxes, etc.

Web offset is used for printing business forms, newspapers, etc.

5.1.4 Gravure[1-3]

This is known as the intaglio process of printing where the image area is recessed below the plate surface.[4] Ink is applied to the plate and forced into the image cavities; the remainder is 'doctored' off the plate surface by a blade. During printing the paper substrate is forced into the cavities picking up the ink (see Figure 70).[6]

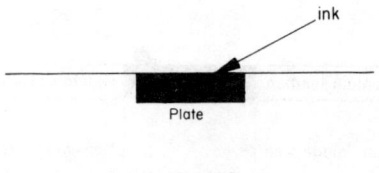

Figure 70 Gravure printing. (Reproduced by permission of Technology Marketing Corporation.)

Gravure printing is excellent for reproducing pictures but the high expense of plate-making usually limits its use to long runs. Gravure can be recognized by the feature that the entire image is screened-type and line drawings as well as halftones may be used. The gravure screen usually contains 150 lines per inch (59 per cm), about the same screen ruling as used in other processes. It is practically invisible to the naked eye. Gravure presses are made both for sheets (sheet-fed gravure) and rolls (rotogravure) of paper. Sunday newspaper magazine sections, colour reprints for newspapers, most premium stamp catalogues, and large mail-order catalogues are common examples of rotogravure printing.

5.1.5 Screen Printing[1-3]

Originally known as silkscreen, this printing method uses a porous stencil.[5] A stainless steel screen or a fine silk such as nylon is mounted under tension on a frame. A photochemically or manually produced stencil is made on the screen by a direct or indirect method or direct/indirect combination in which the non-printing areas are protected by the stencil. The printing is performed by feeding the substrate under the screen, applying ink with a paint-like consistency to the screen, and spreading and forcing it through the fine mesh openings with a rubber squeegee. Screen printing is characteristically recognized by the thicker layer of ink and the screen texture on the printing. The thick wet film of ink may be of the order of ten times as heavy as that of letterpress and can be controlled by the stencil mesh, the ink rheology, and various additives. Production rate has been limited by the ink drying time but currently has been increased by the development of automatic presses and improved dryers (jet drying principle), faster solvent-evaporation inks, and also now by ultraviolet curing.

Screen printing has the great advantage of versatility. Almost any surface can be printed, such as plastic, fabric, metal, wood, paper, glass, etc., in any shape, thickness, or design, and in any size from large posterboards to postage-stamp size.

It is possible with this process to print white and partial colours on dark backgrounds in one printing with complete opacity. Metallic inks such as golds and fluorescent colours are also very amenable to this process owing to the heavy film thickness. Printed circuitry is another large area, as is the wallpaper industry.

5.2 PRINTING INKS[1-2]

A printing ink may be regarded as a dispersion of a pigment or dyestuff in a fluid vehicle.[6] Inks generally contain a polymeric or resinous component that forms, on drying, a continuous film. Pigmentation can be black (carbon), white, or coloured. The colourants can be inorganic or organic compounds and inks can be either opaque or transparent. Dyestuffs can also be used. There are numerous vehicles employed. These may be non-drying oils, drying oils, resin–solvent solutions, resin–oil combinations, or resin–wax mixtures. There are many additives in printing inks, such as driers, wax compounds, greases, lubricants, reducing oils, body gums, binding varnishes, anti-oxidants, anti-skinning agents, or surface-active agents.

Generally, printing inks are prepared by dispersing powdered pigment by mechanical mixing into a vehicle. Powdered pigment consists of agglomerates and aggregates of primary particles. For example, carbon black pigments consist of 30–100 μm diameter agglomerates and aggregates of 0.02–0.03 μm diameter primary particles. The process of dispersion usually involves pre-mixing and mixing steps during which the large pigment agglomerates and aggregates are broken down to a colloidal dispersion of small aggregates and primary particles. A fine degree of dispersion is needed for some types of printing inks, such as litho, as the film is much thinner (1.5–2 μm) than others.[7]

Also, for pigments there is an optimum degree of dispersion or particle size distribution for development of opacity or colour strength.

An idea of some pigmentation levels, viscosity and film thicknesses are shown in Table 9 for inks used in the different printing processes.[8]

Table 9 Pigmentation, film thickness, and viscosity of printing inks

Printing method	Pigment (%)	Viscosity (poise)	Film thickness (wet) (μm)
Letterpress	20–80	10–500	5
Flexography	10–40	1–100	2–3
Newsink	8–12	2–10	1
Lithography	20–80	100–800	1.5–2
Gravure	10–30	0.5–10	12.5
Screen	4–30	18–30	30 (thin film) 60 (thick film)

Printing inks may be divided according to solvent-based and solventless methods of drying.[1–2,10]

5.3 SOLVENT-BASED METHODS OF INK DRYING[1–2,9]

Drying of inks is often defined as the phase change from the liquid to the solid state and may be achieved by a combination of methods. Ink drying comprises two factors, ink 'setting' and 'hardening', usually one following the other without any clear demarcation.

Ink setting means the wet film has adequately set to withstand such stresses as contact with the conveyor and the back of the next sheet in the print pile, giving no or marginal set-off, and in some cases it will withstand contact with the next impression cylinder. These 'set' inks are in a stiff viscous and non-tacky state (or nearly so), which renders it difficult to smear and acceptable for light mechanical handling.

Ink hardening means that a 'set ink' has thoroughly dried to a hard film and is unchangeable in properties.

5.3.1 Absorption[1-2, 9]

Surface tension will cause a liquid in contact with a porous material to be drawn into the openings in the surface structure. Most papers are extremely porous materials because of the network of narrow channels (capillaries) between fibres and particles of mineral loading. Capillary attraction of an ink into a paper surface gives the simplest physical drying method. Drying in this case is not in the sense of the liquid being transformed into a solid; instead it becomes an integral part of the solid paper.

Newsinks, some letterpress inks, and certain web offset inks when applied to absorbent stock, dry only by penetration. In combination with other drying methods, absorption is an important contribution to all printing ink drying, except for non-porous substrates such as metals and plastics.

Efficient absorption drying requires a balance between the paper oil absorbency and the ink viscosity. In the case of insufficient penetration, absorption drying involves the inert vehicle soaking into the porous paper, leaving a pigment plug on the surface. Equilibrium finally occurs between the capillary forces in the ink plug pigment and the medium, and absorption forces caused by the affinity of polar paper fibres for non-polar mineral oil.

Eventually absorption drying causes pigment to be bound to the surface by oil and held there by mechanical adherence. On non-porous substrates these kind of inks would never appear to dry.

Absorption drying may be regarded as having a three-stage mechanism with no clear demarcation limits.[9]

Stage I Ink Penetration into the Paper Valleys and Large Capillaries

This is dependent on the following factors:

1. The bulk entry of the printing ink into the paper. About half of an absorbent paper is made up of air, distributed in large voids around a cellulosic fibre mesh. Penetration is controlled by the number and size of the air voids. Generally, absorbent papers are fairly rough and have valleys in the surface which fill with ink.
2. The ink/paper wettability. The ink film should have good wetting affinity for the stock and not have a contact angle. Therefore, a relatively low surface tension is needed by the ink.
3. High fluidity of the printing ink. At high temperatures, penetration is normally greater as the ink has a lower viscosity and therefore is more mobile into the porous substrate.

Stage II Partial Medium Drainage from the Pigment

Fine capillaries exert suction forces proportional to:

$$\frac{1}{(\text{contact angle}) \times (\text{average diameter of capillaries})}$$

This force is maximized when the contact angle is zero and the capillaries are very narrow. Some vehicle will be pulled into the paper, and the ink film contracts. The pigment particles come into closer contact and develop higher attractive forces. At the stage where the ink and paper exert equal forces, the vehicle drainage stops.

For a high vehicle drainage rate the ratio

$$\frac{\text{no. of small capillaries}}{\text{no. of large capillaries}}$$

must be a maximum and the rate of vehicle drainage is proportional to fluidity of medium/ink fluidity.

Stage III Diffusion of Medium through Stock

If medium diffusion is inadequate and large amounts of solid forme have been printed, then such prints may not dry readily. Ink remaining on the paper surface can not dry quickly enough to avoid set-off onto the next sheet. In some cases this ink can act as an adhesive between the printed surface and the back of the next sheet, causing the two to stick to each other. This phenomenon is known as blocking.

If too much of the medium is absorbed into the paper, unprotected pigment remains on the surface and this can rub off or set off on to the following sheet. This phenomenon is called powdering and is more common with coated papers than with papers possessing a more open surface, as the latter permit a larger amount of the pigment to be transported into the paper structure with the medium. Excessive medium absorption can also cause show-through or strike-through, rendering the printed image visible from the reverse side of the sheet.

Absorption drying takes place very quickly and the fastest press speeds have been obtained with inks that rely only on this method. The chief drawback is that inks that have been absorbed into a paper tend to be dull and lifeless in appearance as very little light can be specularly reflected from its surface.

To remedy this and to obtain an ink with an attractive glossy appearance, and at the same time retain the advantage of rapid absorption drying, the 'quicksetting' inks process was developed. These inks use a two-phase medium. The two components of the medium are:

1. a resin often combined with a drying oil

2. a thinner for the above composed of a thin high-boiling solvent such as a paraffin hydocarbon distillate boiling around 300 °C.

When printed, the thin component is quickly absorbed, leaving the rest of the ink on the surface. This viscous combination of resin, drying oil, and pigment sets and then is slowly converted into a hard dry film as a result of oxidation and polymerization processes.

5.3.2 Oxidation and Polymerization[1, 2, 9–11]

The chemical reaction that causes a phase change from the liquid state to a solid of a drying oil (such as linseed) is a combination of oxidation and polymerization.

Atmospheric oxygen performs a major role in the gradual thickening of linseed oil. Linseed oil dries too slowly in its natural state and is not sufficiently viscous to be used as a printing-ink medium. Various heat treatments overcome this, some under vacuum. The molecules of the drying (linseed) oil link up into longer chains and are resistant to flow. This amount of polymerization is dependent on the time length for which the heat is maintained. These heat-treated drying oils were the traditional letterpress and litho ink media. Natural drying oils such as linseed have since been chemically modified, and also completely synthetic drying oil vehicles have been developed to create quick-drying inks with better properties.

A drying oil varnish on atmospheric exposure undergoes molecular linkage until a hard, flexible and transparent material results. These reactions occur between drying-oil molecules and oxygen at unsaturated points on the carbon chain, creating bridging links across to adjacent molecules.

Driers are incorporated in small quantities as catalysts to accelerate the rate of drying. Driers are metallic soaps which are oxygen carriers, transferring atmospheric oxygen to the molecules of drying oil faster than they can obtain it for themselves.

5.3.3 Evaporation Drying[1–2, 9, 12]

The removal of a solvent or a solvent mixture into the atmosphere is known as evaporation drying and is a pollutive process.

Printing inks of this kind have a medium comprising a hard resin dissolved in a volatile solvent. The solvent evaporates after printing, leaving the resin binding the pigment to the paper surface. Solvent removal can be accelerated by heat and blown air. Evaporation drying is mainly used for gravure and flexographic inks, in particular to non-porous substrates.

Solvent choice dictates to some extent the properties of the ink. The solvent must effectively dissolve a suitable resin binder and also permit an ink to have adequate press stability and drying rate.

The main factor governing the evaporation rate is the solvent vapour pressure. The evaporation rate is governed by the ratio of surface area to volume of ink, that is:

$$\text{Evaporation rate} \propto \frac{\text{ink surface area}}{\text{ink volume}}$$

A greater evaporation rate is therefore obtained from thin films. The more concentrated the resin varnish, the slower the evaporation rate will be, as the resin particles fill some of the liquid interface and reduce the effective area from which the liquid can evaporate. All solvent is therefore difficult to remove from an ink by evaporation and a residual tack tends to develop. Pigments and extenders tend to retard evaporation by obstruction of the surface area, varying in their individual properties such as concentration, specific gravity, size, shape, and absorption.

Volatile solvents are highly flammable, and very fast-drying solvents are often excluded as their flash point is too low. The flash point is the lowest temperature at which a solvent will form a vapour/air mixture which will produce a flash (explode weakly), without continuously burning at atmospheric pressure. Rapid evaporation rates can cause flow-out problems. Odour is another important factor in inks for packaging.

5.3.4 Precipitation[1, 2, 13–14]

Letterpress moisture-set inks rely on this process. These inks are formulated with a vehicle composed of a water-insoluble resin dissolved in a water-miscible solvent. When printed onto paper, water from the paper and neighbouring air dilutes the solvent to the limit where it can no longer retain the resin in solution, and this resin is precipitated around the pigment forming a dry ink film. When insufficient water exists in the paper to cause rapid precipitation, a jet stream can be blown on to the wet print. Glycols are the main solvents employed in moisture- or steam-set inks as these compounds are water miscible, hygroscopic, and have reasonable solvent power for a range of resins.

Glycol-based inks are free from unpleasant odours such as by-products formed during the polymerization of drying oils. They have poor press stability as they absorb water from the atmosphere, which causes the resin to come out of solution on the distributing rollers instead of on the paper.

5.4 SOLVENTLESS METHODS OF INK DRYING[1, 2, 9]

The elimination of polluting solvents has centred on three main approaches:[15]

1. Burning of solvents given off from solvent-based inks.
2. Solvent recovery.
3. Solventless inks.

The choice of the above three will be governed by energy shortages (natural gas or fuel oil) and raw-material shortages (solvents).

Burning of solvents contained in heat-set inks can be at about 1500 °C to form carbon dioxide and water, but a supply of natural gas or fuel oil is needed for this and an energy shortage would limit this approach.

Solvent recovery is possible for gravure and flexographic printing. The high concentrations of low-boiling solvents are usually diluted further in the press room and then these solvents can be absorbed on activated charcoal. The charcoal on saturation has the solvents removed by steam distillation. These recovered solvents can be re-used or sold, although it is not practicable to fractionate a recovered solvent blend boiling in the same temperature range, reducing its commercial value. Energy shortages leading to difficulty in obtaining the solvents used for dilution may restrict the use of this solvent-recovery approach.

Solventless inks are very promising in the sense of avoiding the problem of solvent emission. There are several types of solventless inks including:

1. Thermally catalysed inks.
2. Water-based inks.
3. Solventless oil-based inks and overcoat.
4. Radiation curing inks.

5.4.1 Thermally catalysed Inks[1-2, 6]

These inks are stable at room temperature but rapidly crosslink when heated. Two main types exist: acid-crosslinkable prepolymers, and prepolymers that become reactive at elevated temperatures.

Acid-crosslinked prepolymers include: polyester-alkyd; urea formaldehyde; and melamine-formaldehyde.

These resin adducts[16-18] are used in combination with an acid catalyst. They are unreactive at room temperature but become reactive at elevated temperatures. p-Toluenesulphonic acid is a typical catalyst of this kind. Acid-crosslinking prepolymers are press stable and are polymerized at temperatures of about 135–160 °C.

Typical prepolymers that become reactive at elevated temperatures are styrenated-alkyd adducts. Lithographic and letterpress inks can be made thermally catalysable. These have the advantage of being curable with current, flame, or hot-air dryers, giving mainly carbon-dioxide and water with small amounts of alcohols and the toxic gas formaldehyde. This type of ink film has good properties, in particular resistance to scuff, abrasion, or scratching. They handle well on the press and have a reasonable shelf-life of about 3–6 months. The chief disadvantage is the high web temperature of 135–160 °C needed for film curing. These high temperatures adversely affect the paper properties. More expensive higher-quality papers must therefore be used. Solvent emission is not completely eliminated (a little solvent is included in formulation) and recycling of paper printed with this type of ink is not currently possible.[19-20]

5.4.2 Water-based Inks[1-2, 6]

These inks do still have organic solvents present, but 'water-based' indicates the bulk of water as the solvent. Water-based flexographic inks can contain up to 20 per cent alcohol.

There are three chief kinds of water-based polymer systems. These are: latex type; water-soluble polymers; and water-solubilized polymers.

A latex is a colloidal dispersion of submicroscopic polymer spheres. The viscosity of a latex is determined by the interactions between the colloidal particles, and is independent of the polymer molecular weight. The polymer concentration in a low-viscosity latex may be 50 per cent or greater. Aqueous solutions of water-soluble polymers are true solutions, being molecular dispersions of polymer molecules. The solution viscosity is dependent upon both the concentration and molecular weight of the polymer. High molecular-weight polymers can be used at the expense of polymer concentration and vice versa. The water-solubilized polymers are intermediate between the latexes and polymer solutions, that is between the colloidal and molecular sizes. The preparation of water-solubilized polymers is often by the neutralization of carboxyl-containing latex particles with base to obtain swelling and partial disintegration of the particles. High solution viscosities are needed to give 'tack' in an ink for printability. The viscosity of most latexes is insufficiently low to produce this tack. Water-soluble or water-solubilized polymers are therefore usually used. When dry, these polymers must become water insoluble, so the majority of these water-based ink vehicles are ammonium hydroxide–neutralized alkaline solutions of carboxyl-containing polymers. When the solution is dried the ammonia evaporates, leaving the polymer in its water-insoluble carboxyl form.[21]

An advantage of these types of inks, apart from low solvent release, is that microwave[22–23] radiation can be used to dry these inks rapidly, owing to the high dielectric constant of water.

The main disadvantage is that water has a heat of vaporization greater than that of commonly used organic solvents, that is 1043 BTU/lb (2426 kJ/kg) compared with about 180 BTU/lb (419 kJ/kg) for butanone (methyl ethyl ketone), toluene, and ethyl acetate. These inks also have limited use in regard to other processes such as letterpress and offset lithography. They dry too rapidly on open roller systems such as in letterpress and cannot be used in lithographic printing as they are miscible with the aqueous fountain solution. The water may swell the paper substrate and give poor register. Paper printed with some water-based inks cannot be recycled with current technology.

Gravure is the process where these inks can be utilized on absorbent substrate as their slow drying speed, low gloss, and limited adhesion properties are weighted against their non-flammability, easy wash-up and dilution with water. They are also used in flexography, sometimes with alcohols and glycols added to the inks to improve printability and drying speed.

5.4.3 Solventless Oil-based Inks plus Overcoat[6]

In this type of system the wet ink film on a printed surface is protected by applying a thin transparent film-forming polymer solution while the ink is still wet.[24] Solventless oil-based inks which take several hours to oxidation-dry are overcoated with a thin transparent coating to protect the drying ink film. The solventless ink is often of the heat-set/quick-set kind.[25] The polymer coatings are oxygen-permeable, abrasion-resistant, alcohol-soluble propionate resins. They can also be soluble polymers.[26]

This system has the advantage that other from the coating solution, solvent emission is eliminated. The printed film exhibits high gloss and good abrasion resistance. Paper printed this way can be recycled without complication. Both alcohol- and water-based coatings can be used.

The main disadvantage is probably economic. This is because of the capital cost of the extra coating operation and the fact that the whole sheet must be coated. Complete drying of the underlying ink film also requires a long time.

5.4.4 Radiation curing inks

Radiated energy can be divided into two categories:

1. Thermal effect radiation
   (a) Infra-red
   (b) Microwave and radio frequency

2. Free radical-forming radiation
   (a) Electron beam-ionizing/radical-forming radiation
   (b) Ultraviolet radiation (electromagnetic radiation of wavelengths less than visible light).

Also there is visible radiation which is non ionizing. It is currently commercially impracticable for ink curing and finds application by photosensitization of dyestuffs in photo emulsion systems.

5.4.4.1 Thermal Effect Radiation

(i) Infra-red[27-28]

Infra-red (IR) is a form of electromagnetic radiation that follows the same laws of propagation, absorption, and emission as UV and visible light. All bodies above 0 °K radiate energy. A body with a constant temperature is therefore absorbing (or receiving another input) energy and also radiating it at equal rates.

An efficiently radiating infra-red body has a high emissivity. This body also possesses an equally high absorptivity. This is demonstrated by a blackened surface which is an efficient radiator and absorber, whereas a highly polished surface is the opposite.

The intensity of radiation from a body may be defined as the loss/unit time/unit area. It is a function of its absolute temperature. The Stefan–Boltzmann law states that:

$$\text{Radiation intensity} = (\text{constant}) \, T^4$$

where T is the absolute temperature.

If the absolute temperature of a body is doubled, the radiation intensity will increase sixteen-fold. An increase in temperature, apart from increasing the total radiation, also causes the peak wavelength of the infra-red emission to shift to a lower wavelength value. This is described quantitatively by Wien's displacement law as

$$\text{Wavelength maximum } (\mu m) = \frac{2885}{T}$$

where T is the absolute temperature.

Many producers of infra-red equipment use this relationship to 'tune' the emission wavelengths of their radiators by regulation of the absolute temperature of the heating elements. Some important wavelength maxima are shown in Table 10.

Table 10 Important wavelength maxima for some infra-red radiators

Temp (°K)	Temp (°C)	Wavelength maximum (μm)	Chemical bonds involved
1697	1424	1.7	C—H
995	722	2.9	O—H
849	576	3.4	C—H
498	225	5.8	C=O

Plotted on log–log paper this relationship is linear (absolute temperature plotted against λ_{max} in μm).

The shape of the radiation curve from IR emission is unfortunately quite broad (i.e. relative energy plotted against λ_{max} at a given temperature).

To absorb the radiated energy efficiently, the surface coating should absorb a major portion of it. This is only feasible if the surface coating has an absorption curve which corresponds to that of the emission curve of the radiator.[29–30]

In the case of a four-colour process set of sheet-fed offset inks, the curve of the black ink shows the required broad absorption with only a few peaks and would be a highly efficient absorber owing mainly to the presence of carbon black in the ink. The colours are problematic as they absorb strongly only in discrete and very narrow wavelength regions. These are (in the regions of interest) 3.4–3.5 μm and 5.7–5.9 μm and are a consequence of the C—H and C=O stretching vibrations from the molecules present in the pigments and vehicles.

This presents some complications in drying different colours as the black ink absorbs energy more efficiently and gets much hotter than a neighbouring area of yellow or red, where most of the infra-red spectrum will not be efficiently absorbed by the ink. This is the opposite of the problem found in UV curing, where the black is slowest curing. This is because it also absorbs UV strongly and carbon black is conjugated and polycyclic and virtually acts as a free-radical scavenging trap. Most coloured pigmented coatings and also many vehicles absorb UV radiation much more efficiently than they absorb infra-red. The differences between curing of colours in UV should not therefore be as large as they are in the case of infra-red.

The wavelength maximum of the source could be placed at about 3.4 μm where most organic molecules absorb strongly though in a very narrow band. As long as the IR source used emits a band of wavelengths there will always be a mismatch with the coloured printing ink. For efficient sources that emitted very narrow bands of radiation (such as an IR laser) that could be tuned to 3.4 μm, all organic coatings would absorb it nearly equally.

Alternatively, an additive could be found for the inks that has a broad IR absorption curve but is not highly coloured like carbon black. This could provide the answer to the mismatch problem.

Infra-red inks find some application in the metal coatings industry.[31] This industry has used radiation for curing since the first hot-air oven was utilized for drying coatings on metal. A 204 °C gas convection oven is also a 6.1 μm radiation oven. All the metal parts inside the oven at 204 °C act as radiation emitters. These radiate 960 BTU's per square foot per hour (10,903 kJ per square metre per hour) at a peak of 6.1 μm of IR.

A production convection oven line runs slower for cure at start-up and the air temperature may read 204 °C; however, the metal parts inside the oven are still cold. After several hours the oven stabilizes at 204 °C and the increased line speed is the consequence of radiation. Hence engineers were led to use direct radiation for curing and drying. The higher the surface temperature, the greater the energy output per unit of area, per unit of time. This led to faster drying and the IR industry was born.

Some energy temperature values are depicted in Table 11 for theoretical black-body radiation.[29, 31–32]

Table 11 Energy/temperature values for theoretical black-body radiation. (Reproduced by permission of the Society of Manufacturing Engineers.)

Surface temp (°F)	(°C)	Peak wavelength (μm)	BTU per sq. ft. per hr.	kJ per sq. metre per hr. ($\times 10^4$)
4000	2222	1.07	680 000	772.26
3000	1649	1.51	245 000	278.24
2000	1093	2.13	64 000	72.68
1500	816	2.67	26 000	29.52
1000	538	3.58	8 000	9.09
800	427	4.14	4 500	5.11
600	315.6	4.91	2 200	2.50
400	204.4	6.10	960	1.09
300	148.9	6.90	580	0.66
200	93.9	7.90	320	0.36

Direct radiation use with IR dryers has encountered varying degrees of success in industry over the years. If a 1000 °F emitter has an output of 8000 BTU's/sq. ft./hr. (9×10^4 kJ/m^2/h), then a 2000 °F emitter with 64 000 BTU's/sq. ft./hr. (73×10^4 kJ/m^2/h) cures faster. A 4000 °F emitter at 680 000 BTU's/sq. ft./hr. (772

× 10^4 kJ/m²/m) would be expected to cure very fast indeed. This has not been observed in practice. It was found that with high-energy, high-temperature, short peak wavelength IR, over-baked dark colours and under-cured light colour coatings resulted, because as previously mentioned different pigments and chemicals possess different absorption characteristics for IR radiation. Short-wave IR is very colour sensitive, long-wave IR (over 2.5 μm) is less colour sensitive, and at above 3.2 μm the radiant energy is almost equally effective on light and dark colours.

The majority of organic chemicals possess little or no absorption below 2.0 μm and good absorption at 3.4 μm.

Table 12 depicts the absorbance of IR for selected materials (for major absorbance lines over 30%).

Table 12 Absorbance/wavelength properties of IR active materials. (Reproduced by permission of the Society of Manufacturing Engineers.)

Material	Wavelength (μm)	Absorbance (%)
Alkyd	3.4	60–85
	5.8	85–95
	7.8–8.2	80
Vinyl chloride	3.4	56
	5.8	86
Epon 828	3.4	70
	6.2	75
	6.6	90
S.S. Nitrocellulose	2.8	31
	6.1	93
Cellulose acetate	2.8	30
	5.6	74
Ethyl cellulose	2.9	40
	3.5	70
	7.3	60
Raw linseed oil	3.4	80
	5.7	78
Tung oil	3.4	93
	5.7	95

Two parameters of importance in the application of IR curing are:[29]

1. The distance of the emitter from the surface coating. Infra-red radiation reduces by the square of the distance, so the emitters should be as near as practical to obtain the most efficient transfer of energy.
2. The substrate. Infra-red does not transfer sensible heat. The heat for curing is generated in the surface of the coating by excitation from the radiant energy, as sensible heat is generated in the printing ink and the substrate acts as a heat sink to ink, drawing the generated heat from the ink by conduction. When the

ink and substrate are in balance then curing begins. This does not mean that the substrate must attain the same temperature as the ink. It is an energy in, energy out, balance; energy into the ink minus energy out of the ink to the substrate. The thicker the substrate the greater the heat-sink effect. In production terms, thicker substrates run slower or require longer IR ovens.

Infra-red lamps with peak radiation in the near IR at about 1200 nm are particularly useful in curing alkyd amino finishes on heat-sensitive substrates such as wood.

(ii) Microwave and radio-frequency drying[34-36]

Microwave radiation is a high-frequency, long-wavelength penetrative radiation producing heat effects by molecular friction. Microwave energy conversion into heat is dependent on the loss factor of the material. The process is therefore most efficient with polar materials such as water. Water-based coatings on substrates that do not readily heat up quickly, such as non-polar plastic films, are a most likely use.

The basic principle of microwave or radio-frequency heating is molecular excitation. Polar molecules, when exposed to an electrostatic field, oscillate at the same rate as the operating frequency. Every molecule responds in varying degrees depending on size, mass, shape, and electrical configuration in attempting to align itself to the field. For ink and coating vehicles, this response occurs through a wide band of frequencies. The microwaves only supply the energy for the ink to dry by whatever mechanism is inherent in its composition, such as oxidation, evaporation, etc. The essential difference between microwave and conventional heating such as infra-red, hot air, etc., is that microwave causes heating throughout the mass. There is therefore no thermal delay when the energy source is turned on or off.

The frequencies available in the UK for microwave heating are restricted by the GPO to a few small bands such as 892, 2450 and 5800 MHz. 2450 MHz is used, as the higher frequency band (> 3000 MHz) sets up eddy currents which dry off the free surface water from paper and film. Also dipole stimulation is created which removes bound water from the fibres themselves, giving a scorched or embrittled web.

Microwave radiation possesses a wavelength of the order 12 cm which lies in the electromagnetic spectrum between visible light whose wavelength is of the order of a millimetre and the long radio waves which can be up to the order of a kilometre in length.

Microwave radiation is composed of both an electric field and a magnetic field. There is total interdependence of these fields or directions of force and they act at 90° to each other. Both these fields within the limits of their own planes of activity are constantly altering their polarity direction at a rate characterized by their frequency. The length of the waves is determined by this rate of change or frequency. The frequency of modern equipment tends to use 2450 MHz which means the fields will be oscillating or changing their directions 2450 million times a second.

Microwave energy is produced by a Magnetron. This is essentially a circular cavity in a powerful magnetic field into which DC power is injected. Electrons injected into the cavity are set into rotation by the magnetic field. The frequency or oscillation rate is determined by the cavity design. For practical and economic considerations the size of the Magnetron is limited to 5 kW. Magnetrons can be arranged in series.

The substrate with its superimposed wet ink film is passed through a closely concentrated field of rapidly oscillating electrical energy – positive and negative charges are changing places at 2450 million times per second and it is the electric rather than the magnetic component of the energy that is used.

If sufficient polar molecules are present in the ink they can try to orientate themselves with this varying field, and a means of generating internal heat by molecular friction is at hand.

It is paramount that the molecules lag a little behind the field variations as this gives more friction and this lag is known as the loss factor.

For a solvent to be effective, certain requirements are necessary:

1. High dielectric constant.

2. High loss tangent (loss factor/dielectric constant).

3. Low boiling point.

4. Low latent heat of vaporization.

5. Low molecular weight.

6. High dissociation constant.

7. Low specific heat.

Pigmentation of the system has to be examined with care as the use of different types of pigments and quantities can produce a thermal gradient in the substrate with a risk of 'cockling'. Metallic inks can short-circuit the waveguide applicator and possibly damage the Magnetron, and electrical flash-overs are possible as they certainly are with the use of foil substrates, leading to fire risks.

High-loss substrates on exposure to microwaves can cause shrinkage and register problems. Embrittlement can result from the removal of water or polar softeners from paper or plastic substrates.

The drying units which emit microwave radiation have found commercial application on web-offset newspaper presses.[37-38] Microwave radiation penetrates the ink film and drives off the solvents without intense direct surface heat. The ink formulation is critical for the efficiency of microwave drying.

Microwave radiation is employed for drying aqueous or highly polar systems such as water based gravure[39,40] or flexographic inks, but with partial success for letterpress[41,42] or lithographic inks.[43,44] Screen inks[45-48] have also been dried by this process.

The higher the loss tangent (loss factor/dielectric constant), the higher the absorbed energy and consequently the greater the heating effect:

$$\tan \delta = \frac{P}{KfE^2e^1}$$

where $\tan \delta$ = loss tangent
P = energy absorbed (W/cm^3)
K = constant for units used
f = field frequency (Hz)
E = field strength (rms V/cm)
e^1 = dielectric constant

The loss factor is dependent on the ratio of energy absorbed by the material to the dielectric constant of the material itself at a given frequency.

Some commonly used materials in the flexographic and gravure areas that have been explored for microwave drying[49–51] are listed in Table 13.

5.4.4.2 Free Radical-forming Radiation

(i) **Electron beam**[52–55]

Ionizing radiation allows direct formation of free radicals in the coating and low-molecular-weight unsaturated polymers, monomers, and oligomers are converted in a fraction of a second to fully cured coatings.

Various sorts of radiation of comparable energy differ from each other in their linear ionization density, which is the number of ions produced by the radiation in the target material per unit of distance. As ionization continues, the radiation loses energy and therefore the depth to which the rays penetrate the material is inversely proportional to the intensity of ionization.

Gamma rays ionize very weakly and can therefore penetrate materials for large distances. The ionizing effect is strongest from α-radiation (helium nuclei) but even in low-density media such as air their range is only a few centimetres. The distance between beam source and the coating surface is generally greater than 5 cm and hence no radiation would reach the coating.

A compromise between the two types of radiation described above is needed. To effectively cure a coating, radiation is required which ionizes intensively, yet penetrates materials for large distances. Electron or β-radiation ionizes intensely but its ionization density is not equal to that of α-radiation. Its ability to penetrate depends on the accelerating voltage to which it has been subjected. Generally for coating curing this is between 300 and 500 kV which causes penetration of 300–400 μm in a material of unit density.

On reaction with high-energy electrons, materials undergo three chief primary processes:[56,57]

Table 13[49] Degrees of absorption shown by ink constituents in 2450 MHz microwave field. (Reproduced by permission of MacNair-Dorland Co.)

Solvent

Good	Average	Poor
Water	Ethyl cellosolve	Trichloro-ethylene
Isophorone	Ethylene glycol	Toluene
Acetone	1:1 ethanol:methylethyl ketone	S.B.P. Solvents (1–5)
Methanol	Methyl cellosolve	Hexane
2-Nitropropane	Ethanol	Heptane
Methylisobutyl ketone	Ethyl acetate	
Methylethyl ketone	Diethylene glycol	
Methyl cyclohexanone	Cellosolve acetate	
	n-Propyl acetate	
	n-Propyl alcohol	
	iso-Propyl acetate	
	iso-Propyl alcohol	
	Butyl acetate	

Resins

Good	Average	Poor
Nitrocellulose	Methyl cellulose	Polyethyl methacrylate
Polysulphides	Cellulose acetate	Polybutyl methacrylate
Polyvinyl acetate	Cellulose acetate propionate	Polymethyl methacrylate
Polyvinyl alcohol copolymers	Cellulose acetate butyrate	Gilsonite
Phenol formaldehyde	Polyvinyl chloride	Polyvinyl acetate
Urea formaldehyde	Epoxy	Poly isobutylene
Melamine formaldehyde	Shellac	Polyethylene
Phenolics	Alkyds	Hydrocarbon resins
Polyamides	Polyesters	Polystyrene
Melamines	Ethyl cellulose	
	Nylon	
	Polyvinyl butyral	
	Polyvinylidine chloride	

Pigments

Good	Average	Poor
Carbon black	Phthalo blue	Benzidine
Titanium dioxide	Phthalo green	Rubines
Chrome	Hansa	Fanals
Iron blue		

1. *Ionization*. An electron is eliminated:

$$AB \rightarrow AB^+ + e^-$$

The ionized molecule then immediately dissociates to form a free radical and a radical ion,

$$AB^+ \rightarrow A\cdot + B^+$$

2. *Excitation*. Sometimes the irradiation energy is not sufficient for ionization, only for excitation:

$$AB \rightarrow AB^*$$

Dissociation into free radicals is the most important reaction of excited molecules:

$$AB^* \rightarrow A\cdot + B\cdot$$

3. *Electron capture*. This is a similar process to ionisation:

$$AB + e^- \rightarrow AB^-$$

This then splits into a radical and a radical ion:

$$AB^- \rightarrow A\cdot + B^-$$

Electron-beam curing is dependent upon the ability to form free radicals.

As a source of electron radiation the β-ray isotopes obtained as waste products from nuclear reactors cannot be used as their β-radiation is always accompanied by strong γ-radiation. The protective screens which must be used against the penetrating γ-radiation, also retain the electrons.

Of the generators available, only linear accelerators and not cyclotrons are available.

Electron-beam curing has a wider range of applications than UV curing, being suitable for both clear and opaque pigmented systems, whereas UV is primarily suitable for clear or semi-transparent systems.

This process allows solventless coatings to be cured in a fraction of a second with several advantages over other radiation curing processes, which include:[58,59]

1. No film thickness limitations.
2. Any substrate can be used without a noticeable temperature increase.
3. Low energy consumption.

There are however, some disadvantages:[60]

1. Generally limited to flat surfaces or those of shallow contour.
2. Capital cost of equipment such as electron accelerators is high.
3. The possible hazard of radiation escape.

Electron-beam equipment[61,62] usually uses a point filament as a source of electrons which are produced in a pencil-like beam which is electromagnetically scanned at right angles to the direction of movement of coated objects under the gun window.

Electron-beam curing inks and coatings are similar in principle to UV coatings with the exclusion of a photo-initiator/sensitizer being needed. Any form of ionizing radiation can initiate vinyl-type polymerizations.[63] These can be neutrons, α-particles, γ-rays, and X-rays, as well as high-energy electrons (β-rays). The initiation mechanism is somewhat different to that of photochemical initiation in that radiation of vinyl monomers gives cations and anions in addition to free radicals.

Electron-beam curing offers higher intensities[64] and therefore faster rates of curing for thick ink or paint films. The impingement of high-energy electrons upon metals causes secondary X-radiation, a more highly penetrating form which needs elaborate shielding for safety. Paper printed with these electron beam–cured inks cannot be recycled at present.

(ii) Ultraviolet curing

There are two prime reasons for the current rapid advance of UV curing technology. These are:

1. Energy and matter conservation.[65-66]

2. Environmental pollution control.

Both of these received an impetus from the USA where the shortage of natural gas in certain states led to the search for alternative energy sources, and the Los Angeles (1966) Pollution Act which resulted from the concern of volatile hydrocarbon solvents being released into the atmosphere.

Ultraviolet drying of printing inks and coatings is effected by electrically derived UV light energy as the driving force to produce instantaneously dried inks. As with all ink drying, this conversion involves a phase change.[67-69] In this case a mobile liquid is converted in a fraction of a second (about 0.02 s) to a hard, rigid, solid network. This process is depicted in Figure 71.

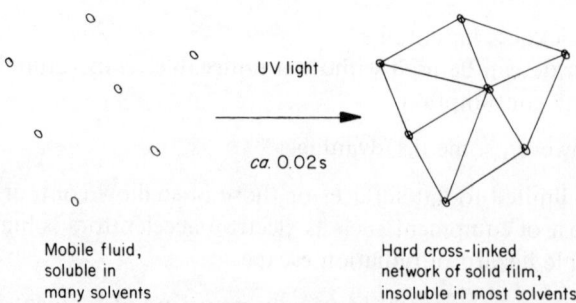

Figure 71 Schematic representation of UV curing of a surface coating.

Energy is conserved as the area of application of the UV light is specific to the coating of ink involved, not to the substrate as occurs with some other methods of drying.

The common types of photo-initiators fall into the classes of:

1. Aromatic ketones and synergistic amines.
2. Alkyl benzoin ethers.

(See Chapter 3, sections 3.4.3 and 3.4.2.2. respectively)

The mechanism of UV curing can simplistically be represented as:

Photo-initiation

$$R-R \xrightarrow{UV\ light} R\cdot + R\cdot$$

$$R\cdot + MH \longrightarrow M\cdot + RH$$

Propagation

$$M\cdot + M_1H \longrightarrow M(M_1H)\cdot$$

$$M(M_1H)\cdot + M_2H \longrightarrow M(M_1H)(M_2H)\cdot$$

$$M(M_1H)(M_2H)\cdot + M_3H \longrightarrow M(M_1H)(M_2H)(M_3H)\cdot \text{ etc.}$$

Or in general:

$$M\cdot + n(MH) \longrightarrow M(MH)_n\cdot$$

Termination

$$M_{197}\cdot + M_{64}\cdot \longrightarrow M_{261}$$

$$M_{32}\cdot + M_{41}\cdot \longrightarrow M_{73} \text{ etc.}$$

Or in general:

$$M_x\cdot + M_y\cdot \longrightarrow M_{(x+y)}$$

R–R is the photo-initiator and MH denotes the monomer or pre-polymer reacting by hydrogen abstraction.

The species denoted with a dot represent highly reactive free radicals.

5.4.4.3 Ultraviolet Coatings

Apart from inks which are described later, many varieties of photopolymerizable coatings occur, including:

1. Wood coatings.
2. Flexible coatings for plastics and textiles.
3. Flexible metal decorating coatings.
4. Clear overprint varnishes for paper and board.
5. Photoresist coatings.

All of the above with the exception of clear overprint varnishes may be pigmented or non-pigmented depending upon the particular application.

(i) Wood coatings[70]

Three distinct coatings are used for wood or wood imitations in the following sequence: filler sealer, undercoat or basecoat, and topcoat.

The filler sealer[71] provides a smooth non-porous surface after a sanding operation and currently these tend to be based upon acrylic systems.[72] A high loading with transparent extender ensures cheapness.

Undercoats or basecoats are opaque for decoration and provide physical strength to the wood coating.

Commercial topcoats must be decorative in the sense of appearing like wood and functional in being able to provide protection and durability to the total film.

Dufour[73] has reviewed the wood-coating area and some of the main points described are outlined below:

Wood finishing is needed to improve its aesthetic appearance and surface mechanical properties. Two chief processes are used to achieve this:

Wet finishing – non-pigmented and pigmented coatings applied directly to the substrate.

Dry finishing – a prefinished sheet of treated paper or plastic bonded to the substrate.

The current requirements for wood finishing include: high productivity to offset labour costs; elimination of pollution by solvent emission; and power-consumption savings because of primary energy source scarcity.

There are many conventional systems employed for wood finishing with the disadvantage of long drying times ingesting large quantities of energy for curing purposes or solvent removal.

Some of these are:

Aqueous systems	– Energy needed is of the order of four times more than organic solvents for water evaporation.
Nitrocellulose systems	– These can possess volatile solvents of the order 80 per cent.
Polyurethan systems	– Possible pot-stability problems and volatile solvents may be up to 60 per cent.
Acid curing systems	– These may possess a maximum of 50 per cent solvent. Irritant formaldehyde is liberated and may persist after curing.

Radiation curing systems permit a solution to the above problems.

In order of choice, the least favourable radiation system is a polyester/styrene ultraviolet photopolymerizable system. Better performance is obtained from an unsaturated acrylic ultraviolet photopolymerizable system. Most favourable of the radiation systems is, however, an unsaturated acrylic system cured by electron beam.

Wood varnishes need a minimum quantity of coating to wet the pores and to obtain good flow from roller application. Also, no micro-air bubbling must occur throughout curing, which must be rapid both at the film surface and within the bulk of the surface coating. Adequate adhesion, hardness, impact and scratch resistance are necessary but with sufficient flexibility to withstand wood dimensional variations. Hardness is an essential quality. Resistance is also needed against abrasions, stains, and solvents.

Cure speeds required are about 18–21 m per minute for two 200 watts per linear inch (80 w/cm) lamps.

The original polyester/styrene ultraviolet photopolymerizable systems developed throughout the period 1960–70 are retained mainly as chipboard sealers and for thick layer finishing. Curing is effected by a lamp combination of weak actinic power lamps TL and 75 watt/linear inch (30 w/cm) HTQ lamps.

The early 1970s permitted, with the advent of the 200 watts/linear inch (79 w/cm) medium-pressure mercury arc UV lamp, the air curing of the polyester in styrene without the need of paraffin wax.

Currently, in France, unsaturated acrylic photopolymerizable systems are gaining ground, and some 20 users are known there, using about 20 dryers (2–3 lamps). The trend is now towards open-pore finishing with acrylic systems. Open-pore finishings in coatings are applied chiefly by rollers at about $8-12\,g/m^2$. Some applications are by spraying. Acrylics are also used for closed-pore finishing in coatings. Application is by curtain coaters at about $80-100\,g/m^2$.

The polyester/styrene system has many disadvantages compared to an unsaturated acrylic photopolymerizable coating. Wetting is not particularly good owing to lumping on the pores. A varnish layer of some $15-20\,g/m^2$ is required followed by sanding to reach a flat surface. This is wasteful in terms of photo-cured varnish. Unsaturated acrylics give good pore wetting and 'lumping' of the coating does not occur. Consequently, thin layers, $8-12\,g/m^2$ only (about 50 per cent reduction compared to polyester/styrene), are needed with minimal sanding and small photo-cured varnish loss.

The polyester/styrene system requires 2–3 lamps (200 watts/linear inch) (80 w/cm) for the base coat and about 5 lamps for the topcoat when curing at speeds of about 18–21 m/min. Acrylics require only about one-third of the lamps. The former system causes the substrate to heat up and subsequent 'microbubbling'. Fillers are often incorporated in polyester/styrene coatings giving a dull finish to counteract, as far as possible, the undesirable non-wetting and microbubbling effects. The use of acrylics can overcome these problems.

The diluent monomer styrene has volatility problems leading to viscosity variations of the coating. This also necessitates the regular cleaning and inspection of the application rollers and also results in the installation of expensive ventilation systems in the workshop. Unsaturated acrylics, on the other hand, have very small volatility and yield a stable viscosity without the cleaning of rollers, and without inspection and associated problems.

To summarize, unsaturated acrylic wood varnishes, although expensive compared to their earlier counterparts, the polyester/styrene systems, offer many advantages. These are mainly the smaller quantity needed per square metre, better quality of coating, less pollution, and considerable reduction in power consumption.[74] They also provide for a smaller working area and high production rates for a low investment.

(ii) Flexible coatings for plastics, and textiles

Flexibility is often required for plastic products such as vinyl flooring[75] and vinyl film.[76] Also, flexible coatings find application in the textile area[77] such as plastisols or coatings to act as a transfer medium for good wash-resistant pigmentation or dyeing of natural fibres such as cotton, where pigment/vehicle adhesion alone is difficult. This is in contrast to synthetic fibres such as polyesters where good adhesion and wash resistance may be obtained without a transfer medium. Sublimatic dyes are used in this textile transfer process.

With vinyl films, good resistance to wear is needed. Generally urethan[78] and isocyanate technology is used to give elastomeric coatings, but this can lead to high-viscosity problems.[75]

Vinyl flooring[75]

These coatings need scrub, stain, abrasion, and scratch resistance. They must also be non-yellowing and not cause the vinyl tile to curl or lift on its borders.

Vinyl film[76,77,79]

Vinyls are often coated to improve surface gloss, burnish and abrasion resistance, stain resistance, or to protect a print pattern.

Vinyls are often laminated to a number of substrates for both decorative and functional purposes. They may be coated prior to or after lamination or as a free film.

Adequate flexibility is necessary to permit embossing and post forming when required.

(iii) Flexible Metal decorating coatings

Metal decorating[2,80-82] is a term describing the application of surface coatings by means such as roller coating, spraying, dipping, and brushing on to all types of metallic substrate. The metal is coated for protective and decorative purposes. There are a variety of application methods according to the physical and chemical needs of the finished product.

For the roller-coating method, the metal generally has one or two coats of a protective lacquer on the inside (internal) and on the outside (external). A decorative system generally consists of a size basecoat, a white enamel coating, then several ink passes to produce the required design. This is followed by a coating of a colourless varnish to provide protection. Every coating and ink pass is then passed through an oven and stoved, although tandem and four-colour passes print the design in one or two passes with an overprint varnish as the final coating, and only one pass through the oven is needed.

Roller-coating decoration is employed for a variety of finished products. These range from bottle tops to aerosols, food cans to children's toys, and from film cassettes to steel drums, etc. The application method for these coating materials is performed on a flat sheet of metal prior to fabrication. All coating materials have to withstand the physical deformation of bending, forming, and stamping and yet must still donate the chemical resistance required.

Sometimes on caps and closures, and some food cans, the coating materials have to withstand the conditions of sterilization in water and steam under pressure without losing adhesion to the metal support. They must also keep their own adhesion properties to some compounds in addition to no loss in chemical resistance.

Ultraviolet curing currently has its largest application in the metal-decorating industry.

Metal-decorating ovens[80]

Two chief kinds of oven are in general use. These are the box and the conveyor or tunnel oven. The box oven currently has little use. The box oven is essentially a heated box into which go the coated sheets on metal racks. This oven incorporates the use of air circulation fans and a fume extraction system. Production is very slow, as fifteen to thirty minutes is needed to permit the oven to reach the temperature required for stoving to begin. One batch of metal may take up to three quarters of an hour to completely stove.

The modern tunnel oven possesses a continuous conveyor. This carries the sheets through the oven, and stoving time is about a quarter of an hour. The sheets are fed directly into the frames on the conveyor which is geared to the varnish machine. A frame receives each sheet as it comes from the belts. After stoving the sheets are automatically stacked and unloaded. Many modern lines employ a continuous system whereby several stacks of metal may be fed into the oven without a break for loading and unloading.

The thermal radiation in the oven is usually achieved from gas or oil heating systems, either direct or indirect. The direct system involves the use of burners in the heating chamber itself. The indirect method heats the air in a separate combustion chamber where it is then piped into the heating section.

Tunnel ovens possess three chief sections:

1. The initial section eliminates the solvents from the coating and raises the sheets up to peak temperature.
2. In the next section stoving occurs because the sheet passes a peak temperature, held throughout for a specified time.
3. In the final section cooling occurs to permit stacking without sticking.

In order to obtain the correct finish from any coating material the oven must always perform these three functions properly. Conditions inside the oven must be kept constant as incorrect stoving may cause serious problems.

Radiation curing with UV light eliminates problems with solvents exhausted to the atmosphere and also enables time in the oven and hence thermal energy costs to be reduced. Ultimate UV systems should eliminate ovens completely.

A detailed description is given below[80] of the substrates and current container and closures used in the metal-decorating industry, now coated with conventional inks and roller coatings from which the applications for UV curing may be gleaned.

Substrates

1. Tinplate
   (a) electrolytic
   (b) hot-dipped
2. Differential tinplate
3. Black plate
4. Hi-top
5. Terneplate
6. Aluminium

Tinplate is most common. The alternatives mentioned above were introduced in order to produce a cheaper metal for containers and closures but with the same properties of strength, ductility and flexibility as tinplate.

1. *Tinplate.* The composition is mild or low-carbon steel that is sheet- or strip-coated on both faces with tin. This comprises in one material the strength and formability of steel and the corrosion resistance, solderability, and good appearance of pure tin. There are two main types:

(a) *Electrolytic tinplate.* This is produced by rolling mild steel into strips or sheets of a finished gauge around 10 thou (254 μm) and coating subsequently on both sides with tin by electro-deposition. The tin is applied to a gauge of 1 thou (25.4 μm) by electroplating from tin-bearing electrodes. The steel base is a continuous strip which passes in one strand through a number of operations; de-coiling, electrolytic cleaning, light pickling, electrolytic tinning, flow brightening of the tin coating, electrochemical treatment, oiling, shearing inspections, and finally recoiling before its ready for use.

Tinplate may be obtained at different tempers and is very ductile. Tin has excellent adhesion qualities to the steel base and flows easily as the metal base is moved or bent during construction of containers and closures. In addition, tin acts as an anti-oxidant coating on top of the steel base.

(b) *Hot-dipped plate.* A greater deposit of tin is applied than that deposited by the electroplating method, although a higher brightness finish is achieved. The metal is usually in sheet form about 75 cm square and is dipped in molten tin, the thickness being monitored as the sheets emerge, by a series of rollers immersed in palm oil. Difficulties may arise with subsequent coating of this metal. These are because some oil remains on the surface even after washing.

The main drawback with this method is the speed at which tinning is carried out. Also a thicker coating is applied than is necessary.

Stabilization of the steel base is necessary. This is because it is very reactive after cold rolling, during processing. Heat is applied to the metal and the temperature is held just below the lower critical change point. Subsequently the steel is finally cooled at a controlled rate. The metal has now returned to its natural stable annealed condition. This means that the highly deformed lattice structure of the plate undergoes 'recrystallization'. There are two methods of annealing: continuous annealing and batch annealing.

(a) *Continuous annealing.* Steel strips on a flow line are heated to a temperature of about 700 °C in 10 seconds and held for 20 to 40 seconds and then cooled quickly but at a controlled rate. This continuous process is the best and most popular used today because a finer-grained metal with better stiffness qualities is obtained and permits the use of a finer tin-coated metal. This is less expensive.

(b) *Batch annealing.* Stacks of steel strips beneath cylindrical covers are heated to about 650 °C and then permitted to soak for a given time prior to cooling at a controlled rate. The total process is performed in a protective gas atmosphere to prevent oxidation. This is a longer and more inefficient process as a large mass of metal has to be heated and cooled before the operation is complete. The average 'batch' size is around 200–300 tonnes of metal coiled in a static base.

2. *Differential tinplate.* This kind of plate has a thicker coating of tin on one side than on the other, primarily for cost reasons rather than any technical advantage. The thicker side is used as the internal side of a container for protective reasons. The thinner side is normally the external part of the container. This gives enough oxidation resistance as needed. The main use of this type of plate is for can work.

3. *Black plate.* This is a mild steel based metal with an oxidised surface. This corrodes easily and all plate of this nature has to be coated. This substrate was introduced as a cheap alternative for tinplate. It is a non solderable metal and an adhesion system is required when this metal is used for can work.

4. *Hi-top.* Essentially this is black plate with a top coating of chromium oxide and was introduced as an alternative to the more expensive tinplate. This metal does not solder readily and where used for can work it needs an adhesive system or a welding technique for sealing, similar to that used for black plate.

5. *Terneplate.* This is a steel base coated with a tin/lead alloy. As with all tinplate substitutes it is made for cost reasons. This has the advantage of being solderable and the need for an adhesive sealing system does not arise.

6. *Aluminium.* Aluminium is not so widely used in the UK as in the USA where electricity is cheap. In the USA this makes production much less expensive than it is in Britain. It is used in the UK for closures of beer tops, e.g. easy-open ring-pull can.

Containers used in metal decoration

General-line cans and containers. For these cans and containers hi-top and tinplate are the normally used substrates. Where sealing is needed an adhesive is required when hi-top is used as it is not solderable. A size is not always applied to these containers as some such as oil cans do not place importance on decoration. For a prestige can such as talc tin, then a size coat is applied. This basecoat aids adhesion of subsequent coatings and gives greater consistency to coating. As a size coat is applied at a very low filmweight, a vinyl-type size is normally used for this type of container as it has good film-forming properties when applied at a solids content of 15–20 per cent.

Enamel coatings used for general-line containers need good resistance to mechanical handling. They must also have good stacking properties. In order to obtain the correct degree of hardness required from the coating, generally an alkyd-based coating with the addition of melamine formaldehyde is used.

Varnishes when applied to any coating are mainly for protective purposes as damage may occur from mechanical handling and during transit. With the general-line type containers a great deal of flexibility is not required of the varnish and a harder varnish system can be used such as a short or medium semi-drying oil alkyd type with melamine formaldehyde added to give the degree of hardness required.

In order to improve the qualities of a colourless finishing varnish, additives are often used. These are lubricants which aid tooling and donate properties of anti-scuff and scratch resistance to the coating.

An internal coating of general-line cans and containers is not always needed. When required it would be in either a decorative or protective capacity or as an aid to fabrication, in particular when using hi-top metal because this is very hard and can damage fabrication tools. An oleoresinous or the more expensive vinyl-type lacquer may be used in the case where internal decoration is needed. If a protective lacquer system is required, an epoxy phenolic system can be employed, but the choice of lacquer is normally dependent on the kind of pack being used. As for coatings to aid tooling, nearly any type would suffice. A lacquer system is therefore not essential and a cheaper vinyl size may be used.

Open-top cans. For these cans, coatings must be hydrophobic (water resistant) in order to withstand retorting conditions. This is the process by which food is cooked in a can at a temperature of 120 °C for about 30 minutes minimum and up to a maximum of 2 hours. Retorting may be either a continuous or batch process. A very high degree of resistance to mechanical handling is also needed from coatings employed in open-top work.

Electrolytic tinplate is mainly employed for open-top cans. A small amount of hot-dipped plate is sometimes used. A vinyl size coat is normally applied on all externally scored cans, such as meat or beaded-type cans.

The type of resins used are of a polyester, acrylic, or non-drying oil alkyd type. This permits the enamel to have satisfactory retorting properties and to be sufficiently hard on drying with good adhesion properties.

The varnish system employed is influenced by the resin system used for the base coat. Usually a non-drying oil with melamine added to give some measure of hardness is adequate in a wet on wet system. As for the rest of the open-top coatings, it must be impervious to water.

Internal coatings may be either of an oleoresinous, phenolic, or epoxy phenolic nature depending on the kind of pack they are needed for. Some examples are:

1. The oleoresinous kind used for fruit such as plums, peaches, strawberries, etc.

2. Phenolic-based coatings employed for meat and vegetables.

3. The epoxy phenolic kind of internals used in place of the phenolic coatings when extra flexibility is needed.

The oleoresinous coating may be pigmented with zinc oxide for use on vegetable-type packs. The zinc oxide absorbs the hydrogen sulphide gas which is evolved in these packs. Absorbing this sulphide eliminates the sulphur odour when the can is opened.

Deep-drawn processed cans. The substrate used is deep stamping tinplate or aluminium (USA mainly). The coatings for these must be flexible and possess good chemical resistance.

All drawn processed cans are sized and this is normally of the vinyl kind as these have improved flexibility properties compared with the epoxy-based type.

Enamel coatings used are either vinyl- or acrylic-based resins as they give better tooling properties during manufacture. The acrylic resin permits better colour retention properties than the vinyl. Both are crosslinked with amino resin but only to the limit that gives the coating satisfactory retorting properties without harming flexibility.

Finishing varnishes may be based on vinyl, acrylic, or even epoxy-ester resins. The vinyl resin is not usually used as it is not suitable for wet-on-wet purposes and is quite costly. The epoxy-ester and acrylic types are therefore used and are normally suitable when crosslinked with melamine and epoxy resins respectively to impart the required tooling and retorting properties. The vinyl-type lacquer systems may be used for internal coatings because they are tasteless, odourless, and offer excellent flexibility and specific chemical resistance. Generally they are used over other coatings where taste considerations are necessary, but they are costly.

Phenolic resins possess good all-round chemical resistance but they do not have the desired flexibility necessary for the production of can ends. They have a limited application value.

Another adequate lacquer system may be created by blending the epoxy and phenolic resins to give adequate flexibility and chemical resistance. In order to overcome sulphur staining, either aluminium pigment or zinc oxide can be added.

The epoxy amino lacquer coatings are an alternative satisfactory internal system for food cans. Use is limited by cost and therefore they usually perform more specialized operations.

Beer and beverage cans. Mainly electrolytic tinplate is used. Hi-top metal can be used. These cans are never sized and all decoration is printed.

A finishing varnish for this kind of work has to have good pasteurizing and retorting qualities in addition to suitable hardness and gloss qualities. The beer can is pasteurized whereas the beverage can is retorted. The varnish must also be free from taste and colour. An alkyd-based resin is mainly used as it is the only non-drying oil alkyd and gives the high degree of freedom from odour and taste that is required. This will normally be cross-linked with a melamine resin for hardness.

Lubrication of beer and beverage can varnishes is important to give good resistance to fabrication and to maintain line mobility during the filling operations.

Internal systems for these cans may be oleoresinous types although in the USA a system known as Budium is used for 90 per cent of all can work. The epoxy urea system is preferred for the beverage can. An epoxy phenolic (self-colouring) kind is also widely used.

Beer cans all possess a treble coating, a primer side seam spray and a vinyl internal spray. The side seam is sprayed with a fast-curing lacquer system. This curing is performed by heat from the soldering process. The insides are sprayed with a tastless odour-free solution and then restoved. The can ends obtain the same treatment but coating is executed in sheet form by the roller-coating method. For ring-pull cans, which are filled from the bottom and sealed, the ends have a coating of extremely flexible lacquer as the draw for such a single area is very extreme. Decoration on this kind of can is printed and small use of enamel systems is employed in the UK.

Waiter trays and display tablets. Every substrate may be used including aluminium. External coatings on these trays include a vinyl size coat, after which a white background coating is generally applied. This can be an enamel or print. The finishing varnish usually is based on a semi-drying oil alkyd which gives the required hardness. A semi-drying oil epoxy ester can also be used which has a high alcohol resistance. A two-coat system can therefore be used combining both properties and still permitting satisfactory flexibility for fabrication.

Internal systems consist of oleoresinous resin types or epoxy-phenolic resin. The epoxy phenolic resin gives better damage resistance of the two.

Extruded tubes. For these aluminium is the most common metal employed. In the case of containing glues and adhesives, lead tubes are used and tin-coated lead tubes for pharmaceutical applications.

A size is not normally applied. Special enamel systems are employed, the epoxy ester-based resin or a vinyl kind being typical.

Internal coatings exist for tubes and these are based on epoxy phenolic resins or a two-pack system of epoxy/polyamide resins. The lacquer is applied by the spray method.

Special inks need to be used as tubes are not usually varnished.

Closures used in metal decoration

Screw and lugged caps. Screw caps are generally formed from electrolytic tinplate or hi-top. Aluminium may be used but is expensive. The metal is always sized to provide improved adhesion of subsequent coatings, better flexibility and higher opacity of subsequent coatings.

The enamel system is often of the stryrenated alkyd type. The varnish coating is generally made from a semi-drying oil alkyd-based resin, crosslinked with some melamine for hardness.

The insides of these closures have either the vinyls or some epoxy phenolic systems as both are adequate for flexibility. The epoxy phenolics are those with a high epoxy content because they are more flexible. Sometimes a specially formulated system is needed.

The lugged caps are always formed from tinplate and all coatings must have good flexibility properties. They must also be able to withstand retorting and possess good colour retention qualities because the internal coating is applied last to prevent scratching on the wickets. Therefore the higher temperature stove can affect the external design.

The size coat is usually a vinyl. Either a vinyl, acrylic, or a saturated non-drying oil epoxy-ester type may be used for the enamel system. The vinyl system, with a low solids content, is expensive and colour retention on reverse stoving is not as good as the rest. The acrylic-based coating possesses greater latitude than the vinyls and epoxy-ester types. These give improved tooling properties as well as thermoplasticity. Epoxy-ester based coatings provide good colour retention properties and are crosslinked with some melamine resin to give hardness and flexibility.

The varnish system may be of the vinyl type. The acrylic or epoxy-based resin systems are, however, more popular.

There are two main types of internal coatings employed and both are dependent on the kind of pack they are to be applied to such as jams, pickles, etc. The acrylic kind is used for jams and a two-coat epoxy phenolic and vinyl system is generally used for pickles.

Crowns. Tinplate is primarily used for crown work. Coatings used have to be flexible, and abrasion and pasteurizing resistant, occasionally under alkaline conditions.

A size may be used but depends on the flexibility of inks and varnishes applied although it is normal to size with a vinyl system. Enamel systems are not often used on crowns. If required the standard general line kind would be adequate.

The varnish system may be based on medium or short oil alkyd or a semi-drying oil epoxy ester. Crosslinking with melamine formaldehyde resin for both cases is required to impart good abrasion resistance. For alkali resistance during pasteurization, the epoxy ester is the preferred of the two.

Pasteurization is performed under water at a temperature up to 100 °C but normally at 80 °C. Retorting is carried out with the action of steam.

Lubrication on finishing varnishes is very important as the sharp edges on crowns come into contact with the crown face in the hopper. All crown varnishes are nearly all lubricated and applied wet-on-wet.

Internal coatings are dependent on the kind of liner present. Cork-lined crowns often use the epoxy phenolic type. This is suitable for all packs including aggressive acid packs; on the vinyl-type lacquers which possess universal properties; or where a non-aggressive pack is used, an oleoresinous type.

In the case of PVC-lined crowns either a vinyl solution type lacquer system or the more popular organosol system can be used.

Roll-on closures. Aluminium is almost exclusively used and with this kind of closure the advantage over the rest is that the caps are literally made to measure as the thread is put on every cap individually in contact with the glass bottle.

Size coats are principally of the vinyl resin solution kind. There are occasions where a sizeless system can be employed. This is where the decoration is printed on base metal and then a roller coating lubrication lacquer system applied on top.

Coatings for pilfer-proof caps need to meet the ensuing requirements:

1. Outstanding flexibility and adhesion qualities.
2. A high uniform gloss-retention capability.
3. They should be capable of withstanding normal hoppering conditions.
4. They should be capable of withstanding the roll seal at closing stage.

The resins employed for enamel systems on these closures need to show a high degree of linearity, i.e. without a tendency to peripheral cracks, in particular on deep-drawn caps. The vinyl resin is probably the prime example of this. Owing to high material cost and low solubility of the resin, other kinds including acrylics and modified alkyds are growing in importance.

Some decorations have a printed gold with the design printed on top. In this case a size coat is applied initially followed by the finishing varnish as the final coat.

For a wet-on-wet varnish system the modified alkyd resin provides best results. Adequate lubrication is important with the varnish system employed as an aid to tooling and for appearances of the cap after it has been drawn.

In the case of the roll-on screw caps it is necessary that the finishing varnish withstands gasket stoving. It must therefore be of a harder nature, although size and enamel systems are the same as for ordinary closures.

For pilfer-proof closures a lacquer system primarily for decoration is needed which will also serve as an aid to fabrication. A vinyl solution toned with a dye for shade satisfies this requisite. The level of any lubrication in this lacquer system should be monitored to provide a reasonable level of flexibility and as a tooling aid. A roll-on screw cap lacquer system could employ the organosol, but as gasket adhesion is the prime requirement and not pasteurization resistance, an acrylic lacquer system is usually preferred.

As outlined, coatings find a wide range of applications in the metal-decorating field, such as nameplates, beer and soft-drink beverage cans, and also in the electronics industry.

Apart from inks, described later, there are two chief types for metal decoration. These are opaque basecoats and clear coatings.

As most metal-coating operations involve a certain amount of post forming, flexibility is required of the coating. It must also be adequate to withstand a post-cure bake which is generally used to thermally cure internal conventional coatings of beverage cans.[83] Photopolymerizable coatings are not likely to be used for these internal applications owing to possible toxicity problems, especially when in contact with foodstuffs.

Photopolymerizable opaque basecoats for metal decoration

As titanium dioxide absorbs strongly in the UV region and scatters visible radiation, photopolymerization efficiency suffers, leading to some adhesion problems.

Adhesion is a consequence of both mechanical and chemical interactions.[84]

Vinyl polymerization occurs with significant volume contraction.[85] Stress reduction alone may be effected by using epoxy-based materials to improve the adhesion as ring opening photopolymerization occurs with less volume contraction and in some cases with expansion.[86]

A thermal post-cure assist often improves adhesion because of:

1. Further addition polymerization of the photopolymerizable coating.
2. Stress relief.
3. Improvement of the metal/photocoating interface surface as a consequence of migration of impurities from the metal surface.

The coating must resist yellowing under the thermal bake conditions. A white basecoat must generally cope with post-forming.

Photopolymerizable clear coatings for metal decoration

Adhesion is important for clears as:[87]

1. The clear must adhere to inks and basecoats and in some cases to uncoated bare metal.
2. The transparent coating must survive manufacturing and handling conditions such as fabrication (in the closure field), pasteurizing, abrasion, steam processing, etc.

(iv) Photopolymerizable clear overprint varnishes for paper and board

These are mainly for the packaging industry where aesthetic and functional considerations are important.[88-90]

Overprint varnishes are used for protecting underlying ink (UV curing or conventional) from scuffing and pick-off.[91,92] They give a high gloss surface with low coefficient of friction and sometimes are used as a moisture barrier.

After curing, photopolymerized transparent coatings perform the function of the starch spray powder used in conventional overprint varnishes.[93]

(v) Photoresist coatings

See Chapter 6.

5.4.4.4 Ultraviolet Printing Inks

Ultraviolet inks currently have their largest application in the lithographic field.[94] The three chief areas are:

1. Sheet-fed lithographic printing for paper and board and plastics.
2. Web offset.
3. Metal decorating (sheet-fed and roller coating).

(i) Sheet-Fed lithographic UV ink printing

Folding cartons are a large and growing area[95] where UV has many advantages. For example, the elimination of offset spray powders has been achieved by UV and as a consequence, led to increased production and better quality. The sheets are dried instantly and are abrasion-resistant on delivery. 'Set-off' is therefore not a problem. Handling for additional processing such as immediate die-cutting is therefore made easy. In-line die-cutting is possible. Ultraviolet-dried prints have, in general, good scuff, scratch, and chemical and solvent resistance. Also UV inks are stable on the press and in the duct as they are involatile and are not subject to oxidation by the atmosphere. This is in contrast to their conventional counterparts which 'skin'. The inks can, therefore, be left 'open' on the press overnight or over a weekend if required with no skinning problems. The need for tedious and expensive wash-ups is, therefore, eliminated.

Ultraviolet curing has allowed for the first time the ability to dry between stations on a lithographic press. This wet-on-dry trapping or inter-colour drying can increase print accuracy and eliminate problems such as doubling or ghosting.

An often-encountered problem in conventional ink sheet-fed printing is 'dry-back' where the dried ink colour is not the same as it was when freshly applied, which is often due to paper and board variables. These are costly in the folding carton industry. Ultraviolet curing helps to eliminate this problem and consequently results in reducing paper waste.

Also, most of the advantages depicted above such as ink stability, wet-on-dry trapping, and instant drying are likely to lead to more efficient lithographic press design.

The odour level is often reduced with UV formulations, due to raw-material suppliers keeping the 'free' acrylic acid content in their monomers and pre-polymers to a minimum.

(ii) Web offset printing[96]

Apart from most of those mentioned above, there are many advantages of the UV process for web offset publishing. Afterburners are eliminated as the inks are solvent-free and give no emission of hydrocarbons to the atmosphere. Gaseous oxides from the combustion of natural gas are also eliminated as the energy source is electrical in nature.

The wet-on-dry trapping principle would be expected to eliminate dot-doubling associated with wet trapping.

The gravure printing process has for a long time had this advantage at the expense of offset printing, and UV helps to tip the balance towards lithography.

The resistance of the UV ink film to skin oils, which is significant for magazine covers and other publications that have excessive handling, is of importance for the publication printer.

Web temperature reduction due to UV radiation cure is also important as blistering and chill rollers are eliminated, as well as the possibility of using thinner, lower-cost, paper stocks with the reduced web temperature.

(iii) Metal decorating[87,97-100]

Ultraviolet curing probably has the greatest growth potential in the metal-decorating field. Curing by UV radiation permits the flat-sheet metal printer to go through the production line at one pass instead of the multi-pass system otherwise used. For conventional inks, metal decorators often print two colours first, thermally cure (generally at 150–175 °C for about 12 minutes) and rack, then print two more colours, a wet varnish, and again thermally cure. This is costly and slow and the advantages of a one-pass system are clearly seen as is the elimination of the enormous cost of installing energy-consuming, long thermal ovens.[65]

Ultraviolet curing has found application in large US companies such as Continental and American Can. In the UK, UV printing inks are used for metal decorating in several companies such as Metal Box and Crown Cork.

Hybrid systems are currently being used with two colours of UV ink, two colours of conventional ink, plus a conventional wet varnish. Complete UV systems without thermal energy are now emerging, although adhesion often leaves much to be desired.

The general film properties of UV-cured metal inks are substantially better than conventional systems. Metal-decorating UV is not only applicable in flat-sheet but is also viable in two-piece can decorating, where the advantages of space and reduced oven maintenance are of paramount importance.

Interdeck UV curing[66] on multicolour tin-printing lines is salient, as, unlike paper and board printing where up to six colours can be printed in a single pass (owing to the substrate usually possessing some absorbency), sheet-metal printing has been limited to two or three impressions per pass followed by stoving and then printing of any further colours. Ultraviolet curing has enabled four colours to be printed in one pass, the first two inks being set by radiation and then overprinted with conventional oxidation-drying inks prior to varnishing and stoving.[6]

The inks are generally only 'set' (surface cured) sufficiently to enable them to be overprinted and finally cured and hardened in the Wicket ovens. The quantity of UV radiation and the formulation of the inks have to be adjusted to ensure the print has sufficient adhesion to plain metal so that it is not picked off by the following printing blanket, while at the same time it must develop sufficient interfilm adhesion with the overlying conventional inks and varnish.

For total UV lines where the ovens are replaced by a UV dryer which is cheaper, the line will occupy much less space and require less energy. The print will have to be cured sufficiently to avoid scratching and, more particularly, set off when the sheets are stacked in the unloader. More radiation (larger number of lamps) is likely to be required than for interdeck setting.

Four major metal-decorating UV ink areas can be depicted in the technical objectives and approaches itemized below:

1. Flat-sheet UV inks for all the metal-decorating industry substrate requirements to be printed dry offset and litho (wet) offset with both conventional and Dahlgren dampening (up to 25% isopropanol present). Untreated tinplate is problematic owing to the presence of rolling oils (palm oil) and is probably best approached from the dry-offset ink type.

2. Ultraviolet inks for the in-round printing process for all metal substrate requirements, both varnish and non-varnish. For two-piece cans, coated or uncoated metals are not problematic as long as metal is pretreated properly or an ink-receptive coating is present.

3. Ultraviolet fabricating ink for caps and closures. These are difficult owing to the extreme flexibility properties needed for tooling.

4. Inks for speciality metal-printing needs (signs, labels, drums, etc.), within the limits of reasonable demands.

The approaches for these include finding novel adhesion promoters, developing new gloss and slip agents, substituting for initiators and accelerators, using new combinations of monomers and prepolymers for the requirements above, as well as researching conventional materials compatible with these systems.

(iv) Plastics[100]

Dry-offset UV seems currently one of the most used processes (especially in the USA) for printing preformed plastic containers and closures (such as tapered pots, cups, tubes, lids, etc.). Ultraviolet curing is of specific relevance when tapered pots and lids are nested, as this helps from the point of view of food safety regulations because any contamination on the inside (for foodstuffs) is extremely undesirable.

Plastic surfaces are generally chemically inert, leading to poor adhesion with inks. Surface oxidation of the plastic is therefore a means of rectifying this problem. Good ink adhesion may be obtained by treating a polyolefine plastic surface with a gas flame, chlorination, or electron discharge to render it adequately polar.

Inks of the conventional type require thermal energy to cure, which distorts the shape of the container. Many glycol-based printing inks used do not ultimately give adequate resistance to many of the materials being packaged, for example, anti-freeze solution and oils for motor vehicles. Ultraviolet-curable inks can help to overcome all of these problems.

5.4.4.5 Ultraviolet Inks in the Field

(i) Equipment compatibility

Ink rollers and blankets

The best rollers and blankets for use with UV inks are of a Buna-n-rubber type.

Rollers and blankets of a urethan and/or PVC composition can cause problems. Nitrile based rollers are to be avoided.

Fountain etch

Runs in the field are with a variety of acid-type fountain etches. In cases where compatibility of a particular fountain etch of UV inks is questioned, then a Rosos-type etch tailored to a pH of about 4 will probably suffice.

Plates

Ultraviolet inks are successful on many plates. These include deep etch, bi-metallic, tri-metallic, and photopolymer. Presensitized and wipe-on types may have problems.

Press conditioning

The press has to be conditioned prior to running UV-curing inks. This conditioning allows the rollers and blankets of the press to reach an equilibrium with the media used in UV inks.

A press previously running conventional inks must be conditioned prior to running UV inks. This removes any conventional ink vehicles from the rollers and blankets and eliminates the possibility of incompatibility between conventional and UV ink media occurring on the press.

The press may be preconditioned as follows. First remove all excess ink then wash press thoroughly with conventional press wash. Rewash press with presswash then apply press conditioner to rollers and blankets and permit press to idle for about half an hour. Finally wash press with presswash. The press is then ready and conditioned to run.

For sheet-fed presses, caution must be observed to ensure the elimination of all spray powder from the delivery system, in particular, the gripper bars. If any powder remains in the grippers, the springs may become inoperable on exposure to UV radiation, resulting in the premature release of sheets.

Wash-up of UV inks

The press is cleaned and washed as for conventional inks. Two press washes are required, one for the rollers and blanket and a milder one for plates.

(ii) Printing problems[1,2,9]

(a) Scumming/tinting (*contamination of non-image areas*)

If a scum or faint tint of ink is contaminating the non-image area of the stock and/or plate and cannot be eliminated by a proper ink/water balance, then:

Possible causes are;

1. Ink not adequately hydrophobic.
2. Ink too soft or too low in tack.
3. Ink inadequately ground.
4. pH of ink too high.

Probable remedies are;

1. Mix hydrophobic varnish with the ink.
2. Add Bentone or bodying compound to the ink to shorten the body or reformulate ink to obtain fuller body.
3. Recheck grind of ink; if necessary regrind.
4. Drop pH of ink.

(b) Piling

Ink builds up on areas of the rollers, blanket and/or plate:

Possible causes are;

1. Ink too water-receptive, becomes waterlogged and piles.
2. Ink badly ground.

3. Additives such as Bentone, wax, photo-sensitizer, etc., inadequately dispersed.
4. Rollers and blanket improperly preconditioned.

Probable remedies are;

1. Mix hydrophobic varnish with ink and reduce dampening to a minimum.
2. Recheck grind of ink; if necessary regrind.
3. Ensure additives are used in finely dispersed compound form.
4. Clean and again precondition rollers and blankets.

(c) *Linting/picking*

The linting or pulling on uncoated stocks and the picking or lifting of the coating on coated stocks has the following possible causes

1. Ink is too tacky for stock being used.
2. Stock has insufficient lint or pick resistance.

Probable remedies are;

1. Lower tack of ink with reducer.
2. Convert to more lint or pick resistant stock.

(d) *Misting or flying*

Fine droplets or filaments of ink formed on roller train during film splitting and sprayed onto surrounding areas.
Possible causes are;

1. Ink too long.
2. Excessive ink carried to obtain desired density.

Probable remedies are;

1. Mix Bentone or other bodying compound to shorten body of ink.
2. Replace with stronger ink and run thinner film.

(e) *Greasing* (*dot spread*)

The halftone dots increasing in size causing the printed signature to lack sharpness.
Possible causes are;

1. Ink too long in body or too low in tack.
2. Ink too receptive to water.

Probable remedies are;

Reformulate ink for improved body and higher tack and if possible use more hydrophobic varnish.

(f) *Poor trapping*

Superimposed inks not trapping or laying well and giving undesirable results.

Possible causes are;

1. Tack of inks out of sequence.
2. Strength of inks improperly balanced. This can result in running one colour full and the following colour spare, in which case the colour-run spare may not trap.
3. Poor ink/water balance.
4. Excessive wax in ink resulting in refusal of subsequent layers of ink.

Probable remedies are;

1. Make sure inks are of the correct, successively decreasing tacks.
2. Lower strength of ink being run spare and carry heavier film.
3. Obtain and keep proper ink/water balance.
4. Carry minimum quantity of wax to obtain desired film properties.

(g) *Ink does not dry adequately*

Printed ink film wet or tacky after exposure to curing unit of press.

Possible causes are;

1. Curing unit malfunctioning.
2. Ink improperly controlled during manufacturing.
3. Ink improperly formulated for desired end result.
4. Ink taking on water.

Probable remedies are;

1. Ensure curing unit is switched on and operating properly.
2. Recheck batch sample versus standard for correct drying, if necessary remake ink.
3. Reformulate ink for desired end result.
4. Reformulate more hydrophobic ink.

(h) *Poor rub and scratch resistance*

Printed ink film appears dry but exhibits poor rub and/or scratch resistance.

Possible causes are;

1. Curing unit malfunctioning.
2. Ink contains insufficient:
    (a) photopolymerizing varnish or monomer;
    (b) photo-sensitizer compound;
    (c) wax compound.

Probable remedies are;

1. Ensure all lamps are on and functioning correctly.
2. Fortify drying with:
   (a) faster and more monomers;
   (b) photo-sensitizer compound;
   (c) wax compound.

5.4.4.6 Ultraviolet Gravure Application[101,102]

From an economic viewpoint, any non-volatile liquid ink is at a direct cost disadvantage compared to a solvent-containing ink, but there are indications that the much higher mileage obtainable because of shallower etch requirements and some of the unusual properties of UV inks for certain uses, can offer a counterbalance.

Speciality gravure and some areas of packaging gravure are the best potential markets for use of UV curing for the following reasons:

1. Ultraviolet inks can give highly resistant film properties owing to their crosslinked state and the higher binder–pigment ratios which always exist in non-volatile ink films.
2. Possible better adhesion can be obtained to the varied and difficult substrates used in packagings.
3. For packaging gravure operations, solvent recovery generally produces a highly complex solvent mix which is of limited resale value and usually of little value for direct re-use in inks, so that afterburners appear to be the indicated current control method, despite the expense of operation. Ultraviolet inks produce no effluents and so eliminate the use of any control device.
4. The ink economics would seem more favourable than in the case of either publication gravure or high-volume packaging gravure.
5. Offset gravure would be well suited to the use of a non-volatile system, and especially as offset gravure is growing in use in the packaging and speciality gravure fields. The variables caused by drying of ink on the transfer blanket would not therefore be a problem with UV inks.

There have been many current problem areas in making low-viscosity UV curing gravure inks:

1. Most multifunctional monomers have high viscosities but the use of these is desirable to achieve more rapid cure speeds.
2. Thin-viscosity systems are generally slower curing than paste ink vehicles as they are farther from being a solid and hence need more dwell-time under the UV lamps to build up a crosslinked solid film.
3. Unless the rheology is correct, screening or patterning of solids owing to lack of flow-out can be problematic, and formulation must therefore make ink flow extremely high to obtain smooth printing in solid areas.

4. Doctoring or wiping must be very efficient, otherwise any residual ink in the non-printing areas is more likely to transfer to the stock owing to the non-volatility of UV inks.
5. Gravure UV inks require higher pigment loadings to obtain proper strength, as shallower engravings and thinner films are required with non-solvent inks for economical and practical operation. Viscosity is also likely to be increased because of these higher pigment loadings.
6. Currently employed etch depths are likely to be too great for non-volatile systems as they are designed for running solvent-based inks having generally less than 30 per cent solids and as little as 10 per cent in some cases.

Viscosity control for gravure ink formulations can be executed by using carefully selected blends of non-volatile monomers and prepolymers, while cure speeds can be maintained at a high level with the proper use of photo-initiators.

Fountains could be modified to maintain slightly elevated temperatures and are another possible solution to obtaining lower viscosities, as the inks are essentially non-volatile and non-heat reactive at these temperature levels (30–60 °C).

Flow-out and wiping can be improved by using additives designed for particular compatibility with UV-reactive varnishes. Reverse angle, non-metallic doctor blades may have use in improving doctoring at high speeds with these non-volatile systems.

For solids it would seem that depths in the 10–20 μm region are most practical.

5.4.4.7 Ultraviolet Flexographic Ink Application[103]

Large-volume flexo markets are not likely to undergo cost saving, and UV is only likely to occur in the speciality areas–those markets that are not satisfied with the present products.

Pollution control cost

The volatile solvents used in flexography will probably be regulated in time, as will any printing system photo-active based on drying by solvent evaporation.

Four different ways to control the release of volatile solvents are:

1. Install afterburners.
2. Introduce recovery systems.
3. Use water-based inks.
4. Use UV curing inks.

The relative costs of the above are major factors.

Afterburners and catalytic oxidation have high capital requirements and very high additional fuel requirements. Solvent recovery has very high capital cost and certain technical problems with flexographic solvent mixtures. The drying of water-based inks on non-absorbent plastic films needs higher heat input than the drying of solvent inks. Obviously, this factor may vary with the solvent blend but several times the amount of energy is needed to produce the same printing speed. Ultraviolet inks are much more expensive than any of the conventional solvent systems.

Most synthetic organic chemicals come from crude oil and over the last years fuel costs have soared, leading to difficulties in obtaining sufficient solvent to operate at former levels.

If the fall in availability of crude oil continues, the petrochemical industry will get a high priority probably at the expense of fuel needs as a shortage in petrochemicals would have a greater effect on the economy.

Effective conservation is more likely in the fuel sections where there is more room for economy and there are alternatives to oil for heat energy.

Solvent prices will therefore rise and it will be paramount to realize that the solvent contributes nothing to an ink film as it is used only to provide liquidity during printing and is driven off during the drying and lost.

Most flexo inks are air dried, using natural gas for heating. Compared with other manufacturing, the cost of natural gas for drying flexo inks is insignificant and the industry could afford the rapidly rising costs if continuing supply were guaranteed; but this cannot be assumed.

Solvent recovery systems need heat energy in addition to drying requirements.

The solvents used for flexography have very low flashpoints and must be used with care to avoid fires and explosions. Explosion-proof equipment and fixtures in the pressroom and in ink-handling areas are a considerable part of the cost of using red-label inks. Insurance costs are naturally higher for plants handling red-label materials.

The flammability of UV inks will be much lower than regular flexo solvents. Ultraviolet inks are much safer to handle and process than solvent-bearing inks and safety equipment costs and insurance rates should all benefit from the change.

Costing of UV flexo inks would indicate a ratio of about 2:1 with conventional inks but this difference will steadily decline as the cost of solvent increases, and no advantage is taken of stronger inks and thinner printed films.

Other less tangible factors will make UV even more attractive, such as smaller wash-up waste, no need for circulating or viscosity control equipment, lower cost for press drying equipment and presses without explosion-proof motors. Insurance rates would certainly be lower.

If the cost for UV flexo inks drops to 50 per cent more than conventional, then UV will come into its own. It will make a flexographer independent of solvent supply and eliminate all air-pollution problems and dry ink regardless of gas availability. These factors will have a bearing on UV press conversions as the cost difference narrows between UV and solvent ink.

Flexo printers have a continuing need to improve quality for various specialized applications. They will probably want better print quality or perhaps greater chemical resistance of the dried ink film. Economics is not likely to deter this as the problems can be solved with the new technology.

Unforeseen factors would accelerate or slow down this technology for the future. Energy legislation, pollution laws, growth of nuclear power facilities, new oil discoveries, economic policy, or world politics could alter the lines.

5.4.4.8 Ultraviolet Screen Printing Application[104–107]

Solvent-based screen ink printing has several disadvantages compared to lithographic, solvent-based printing. These include:

1. Poor ink mileage.
2. Expensive drying cost.
3. Poor print resolution.
4. Slow printing speed.
5. Large percentage of solvent present.

Screen printing does, however, have certain exclusive features with which lithography or other processes cannot currently complete. These features include:

1. Good ink adhesion and other chemical performance properties on a wide variety of substrates.
2. High gloss and brilliant uniform colours.
3. Good lightfastness and weather resistance.
4. Brilliant fluorescent colours with reasonable lightfastness.
5. High film strength for dry and water slide transfers.
6. Good price competitiveness on relatively short-run work, as equipment is generally simple.

Wet-screen ink films are of the order 20–25 μm thick whereas lithographic films are of the order 2 μm. Lithography therefore has about a 10:1 advantage over screen. A micrometre is equivalent to a millilitre per square metre, 1000 square metres per litre, 40 720 square feet per US gallon, or $\frac{1}{25}$ mil.

Carton, paper, and board litho printing have no limiting drying costs as the printing is deep stacked directly at the press delivery. The print detail (resolution) can be compared by means of halftone printing in which lithography with halftone screens of 100–150 lines per inch (39–59 per cm) is roughly comparable in dot quality to screen at about 55–65 lines per inch (22–26) per cm, that is, lithography has about a 200 per cent better print detail resolution.

This ink mileage limitation and drying would price screen ink printing out of the printing market apart from the exclusive advantages of screen over lithography described above. Poor print resolution excludes screen printing from the vast market for fine-line four-colour halftone printing that is dominated by lithography.

On a print area basis, the relative total sizes of the litho and screen markets in the UK becomes about 100:1 (litho/screen). Some 90 per cent of the lithographic market needs fine-line halftone printing or fine text reproduction which screen cannot produce. The screenprint market could approximately double if it captured as little as 1 per cent of the lithographic print market.

Long-run work only becomes a serious limitation owing to the speed of screen printing machines but rotary screen machines can run at speeds of the order 120 m per minute.

Three of the limitations mentioned, poor ink mileage, high drying costs and poor print quality, probably arise from one cause, that is solvent-based inks.

Evaporation-drying solvent-based inks have some 50–70 per cent of solvent by volume on thinning to printing viscosity. If a 90 mesh screen is used to apply solvent-based screen ink, roughly a 20 μm wet film thickness will be applied but will dry to only about 8 μm dry thickness. As these solvents are derived from coal and oil resources, this is an inefficient process. The solvents also represent in many cases a fire and explosion hazard as well as a health hazard if certain threshold values are exceeded.

Solvent-based screen inks lead to screen instability and are slow drying, each of which adversely affects the other. Ink evaporation on the screen leads to blocking of mesh apertures in non-stencil areas and either a print quality loss such as dried-in edge occurs, or a total print reject. To clear a screen, wash-ups, extra thinning, retarding and running water points are needed as daily practices in a screen printing plant. High ambient temperatures increase screen instability and adverse conditions are found in hot weather when there is no air conditioning, causing work loss. The main factor in limiting print resolution is screen instability. For finer meshes (such as 150 mesh per cm), the screen instability problem is increased. A rewetting problem occurs if more retarder, that is a solvent with a lower evaporation rate, is used because a second colour overprinted onto a first colour renders this tacky and so causes bad screen 'snap' and loss of print quality or even bleeding together of the colours on the screen.

Drying equipment is costly for solvent-based screen inks both for machinery, space, and power consumption. A 2 m printing press could require an 8 m dryer and therefore some 80 per cent of a floor area of a screen printing plant can be occupied by drying equipment. Printers often find that their drying capacity is inadequate for maximum press speed. The time factor involved for complete drying, particularly on plastics and two-sided work, can slow down the passage of work through a plant and prevent in-line printing, finishing, and packing.

For the screen ink industry there are several types of inks which could replace the solvent-based types. These include:

1. Water-based inks.
2. Plastisols.
3. Ultraviolet inks.

Water-based inks are used in flexographic printing, the solvents being replaced by water. In screen inks, as in flexographic inks, this would eliminate solvent costs and lower unit ink costs and eliminate solvent hazards. Ink mileage, however, is unlikely to be increased. Water is slower drying than the usual screen solvents and would therefore increase drying costs. Screen instability is likely to worsen as some water-based polymers are emulsions of colloidal systems which on drying are not easily redispersible with water, leading in many cases to a blocked mesh, that needs a specific wash up mixture. Print quality is therefore likely to become worse with water systems.

Also a new range of problems could occur with water-based screen units, such as:
(a) foam on the screen;
(b) wet rub resistance;
(c) weather resistance;
(d) paper cockling;
(e) ink freezing in cold weather.

Banning of solvents by government legislation is the most likely means of causing water-based inks to displace solvent systems.

Plastisol inks which are based on a dispersion of finely divided polyvinylchloride powder in liquid plasticizer are used for direct textile printing and for some textile transfers for cotton fabrics. Unfortunately plastisols need, during drying, to heat momentarily to about 180 °C to optimize ink adhesion and hardness. The polymer powder size is also too coarse for very fine meshes.

Ultraviolet screen printing inks look the most viable to overcome the chief limitations mentioned, and opacity is achievable to match solvent-based inks. Ink mileage and resolution are able to be increased and for the first time, screen printing could be able to rival lithography as a cheap alternative to produce fine-line halftone. Drying costs are less than those for a solvent system.

As previously mentioned, a wet screen ink film through say a 90 mesh/cm screen would be deposited as some 25 μm thickness and with say two-thirds solvents being present, would yield a dry film of about 8–9 μm. A UV ink deposited through a finer mesh such as a 200 mesh/cm screen would yield about 8–9 μm wet and about the same dry after cure as there is no solvent loss. There is therefore a mileage factor of some 3:1 in favour of UV inks, depicted in Figure 72.

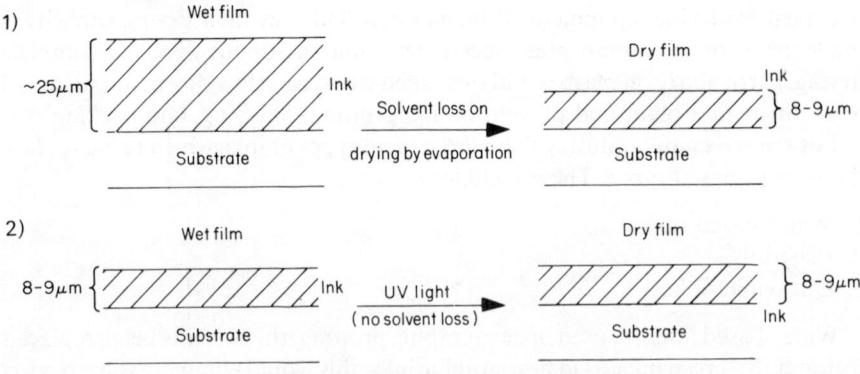

Figure 72 UV screen ink mileage.
1) 90 mesh/cm, solvent based screen ink
2) 200 mesh/cm, UV solvent less ink

5.4.4.9 Advantages and Disadvantages of UV Curing[108-116]

To summarize, some of the chief advantages of UV curing inks for the various processes are:

1. They use solvent-free formulations, eliminating afterburners, ovens, and other emission-control devices.
2. They remove dependence (USA) on natural gas.
3. UV lamps and equipment take little space (unlike ovens).
4. Production lines and throughput can be accelerated.
5. They allow in-line decorating.
6. Wet-on-dry trapping (interdeck curing) is obtainable.
7. Improved film properties are obtained.
8. Press stability (ability of ink to be left open) is increased.
9. Wax/starch spray (paper and board) is eliminated.
10. Instant drying (elimination of dry rack) (folding cartons).
11. In-line die cutting.
12. Elimination of thermal distortion (plastics).
13. Improved product resistance (plastics).
14. Better colour reproduction (plastics).
15. Fire risks (gravure) are reduced.
16. Mileage is reputed to be equal or about the 15 per cent level greater than conventional inks and the elimination of dry-back should enlarge the effective mileage.

There are, of course, disadvantages such as:

1. Ultraviolet equipment costs are high to install.
2. Ultraviolet materials are costly.
3. There are possible adverse toxicology properties.
4. Certain highly pigmented systems that are highly opaque are difficult to dry with UV. Also a pure metallic pigment ink will reflect the UV and resist cure. If surface cure is obtained, this can retard or prohibit 'leafing' occurring.

Many of the above advantages have been claimed for UV curing but as to how well it has lived up to expectation is open to controversy. Some claimed advantages of UV curing are shown below with possible non-fulfilments:

1. Inks claimed to be 100 per cent involatile – often not true in practice i.e. free acrylic acid present in some systems, monomers and diluents often used with slight volatility. On stoving many metal decorating systems lose a large amount of volatile material (up to 20%) to the atmosphere and do not effectively reduce pollution as is often claimed. Photo-initiators such as benzophenone are volatile at these temperatures and can be responsible for part of this.
2. Instantaneous drying – not always of the through-cure scratchproof type, sometimes of the surface-cure type which may lead to smudging.

3. Excellent setting on a wide range of substrates (including non-absorbent substrates) – inks often fail to have good true adhesion characteristics on the latter types such as PVC, polythene, untreated tinplate, aluminium, etc.
4. Excellent rub resistance – so dependent on choice of monomer/prepolymer in system as brittleness can result on overcure with, for example, epoxyacrylates.
5. Extremely good stability on the press – not always borne out in practice. Often a tack build-up is observed causing 'picking', etc. This is due primarily to absorption into the rollers of either monomers and or photoinitiators, benzophenone in particular possibly being a culprit. Also molecular weight control of prepolymer is probably important with this problem, as also should be pre-run press conditioning (i.e. saturate rollers with UV materials).
6. Elimination of atmospheric pollution caused by evaporation of solvents – not strictly true, see (1) above.
7. Very low-odour prints – so dependable on monomer and photoinitiator choice. Acrylic systems' odour tends to linger, as monomer may not be completely converted to polymer and remains trapped in the film. Also any small amount of acrylic acid in the system will cause a characteristic odour.
8. Greater mileage – the inks are much dearer than conventional types and in the litho field seem to show no real evidence of greater mileage.
9. Paper recovery – de-inking of paper is generally more difficult than that of conventional inks.
10. Economic advantages over conventional inks – not always justified (cf. infrared); perhaps only really justified in the metal-decorating area where very expensive Wicket ovens are replaced by cheaper lamps.

The cost of a UV system can be tabulated in the manner below:

1. Installation cost.
2. Depreciation.
3. Replacement of lamps.
4. Consumption of electricity.
5. Increase in price of printing inks.

The advantages to be obtained from the system in the printing shop and dependent upon the particular local conditions of the printing shop are:

1. Increase in productivity.
2. Reduction in waste.
3. Reduction in stock.
4. Improvement in quality.
5. Elimination of anti-set off spray powder and difficulties which use of the powder bring.
6. Possible use of cheaper substrates.
7. Diminution of pollution.

The above factors should come into the economic balance and should be costed by individual printers. Only by a precise study can one judge whether the costs are compatible with the improvement in output and whether such improvement is in terms of quality and quantity.

5.5 REFERENCES

1. Young, L. C., *Materials in Printing Processes*, Focal Press, London and New York, (1973).
2. Askew, F. A., *Printing Ink Manual*, Heffer & Sons Ltd., Cambridge, (1969).
3. *The Universal Encyclopaedia of Machines, How Things Work*, Paladin, (1972).
4. Cerutti, C., *Printing Technology*, **8**, 42, (1964).
5. Autotype International Ltd., *Stencil Techniques*, (1977).
6. Van der Hoff, J. W., *J. Radiation Curing*, **1**, 4, 7–21, (Oct. 1974).
7. Mitton, P. D., 'Opacity, Hiding Power and Tinting strength', in *Printing Ink Handbook*, Vol. 111 (Ed. Patton, T. C.), J. Wiley & Sons, New York, pp. 289–339,(1973).
8. Zettlemoyer, A. C. and Myers, R. R., 'The Rheology of Printing Inks', in *Rheology*, Vol. 3 (Ed. Eirich, F. R.), Academic Press, New York, p. 146, (1960).
9. Apps, E. A., *Inks for the Major Printing Processes*, Leonard Hill, London, (1963).
10. PIRA, Evaluation of Accelerated Ink Drying Systems for Rotary Presses, *Prod. J.*, 48, 20–23, (July 1970).
11. Fierz, H., 'Drying Printing Inks with a supply of Energy', *Ugra Mitt*, 24, p. 583–588, (March 1972).
12. Surgeon, W. R., 'Two New Ink Systems', *Graph. Arts Mon.*, **43** (9), 42–44, (Sept. 1971).
13. Anon, 'Inks Drying on the Paper and not on the Press', *Lithoprint*, 29–34, (May 1972).
14. Bassl, A. and Bohm-Kasper, K. H., *Farbe und Lacke*, **73**(10) 916–923, (Oct. 1967).
15. Nass, G., 'Solventless Printing Inks', *Am. Ink. Mkr.*, **49**(1), 25, 63–64, (Jan. 1971).
16. Bassemir, R. W., Preprints, 8th TAPPI (Graphic Arts Conference, Miami, Fl.), 19–22 Oct., 21, (1971).
17. Bruno, M. H., 'State of the Art of Printing in the USA', paper presented at the 11th International Conference of IARIGAI, Rochester, New York, (12–19 May, 1971).
18. Morano, A., 'Catalytic Thermo-Cured Web Inks', a talk presented at the New York Printing Ink Production Club Meeting, (20 Sept., 1972).
19. Vanderhoff, J. W., *American Ink Maker*, **51**(4), 38, (1973).
20. Matheson, J. W., '*Tech Talk No. 5*', American Paper Institute, (1971).
21. Dunn, H., Paper presented at the 8th TAPPI Graphic Arts Conference, Miami, Fla. (19–22 Oct., 1971); published in *Penrose Annual*.
22. Bielenda, A. S., Gravure, p. 13–15, Feb. 1968; pp. 24–25, 54, 56, March 1968.
23. Inokuma, S., *Graphic Arts Japan*, p. 65–67, (1969–70).
24. Costello, J. P. (to Fred'k H. Levey Co., Inc.), *U.S. Patent 2,696,168*, (7 Dec. 1954).
25. Rocap, W. A., Jr., Meredith Printing Co., Des. Moines, Iowa, private communication, (Oct. 1972).
26. Eastman Chemical Products Publication X-214, *Alcohol-Soluble Propionate (ASP) in Printing Inks*, and Customer Service Report 217–1, *Alcohol-Soluble Propionate in Flexographic and Gravure Printing Inks*.
27. Barrow, G. M., *The Structure of Molecules*, W. A. Benjamin, Inc. New York, (1964).
28. Walker, S. and Straw, H., *Spectroscopy*, Vol. II, Chapman & Hall, (1962).
29. *Infra-red Spectroscopy – its use in the Coatings Industry*, Fed. of Societies for Paint Technology – Philadelphia, (1968).
30. Early, H. C. and Miller, D. B., U.S. Patent 3,159,464, *Method of Drying Printed Web*, (1 Dec. 1964).

31. Pray, R. W., Thermogenics of New York, Inc., New York, entitled *Radiation Curing of Coatings to Metal Substrates*. Paper originally published 1974, by the Society of Manufacturing Engineers, 20501, Ford Road, Dearborn, Michigan.
32. Comarc Engineering, Technical Data Sheet, (April 1965).
33. *Infra-red Spectroscopy*, Chicago Society for Paint Technology, (31 Oct. 1961).
34. Lendle, E., *Druck/Print*. pp. 736, 738, (Nov. 1972).
35. Weller, G., *Polygraph Jahrbuch*, pp. 183–185, (1969).
36. Trembley, J. F. and Loring, C. M., Jr., *TAPPI*, 1847–1850, (Oct. 1969).
37. Wheaton, R. E., *ANPA Research Institute Bulletin*, No. 942, 398–402, (26 Dec. 1967).
38. Finch, P. A. W., *Production Journal*, 21–25, (Oct. 1967).
39. Gilliatt, C. L. and Kihn, W. W., *Gravure Technical Assoc. Bulletin*, 95–97, (Autumn 1969).
40. Gerling, J. E., *Gravure Technical Assoc. Bulletin*, 92–94, (Winter 1968).
41. Jaffe, E., *ANPA Research Institute Bulletin*, No. 1028, 357–359, (18 Sept. 1970).
42. *ANPA Research Institute Bulletin*, No. 972, 1–2, 15 Jan. 1969.
43. Brewer, R., *British Printer*, 122–125, (Sept. 1969).
44. Ritchie, R., *Penrose Annual*, p. 193–196, (1970).
45. Camus, F., *Siebdruck*, 466, 468, (Oct. 1971).
46. Hoch, H. J., *Adhesion*, 72–75, (Feb. 1969).
47. *Siebdruck*, 312–314, (June 1970).
48. Lendle, E., *Siebdruck*, 584–586, 588–590, 592–593, (Dec. 1968).
49. Woods, D. H. and Hartshorn, I., *Am. Ink. Maker*, 39–40, 42, 44, 59–61, (Dec. 1970).
50. Gerling, J. E., *Am. Ink. Maker*, 32–34, 36, 38, 40, 64, (April 1969).
51. Hutchinson, G. H. (pp. 2/1–2/3) and Ritchie, R. (pp. 18/1–18/5), Colloquium on R. F. and Microwave Industrial Heating by Institution of Electrical Engineers, Bradford University, 27–28 Oct. 1971.
52. Tawn, A. R. H., *J.O.C.C.A.*, **51**, 782–791, (1968).
53. Du Plessis, T. A., and De Hollain, G., *J.O.C.C.A.*, **62**, 239–245, (1979).
54. Ranney, M. W., *Irradiation in Chemical Processes, recent developments*, Noyes Data Corp, London, pp. 114–234, (1975).
55. Van Den Broeck, P., *Revue IRE, Tijdschrift*, **2**, 12, (1977).
56. Spinks, J. W. T. and Woods, R. J., *Introduction to Radiation Chemistry* (2nd edn.), J. Wiley & Sons, New York, p. 122, (1976).
57. Deninger, W. and Patheiger, M., *J.O.C.C.A.*, **52**, 930–945, (1969).
58. Haring, E., *Durussement par electrons des peintures et vernis lors de la fabrication industrielle des portes*, Cahier 105, Ed. EURISOTOP, Brussels, (1976).
59. Morgenstern, K. H., *The Technique of Electron Irradiation Curing of Coatings*, Soc. Automotive Engineers Congress, Detroit, (Jan. 1967).
60. Laizier, J., *Rad. Processing*, **4**, 13, (1977).
61. Nablo, S. V., Quintel, B. S. and Fussa, A. D., paper presented at the Annual Meeting, Federation of Societies for Paint Technology, Chicago III, (Nov. 1973).
62. Davison, W. H. T., *J.O.C.C.A.*, **52**, 1946, (1969).
63. Hoffman, A. S., Jameson, J. T., Salmon, W. A., Smith, D. E. and Trageser, D. A., *Industr. Eng. Chem. Prod. Res. Develop.*, **9**(2), 158, 1970.
64. Burlant, W. and Hinsch, J., *J. Polym. Sci.*, 2135, (1964).
65. Parrish, M. A., *J.O.C.C.A.*, **60**, 474–478, (1977).
66. Lott, A. D., *J.O.C.C.A.*, **59**, 141–145, (1976).
67. Carlick, D. J., 'U.V. Curing of Inks & Coatings', paper presented at Miami, Florida, (Oct. 1973).
68. Carlick, D. J., 'The Suncure System', *The Penrose Annual 64*, (1971).
69. Carlick, D. J., 'U.V. Curing of Inks', *Modern Packaging*, 64–67, (Dec. 1972).
70. Gruber, G. W. 'UV Curing of Coatings', in *U.V. Curing: Science & Technology* (Ed. S. Peter Pappas), Technology Marketing Corporation, (1978).
71. Harris, W. E., *Industrial Finishing*, **46**(11), 44, (1970).

72. Carder, C. H., *Paint & Varnish Production*, **64**(8), 19, (1974).
73. Dufour, P., *J.O.C.C.A.*, 62, 55–58, (1979).
74. Koch, R. L., *Amer. Paint Journal*, **7**, 10, (1971).
75. Prane, J. W. and Bluestein, C., *SME Technical Paper F.C.76-527*, (1976).
76. *PPG Coating & Resins Magazine*, **8**(1), 16 (1976).
77. Walsh, W. K. et al., *SME Technical Paper F.C.76-520*, (1976).
78. Howard, D. H. and Martin, B., *Radiation Curing*, **4**(2), 8, (1977).
79. Anon., *Radiation Curing*, **1**(1), 4, (1974).
80. Nairn, N., *Report on Metal Decoration* for The Andrew Holmes Memorial Fund Scholarship.
81. Hitchins, B., presentation at a meeting in London, published in *Professional Printer*, **19**(5), 11 March 1975.
82. Selection & Industrial Training Administration Ltd., *Printing Ink Technology – Roller Coatings*, Modular Training Manual, (1971).
83. Libutti, D., *Radiation Curing*, **2**(2), 16, (1975).
84. Blomquist, R. F. 'Adhesives' in *Kirk Othmar Encyclopaedia of Chemical Technology*, (2nd edn.), Vol. 1, John Wiley & Sons, Inc., New York, pp. 371–405, 1963.
85. McGuiniss, V. D. and Dusek, D. M., *J. Paint Technology*, **46**(589), 23, (1974).
86. Bailley, W. J., Iwama, H. and Tsushima, R., *J. Polymer Sci.*, Symposium No. 56, 117, (1976).
87. Aloye, J. A., Jnr., *SME Technical Paper F.C. 76-504*, (1976).
88. Coppinger, C., *Paperboard Packaging*, **62**(10), 108, (1977).
89. Carter, R. G., *SME Technical Paper FC.76-533*, (1976).
90. Nowak, M. T. and Rybny, C. B., *American Inkmaker*, **55**(12), 25, (1977).
91. Magdelinkas, M. M. & Wolinski, L. E., *SME Technical Paper F.C. 76-532*, (1976).
92. Martin, M. J., *Paperboard Packaging*, **62**(2), 40, (1977).
93. Coppinger, C., *Radiation Curing*, **3**(3), 3, (1976).
94. Bennett, J., 'Convertible Systems', *Aust. Lithogr.*, **6**(35), 28–30, (May–June 1971).
95. Carlick, D. J., *Printing Magazine*, 36–37, 42, 53, (Jan. 1972).
96. Covell, P. L., 'Pre-Print-Ink Requirements for the 70's', *Prod. J.*, 31, (Nov. 1970).
97. Sallee, E. D., *Radiation Curing*, **3**(1), 13, (1976).
98. Mattingly, G. S. and Buchoveckey, A., *Radiation Curing*, **1**(4), 29, (1974).
99. Fefferman, G. B., *Electronic Packaging & Production*, **15**(6), 29, (1975).
100. Carlick, D. J., UV Presentation, Caracas, Venezuela, (Feb. 1974).
101. Williams, C. H., 'Progress in Inks & Lacquers', *Packag. Rev.*, **91**(6), 53–54, (June 1971).
102. Carlick, D. J., 'U.V. Inks Projected into Gravure', *G.T.A. Bulletin*, **12-2**, (Summer 1972).
103. Anderson, A., *South African Printer*, 16, 19, (June, 1974).
104. Reed, K. J., Managing Director of E. T. Marler, 'U.V. Screen Printing', *Point of Sale News*.
105. Lavell, H., *Point of Sale News*, 40–41, (1976).
106. Reed, K. J., *Uviscreen – The Future of Screen Printing*', Technical release, E. T. Marler Ltd.
107. Reed, K. J., *The Wonder Cure*', Screen Print Technical release, (1976).
108. Bean, A. J. and Bassemir, R. W., 'U.V. Curing of Printing Inks', in *U.V. Curing: Science & Technology* (Ed. Peter Pappas), Technology Marketing Corporation, USA, (1978).
109. Nass, G., 'The Economics of U.V. Curing Inks', *American Ink Maker*, 25, (June 1975).
110. Bassemir, R. W. 'The Gloss of U.V. Cured Press Applied Films', *Radiation Curing*, 10, (May 1976).
111. Vanderhoff, J. W. *De-Inking of Wastepaper Printed with Solventless Inks*, Tappi De-Inking Conference, (1975).

112. Bock, R. F. and Rosenberg, S. H., 'Models for Economic Evaluation of U.V. Ink Drying Systems', Graphic Arts Technical Foundation. *Research Project Report No. 89*, (March 1972).
113. Ecology & The Graphic Arts: 'Air Pollution from Ink Solvents', *Graphic Arts Monthly*, 38–40, (July 1972).
114. Parish, M. A., 'U.V. Light Curing: some benefits and recent advances', *J.O.C.C.A.*, **60,** 474–478, (1977).
115. Pierce, R. J., *Printing Technology*, **17**(3), (1973).
116. Poulter, S. R. C., *Technical and Economical Aspects of Using U.V. Radiation for Rapid Ink Drying*, Leatherhead, PIRA, p. 33, (Nov. 1973).

Note. Parts of the articles in question are reprinted from *Journal of Radiation Curing®*, Vol. 1, No. 4, 1974, published by Technology Marketing Corporation, 17 Park Street, Norwalk, CT 06851, USA. Copyright © 1974 Technology Marketing Corporation.

6

Photoresist technology

6.1 DEFINITION, HISTORY and GENERAL APPLICATION

A comprehensive review of photoresist technology has been made by Hepher[1] and material from this paper is reproduced with the joint permission of the Royal Photographic Society and Kodak Ltd. The term 'resist' may be defined as a protective surface coating applied to an underlying material. Protection is chiefly against physical, chemical, and electrical effects. Photoresists are generally up to 5 μm in thickness.

There are several important areas of application for photoresists.[2,3] Early ones were centred mainly with the engraver and lithographer. Currently this now includes electronic circuits and components, printed illustrations, textile patterns, nameplates, dials, decorative panels, glass scales and graticules, mechanical parts such as electric razor heads, shutter blades for cameras, and photo-milled structures for aircraft.

Photoresists[4,5] are essentially systems where a layer of material that is sensitive to light is coated onto a surface and then exposed to actinic radiation generally in the visible and near-UV regions of the electromagnetic spectrum through a master-image or pattern. This causes a solubility change of the layer. Normally, insolubility of the photo-irradiated portion of material occurs but the reverse process can also be utilized. The more soluble parts can be removed on treating the layer with a solvent. This leaves an image-wise stencil on the support (see Figure 73).

The image formed may then perform as a resist, often against chemicals such as acid or alkali, although the same process is also used for solvent resistance, for the supporting base surface. This type of photoresist may act as a mask for metal electroplating, dye deposition, or even sand blasting. Coloured (dyes) resist images are made for decoration. The photoresist can also be made oleophilic for the manufacture of a lithographic printing image.

There are many classes of systems used for photoresists. These include:

1. Crosslinking of a light-insensitive colloid or synthetic polymer by the addition of a photo-sensitizer that causes decomposition to form an active species that insolubilizes the surrounding vehicle or binder. For example, dichromate/gelatin or azide/rubber.

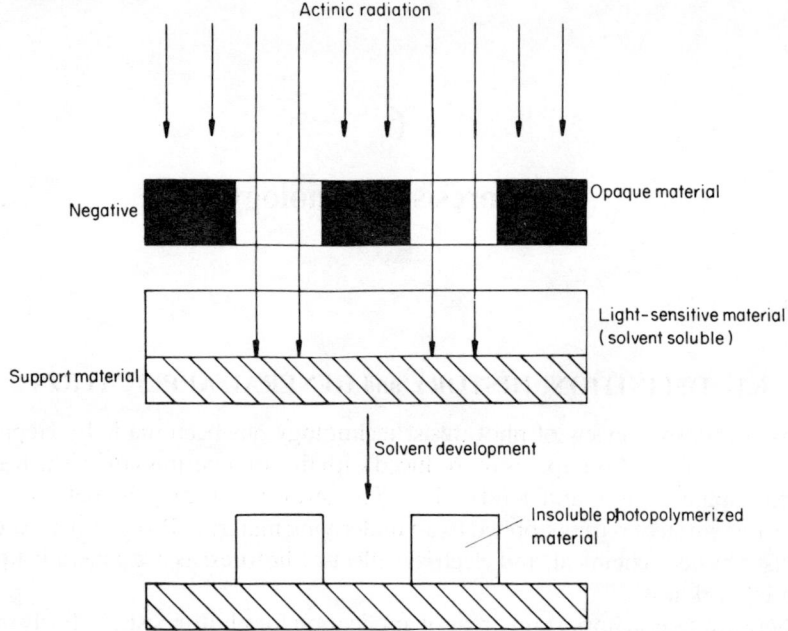

Figure 73 A typical photoresist.[1] (Reproduced by permission of The Royal Photographic Society of Great Britain and Kodak Limited.)

2. Polymers containing photosensitive groups built *in situ* of the structures, forming crosslinks between their molecules on exposure to actinic radiation, to significantly lower their solubility, such as polyvinyl cinnamate.

3. Photopolymerization providing the means of insolubilization, whereby monomeric compounds are made to polymerize via actinic radiation with or without a photo-sensitizer/initiator present.

Probably the earliest recorded photoresist system was the process of Joseph Nicephore Nièpce around 1826. He found that bitumen layers derived from Judea asphalt on prolonged exposure to light (several hours) became insoluble to a mixture of lavender oil and mineral spirit. Images were made by him on silvered glass, the first photographs. Nièpce also developed the system for copper engraving, producing required designs and also stone lithographic printing images. He named his photoengraving a 'heliograph', and this required about 12 hours outside exposure.

Bitumen is a complex hydrocarbon polymer mixture with some unsaturation present. Exposure to short wavelengths probably destroys some of this, causing a change in solubility due to the crosslinking formed within the structure. The change in solubility of photo-crosslinking systems is as a rule slight and needs to be discerned by careful selection of solvents for dissolving the unexposed material.

Not all bitumen fractions are light sensitive. The most active appear to contain more sulphur than the rest and this may involve a process similar to the vulcanization of rubber.[6]

Nièpce's photoresist gave a hydrophobic layer. Aqueous acid or alkali etching fluids would not easily destroy or swell it.

The disadvantages of bitumens were their long exposure time needs and non-uniform composition.

From the above it can be deduced that desirable features for an acid/alkali photoresist include:

1. Good hydrophobic character.
2. Rapid light sensitivity (minutes as opposed to hours).
3. Good solubility differential.

Many natural organic colloids have been used, including proteins such as albumin, animal glues, gelatin, casein, etc. Also, carbohydrates have been used, such as starch, gum arabic, etc., but these all have the disadvantage of short shelf-life and non-uniform composition.

One of the most important photoresists to be developed was that using dichromates[7–9] as a photosensitizer. In this system, the photomechanical arts have used chromium salts to cause colloid or resin insolubilization. Perhaps the most popular to have had any 'in-depth' mechanistic studies was the combination of dichromate and polyvinyl alcohol and its derivatives as described below.

6.1.1. THE DICHROMATE/POLYVINYL ALCOHOL SYSTEM

6.1.1.1 Photochemical Curing Mechanism

The photosensitization reaction has been interpreted as being caused by the reduction of the dichromate ion Cr(VI) to produce the chromium Cr(III) ion, with the simultaneous oxidation of the natural colloid or synthetic resin present.[10] The latter compounds either become crosslinked by the chromic ion or by the oxidation or a combination of the two reactions.

The mechanism is obscure[11–13] but it is thought that it depends initially on the transition of chromium from a high to a low state of oxidation, followed by an interaction process with the colloid to form a chromium/colloid complex in which binding between the colloid and photo-released chromium compound involves either primary valency forces or physical forces of absorption.[14–17] The resulting complex causes a solubility decrease. As water-soluble colloids are used, a solubility differential is readily obtained by aqueous system development. The formation of the chromate ion is thought to be an important initial step,

i.e. $$Cr_2O_7^{2-} \xrightarrow{h\nu} CrO_3 + CrO_4^{2-}$$

The absorption spectrum of an aqueous solution of potassium dichromate is concentration-dependent, owing to two chief parameters:[18]

1. Hydrolysis reaction: $Cr_2O_7^{2-} + H_2O \rightleftharpoons 2HCrO_4^-$.

2. pH. This is owing to the dissociation of the acid chromate ion (pH ~ 6.5).
$HCrO_4^- \rightleftharpoons H^+ + CrO_4^{2-}$.

Smethurst[19] has reviewed the photoreactivity of dichromate/hydrophilic colloids but obscurity of a definite mechanism is apparent. Kläning et al.,[20-21] and Wiberg[22] have shown that the thermal oxidation of secondary alcohols by the $HCrO_4^-$ ion in weak acid solution occurs via the acid-catalysed creation of an acid chromate ester. Cleavage of the carbon–hydrogen bond of the hydroxylated carbon atom is found to be the rate-determining step. The chromate ester has been found to be the photo-responsive entity and the creation and decomposition of this moiety may be depicted as:[23]

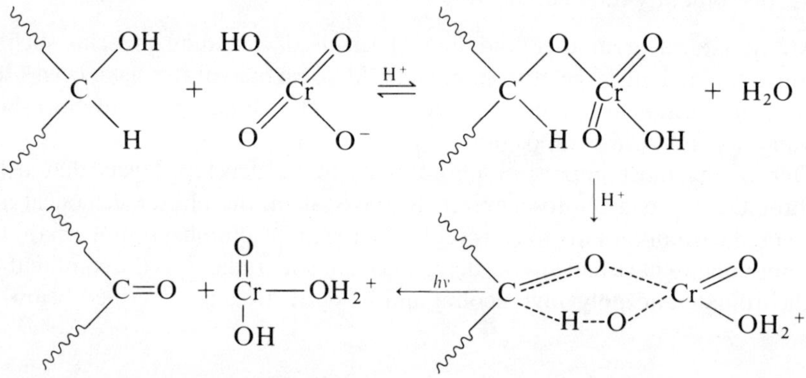

The primary photolysis product would be one carbonyl group which would result in two being formed by the subsequent reaction of the species chromium(V) formed. That is, the reaction may be written as:

$$\begin{array}{c} R' \\ R'' \end{array}\!\!\!\!CHOH + Cr(VI) \longrightarrow \begin{array}{c} R' \\ R'' \end{array}\!\!\!\!C=O + Cr(IV)$$

$$Cr(IV) + Cr(VI) \longrightarrow 2Cr(V)$$

$$Cr(V) + \begin{array}{c} R' \\ R'' \end{array}\!\!\!\!CHOH \longrightarrow \begin{array}{c} R' \\ R'' \end{array}\!\!\!\!C=O + Cr(III)$$

Oxygen also partakes in the chromic acid oxidation of secondary alcohols, as found by Driscoll and Mosher.[24] Hasseberger and Mosher[25] confirmed by electron spin-resonance studies the existence of the species Cr(V). These studies allow a set of secondary reactions to be considered as well in the above mechanism and may be extrapolated as a possibility for polyvinyl alcohol which is a secondary alcohol.

$$\text{Cr(IV)} + \begin{array}{c} R' \\ R'' \end{array}\!\!\!\!\!\text{CHOH} \longrightarrow \begin{array}{c} R' \\ R'' \end{array}\!\!\!\!\!\text{CHO} + \text{Cr(III)}$$

$$\downarrow$$

$$R'\text{—CHO} + R''\cdot$$

$$R''\cdot + O_2 \rightarrow R''O_2\cdot \rightarrow \text{autoxidation}$$
$$R''\cdot + \text{Cr(VI)} + H_2O \rightarrow R''OH + H^+ + \text{Cr(V)}$$

During photolysis of the chromic ester and photo-oxidation with dichromate it has been found that deposition of an insoluble brown compound occurs.[26,27] This compound may be:

(a) chromium dioxide, CrO_2;

(b) chromic chromate $(CrO)_2CrO_4$;

(c) basic chromic chromates such as $(Cr(OH)_2)HCrO_4$ and $[Cr(OH)_2]_2(CrO_4)$.

An insoluble infinite network would be expected to be formed by the production of an average of more than one crosslink per molecule in the polyvinyl alcohol system. Similar deductions have been made for the photo-insolubilization of polyvinyl cinnamate.

The crosslinking mechanism has been thought by Duncalf and Dunn[10] to be owing to the co-ordination of the hydroxy groups by polyvinyl alcohol and the chromic ions created in the film by dichromate reduction under non-aqueous conditions (i.e. dried film). Polyvinyl alcohol may be insolubilized without irradiation or oxidation when a hydrated Cr(III) salt incorporated in the film is dehydrated thermally. Schläpfer also concluded this in his studies.[28,29]

Owing to the consumption of hydrogen ions during the oxidation, actinic radiation has the effect of raising the pH of dichromated coatings, which may mean that this crosslink sensitivity to acid is a result of the ligand being the alkoxyl ion co-ordinated to Cr(III) rather than the alcohol group in polyvinyl alcohol.

Elöd and Schachowrsky[30] found that polyvinyl alcohol films containing Cr(III) salts become partially insoluble without heating by raising the pH by exposure to ammonia vapour.

Photochemically induced crosslinking of polyvinyl alcohol with dichromate yields a film which is unstable to acid at about the one molar concentration range. For photoengraving, it is necessary to have this film acid etch resistant and this is achieved by a 'burning-in' process. The coating is generally heated to about 300 °C for a short duration. Zinc plates must not be heated above 230 °C, to prevent detrimental crystal structure changes in the metal. Often this type of plate is 'fixed' by immersion in an aqueous chromic acid solution and then warmed with water before burning in for about 3 minutes. The fixing is thought to increase the carbonyl content of the polyvinyl alcohol and as a result determines its thermal degradation behaviour.[31,23]

The dichromate system has several drawbacks which include:[32]

1. Instability of the coated layer with its tendency to insolubilize in the dark and/or lose its light-sensitive properties.
2. Non-uniform composition of natural organic colloids.
3. Only water-soluble colloids are particularly efficient for dichromate sensitizing.
4. Coatings are fairly susceptible to climatic conditions such as humidity and temperature.
5. The process is slow in cure speed.
6. The dermatitic and toxic nature of dichromates.

These have led to the search for novel systems.

6.1.1.2 Alternative Binders and Sensitizers

Fish glue or gelatin are commonly employed as photo-crosslinkable vehicles for this type of dichromate-sensitized system. An aqueous insoluble image is created, therefore giving good handling latitude. The final stencil is, however, still water permeable and does not possess resistance to acid/alkali etching solutions. There are several means to improve these resistance defects.

For the photoengraving process, it is common, as previously mentioned, to 'burn in' or heat the processed image at a level fractionally below the charring stage, resulting in an 'enamel' layer. Fish-glue images are particularly responsive to this although gelatin can also be thus treated. 'Burn-in' is limited to the metal area as photoresists that are needed on plastic laminates and thin foils would not tolerate these high-temperature conditions.[31,33]

Novel systems have been developed to overcome these kind of limitations, that is to obtain dichromate-sensitized, water-insoluble colloids or resins giving etch-resistant layers without the 'burn-in' stage.

A useful achievement in this area has been the addition of alcoholic shellac solutions to alkali dichromates. The shellac must be partially hydrolysed (via ammonia addition) or else the water-soluble dichromate cannot be used. These dichromate photosensitized shellac resists have the drawback of not possessing resistance to some strong etching mordants.[34]

Polyvinyl acetate and polyvinyl butyral have also been used as water-soluble resins for this type of system.[35] They have limited hydrophobic character, giving the drawback of poor alkali/acid resistance in the final film.

There are many synthetic polymers that have been investigated for this type of system, including polyvinyl alcohol, polyacrylamide, polyvinyl pyrrolidone, and specific polyamide resins. The photoengraving process has usually adopted the use of polyvinyl alcohol/ammonium dichromate giving a satisfactory solubility differential and clearcut images.

Again, after irradiation, the stencil needs an after-treatment such as a 'burn-in' as it is permeable. This 'burn-in' gives acid/alkali resistance and also the temperature necessary for this is significantly lower than that required for fish glue.

The polyvinyl alcohol based photoresist image may be treated with a solution of a polynuclear hydroxy compound such as logwood dye, in order that after a further chromic acid bath, more tanning of the layer occurs.[36]

In the absence of pretreatment with the polynuclear hydroxy compound, the chromic acid is thought to have little extra effect on the original photo-induced crosslinking. The logwood dye absorbs to the polyvinyl alcohol structure, augmenting its molecular weight. Chrome tanning is subsequently possible and the resist image formed needs a relative low burning-in temperature only. This gives complete strong etch solution resistance.

In general, polymers either natural or synthetic possessing pendant polar groups such as hydroxyl, carboxyl, carbonyl, amino, etc., can be used to attempt to produce a light-sensitive dichromated colloid system. This type of system is basically the crosslinking of saturated polymers containing polar groups by an additive crosslinking agent.

A double-layer[37] system may be used to obtain better etch-resistant layers with dichromated colloid resists. The base support is initially coated with a good water-resistant resin layer such as bitumen, nitrocellulose, or bakelite resin. A surface layer of a dichromated colloid is then applied. On exposure and development of the dichromate-sensitized layer, the underlayer of resin is removed in the unprotected areas by solution with an appropriate solubilizing compound. This process is depicted in Figure 74. This latter solvent must be carefully chosen for the underlayer to prevent sideways penetration and image attack.

A red-shift technique, that is, extending the light sensitivity of the dichromated colloid system towards longer wavelengths, is often used, whereby a photo-reduceable dye such as methylene blue and a mild reducing agent are jointly added. The dye becomes reduced to the leuco form when an electron-donating compound is present by absorbing the longer-wavelength radiation. This leuco dye then reacts so as to cause reduction of the dichromate ion to give Cr(III) and this participates in the gelatin or synthetic resin crosslinking.[6,7,38]

Dichromate resists have the disadvantage of being unstable, and once coated and dried they will thermally begin to crosslink in the absence of light. The three prime factors affecting this thermal reaction are moisture, temperature, and pH. Insolubilization often occurs within a few hours of coating, although for some colloids such as dextran a lifetime of up to a week is claimed.[39]

A search for novel compounds to supersede dichromates[40,41] produced the diazo compounds. Peter Greiss in 1858 prepared the first aromatic diazo compound. Some workers produced a hybrid dichromate/diazo compound system.[42] The early 1920s represented, however, the first application of these compounds in the reproductions industry as:

1. diazo paper (or blue print paper);

2. lithographic printing, for which diazo compounds are employed in both:

    (a) negative working systems – that is where light-struck areas become insolubilized, and

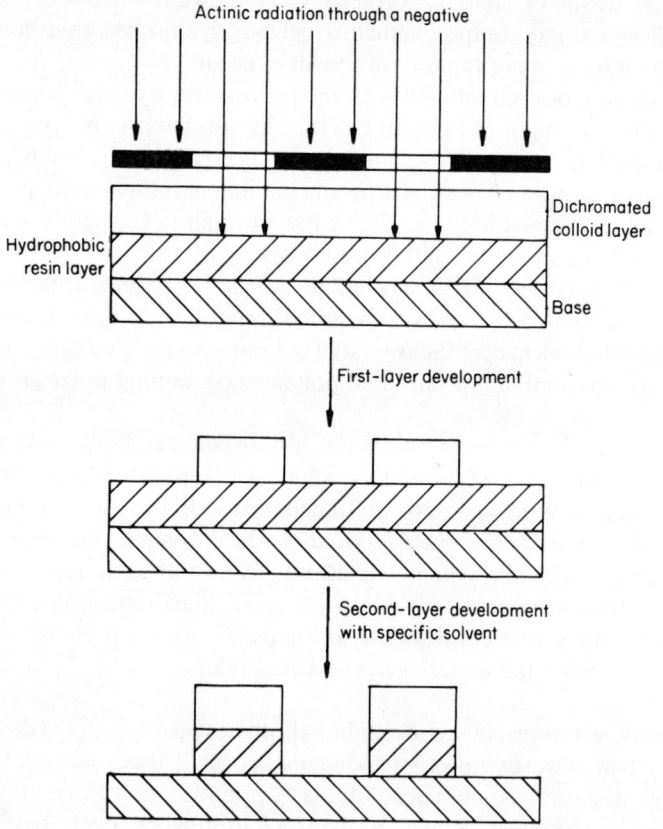

Figure 74 The double-layer photoresist principle.[1] (Reproduced by permission of The Royal Photographic Society of Great Britain and Kodak Limited.)

(b) positive working systems – where the light-struck areas produce a resultant compound that is soluble and removable by a developer.

Diazo salts in aqueous conditions decompose to give nitrogen gas and phenol compounds such as:

$$\text{Ph-N}{\equiv}\text{N}\}^+ \text{X}^- \xrightarrow[\text{H}_2\text{O}]{h\nu} \text{Ph-OH} + \text{N}_2 \uparrow + \text{HX}$$
$$\text{phenol}$$

Under anhydrous conditions such as the diazo-sensitized resin coatings of lithography, the decomposition is thought to proceed via a radical reaction. In this case, biphenyls, are thought to be the products.

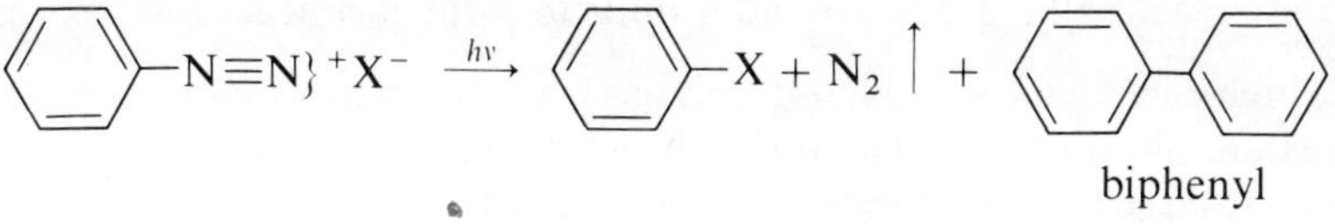

$$\text{Ph-N}\equiv\text{N}^+\text{X}^- \xrightarrow{h\nu} \text{Ph-X} + \text{N}_2\uparrow + \text{biphenyl}$$

An important group to be discovered were the 'diazo resins' which are the condensation products of diazo salts and aliphatic aldehydes such as 1-diazo-4,4-diphenylamine with formaldehyde. Possibly, the formaldehyde forms a methylene bridge amongst aromatic nuclei to give a high molecular weight resinous material similar to the condensation reaction to form phenol formaldehyde resins[1,43–44,73]

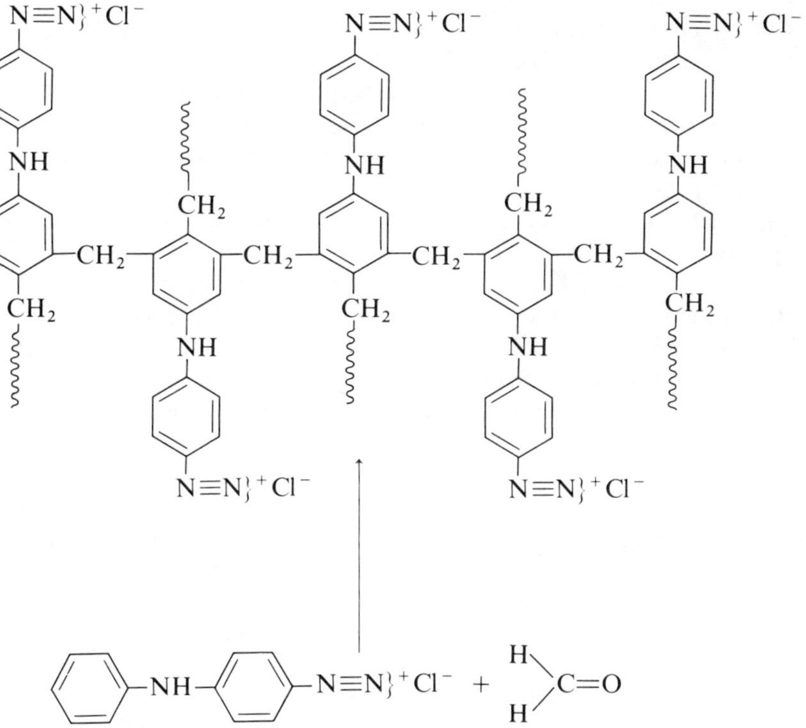

[structure of diazo resin network showing diazonium-substituted aniline units linked via NH to phenyl rings connected by CH₂ methylene bridges]

$$\text{Ph-NH-C}_6\text{H}_4\text{-N}\equiv\text{N}^+\text{Cl}^- + \text{H}_2\text{C=O}$$

With this large structure, the resin is, owing to its polar diazo groups, water soluble. It is readily insolublized on exposure to actinic radiation, as then the diazo groups are destroyed. Presensitized lithographic plate formulations which are only partially stable on ageing are often based on this type of system as they are adequate enough to permit factory coating by the user.[45] Substantial photoresist stencils are not formed from diazo resins themselves but need to be incorporated into a hydrophilic colloid and are subsequently reasonably efficient at insolubilizing the colloid on exposure to actinic radiation. The mechanism of this is unclear, although two chief possibilities are viable:

1. The diazo resin may create a large molecular network that causes immobilization of the colloid binder around it.

2. It may react with specific colloid groups to form chemical crosslinking.

The literature favours the second as it has been discovered that the crosslinked colloids are physically similar to the chromate-hardened ones.[46]

Diazo resin sensitizers have the disadvantage of nitrogen gas release, resulting in 'bubbling' of the final coating, although a system was developed utilizing this property.[47]

The addition of iodoform to bitumen was a novel system attempted for sensitization. For partially crosslinked resins, sensitization with iodoform or lead iodide gave on exposure to actinic radiation a degree of crosslinking satisfactory for a solubility differential. Phenol formaldehyde resins and partially gelled drying oils have been used for this.[48–51]

Current photoresist systems rely on several useful resin systems. One of these is that based upon cinnamic acid.[52] This substance has an inherent photosensitivity and permits a method of forming a synthetic photoresponsive resin. The basic photochemistry of cinnamic acid is dimerization on exposure to actinic radiation, two molecules taking part to form optical isomeric truxinic and truxillic acids,[1,53] see (a) on page 273.

Dicinnamalacetone in conjunction with a resin vehicle was found to undergo this type of photo-crosslinking to produce a dimer or polymeric structure.

The resin vehicle did not take part in the crosslinking and therefore only a partial solubility difference occurred. The system had stability.[54,55]

The synthesis of cinnamic acid esters of polyvinyl alcohol and cellulose expanded the field, as these were polymers with photosensitive groups built *in situ*. Polyvinyl cinnamate is typical, having a vinyl backbone with cinnamate groupings as side chains,[1,23,56–58] see (b) on page 273.

Polyvinyl cinnamate coatings on exposure to actinic radiation become virtually insolubilized in most common solvents giving a very wide development latitude, and extended solvent contact is not detrimental to the exposed resin areas. The system has low sensitivity to actinic radiation but this has been overcome by photosensitizing with ketones, quinones, and organic nitro compounds.[59] The combined system is many magnitudes faster than a dichromate-sensitized colloid system and also has good stability, wide development latitude, and good resistance to aqueous etching mordants.

Other systems using cinnamic-type groupings have been polyester resins in which a cinnamic acid derivative forms part of a polymer chain.[60] Also, in another, an isocyanate linkage is used to join the cinnamate group to the polymer backbone.[61] These type of systems have all had their sensitivities extended to longer wavelengths and speeded up by 'red shift' photosensitizers. These type of photocycloaddition reactions are described in some further detail in section 6.2.2.1.

As well as diazo compounds being used to insolubilize hydrophilic colloids, certain organic azides were found to react similarly.

The azide group is unstable to heat or short-wavelength radiation, often explosively so, but when attached to a large organic nucleus has some stability conferred upon it. Actinic radiation causes the azide group to release nitrogen gas

(a)

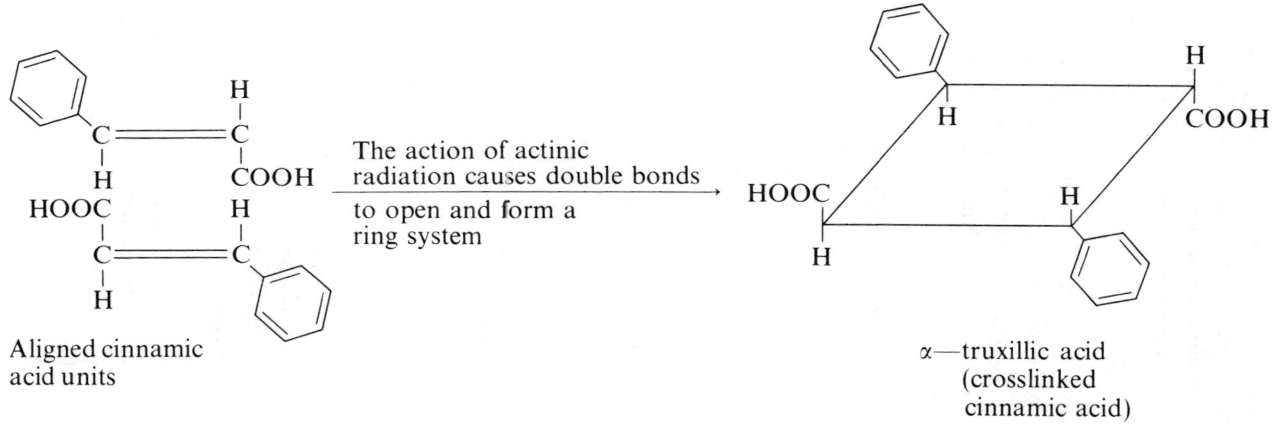

Aligned cinnamic acid units

→ The action of actinic radiation causes double bonds to open and form a ring system →

α—truxillic acid (crosslinked cinnamic acid)

(b)

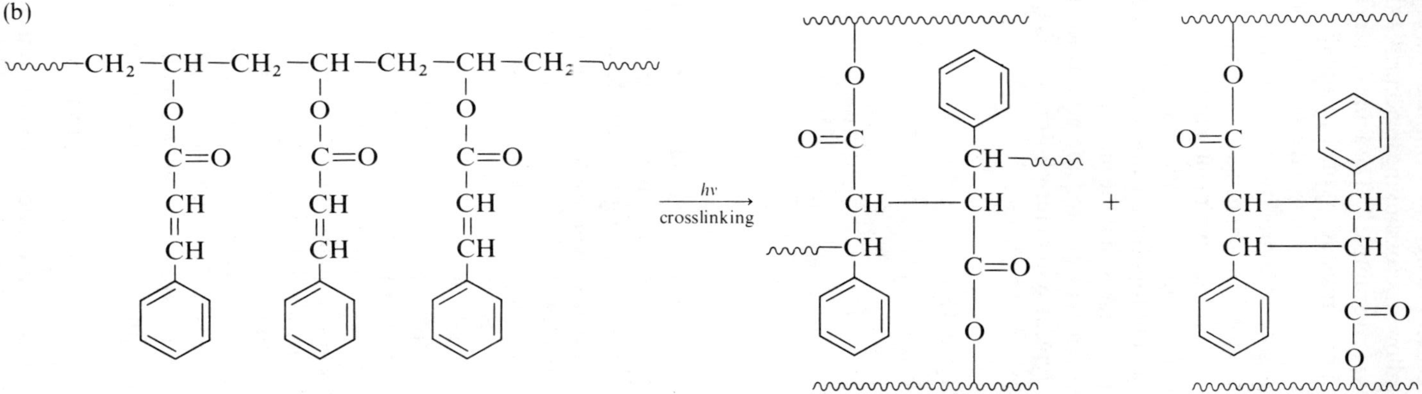

$\xrightarrow[\text{crosslinking}]{h\nu}$

and generate a free radical which will couple with itself or any adjacent molecule, thus allowing crosslinking to occur. 4,4′-Diazido stilbene disulphonic acid/gelatin is a typical example.[62] The azide system has good stability under these conditions but poor resistance of hydrophilic images causes application limitation. Hydrophobic systems have been developed, namely, those based on bis-azides giving solvent solubility instead of water solubility. These were used with certain hydrophobic resins such as hydrocarbons.[63,64] Polymeric compounds having the azide group built *in situ* have now been synthesized.[65] A further description of this type of system is to be found in section 6.2.2.2.

For photopolymerization as a means of forming a resist, a practical disadvantage is that the monomers are more often than not in the liquid phase at room temperature and the non-irradiated coating is inconvenient to handle.

A typical system for photoresists is the formation of polyacrylamide from the monomeric acrylamide:[1]

$$n \begin{array}{c} CH_2=CH \\ | \\ C=O \\ | \\ NH_2 \end{array} \xrightarrow{\text{with or without photoinitiator/sensitizer}} \left[\begin{array}{cccc} CH_2-CH-CH_2-CH-CH_2-CH \\ | & | & | \\ CO & CO & CO \\ | & | & | \\ NH_2 & NH_2 & NH_2 \end{array} \right]_{n-3}$$

acrylamide monomer → polyacrylamide

Relief printing plates are often prepared by a photopolymerization technique. A monomer possessing bi-reactive sites is exposed to actinic radiation in admixture with a photosensitizer/initiator and a polymer. Light absorption results in the self-polymerization of the monomer or to form crosslinking bonds with the surrounding polymer, leading to a reduction in solubility. Examples of this type of system are methylene bis-acrylamide which is doubly unsaturated and is able to create two links with the surrounding carrier polymer. This carrier polymer may be a cellulose-ester, polyamide, or polyvinyl alcohol and its derivatives. Benzoin or diacetyl can be the photo-initiator/sensitizer.[66] The carrier polymer may be modified to permit alkaline solubility which allows the use of aqueous solvent developing. The main area of usage is for the preparation of photopolymer relief printing plates rather than the photoresists field. Reversal photoresist images are often needed in photomechanical work. That is, it is necessary to make positive resist images from positive originals. In order to perform this directly a reversal system is required that on exposure to actinic radiation allows a solubility increase in the resist layer.

The ferric-ferrous/hydrophilic colloid system is an example of this system. Photosensitization with a ferric salt is a fairly water-insoluble coating when dried as a film. Exposure to actinic radiation causes reduction of the ferric ion to the ferrous ion with an increase in binder solubility.

6.2 TYPES OF PHOTORESISTS

Photoresists are often broadly categorized into two main types. These are positive working photoresists and negative working photoresists.

6.2.1 Positive Working Photoresists

These become more soluble in exposed areas by increasing their acidity and consequently become more soluble in dilute aqueous base solutions. They have three main components. These are:

1. A photoresponsive compound which is destroyed on exposure to actinic radiation.
2. An acidic polymer.
3. A solvent as an aqueous basic developer.

Some compounds used for this type of system are quinone diazides.

6.2.1.1 Quinone Diazides

Direct positive systems exist where solvent soluble compounds based on the *ortho*-quinone diazide or diazo-oxide structure become soluble in aqueous solutions on exposure to actinic radiation. Initially the compound contains a diazo group stabilized by the formation of a diazo-1,2-ketone and is water insoluble. Actinic radiation causes the liberation of nitrogen and consequent ring contraction to form a water-soluble carboxyl group in aqueous alkaline solutions.[67-70] This type of reaction[1,71,72] is known as a photo-induced Wolff rearrangement:

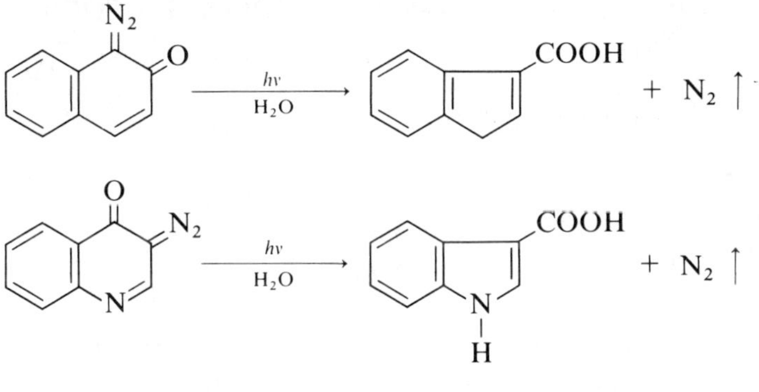

These diazo oxides have been extensively used for lithographic positive working plates.[73-75]

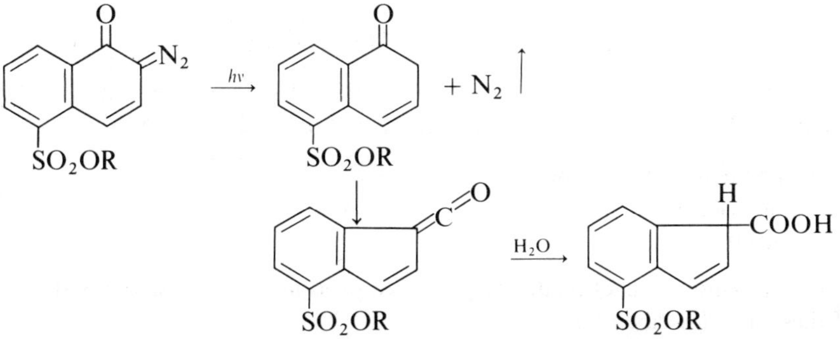

6.2.1.2 Acidic Polymers

Two main types are used in positive photoresists: *Novalaks* and *acrylics*.

Novalaks are phenol/formaldehyde complexes.[76-79] Good contrast and resistance to swelling during development are obtained. This is in contrast to the crosslinked regions of an exposed negative resist which swell whilst the developer is dissolving it. The swelling has the adverse effects of causing relief images to augment, to flow on the substrate, and to fuse with adjacent features.

Acrylics used are generally terpolymers.[2,80] Two vinyl monomers give a compromise of hardness and flexibility. A third provides a pendant solubilizing group. Other methods involve the photo-crosslinking of light-sensitive polyamic esters.

After development, thermal conversion of the image to a polyimide by degradation and volatilization of the crosslinks occurs. This confers upon the resulting image the high temperature stability characteristics of polyimides.[81]

This is depicted below:

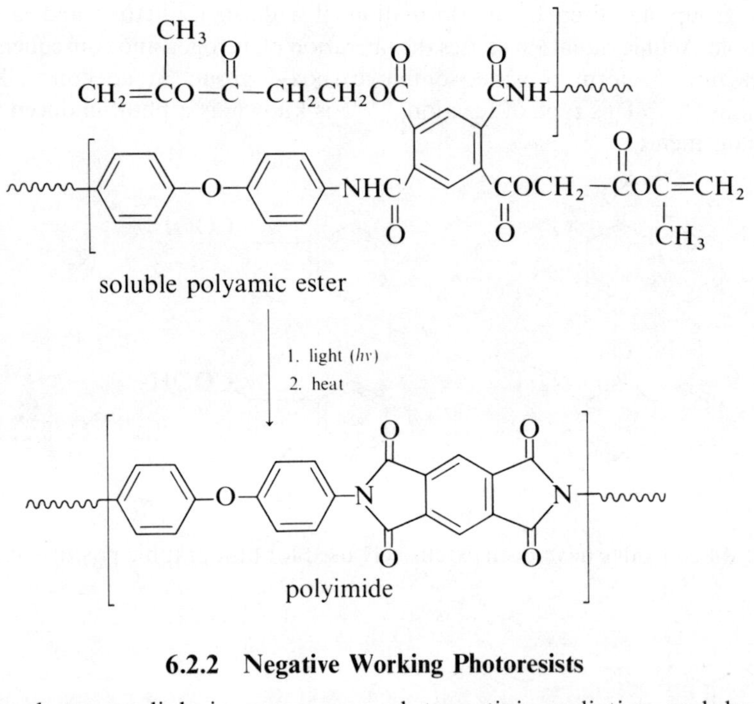

soluble polyamic ester

1. light ($h\nu$)
2. heat

polyimide

6.2.2 Negative Working Photoresists

These photo-crosslink in areas exposed to actinic radiation and become insoluble. There are three chief ingredients for these. These are:

1. A chemically reactive polymer.
2. A photosensitizer.
3. A solvent.

Several kinds of reaction are employed to produce these types of resists. Some of these are described below.

6.2.2.1 Photocycloaddition Reactions

(i) Cinnamate esters

Photocyclodimerization of cinnamic acid and its alkyl esters occurs.[82] This is not immediately shown by the *cis* isomer which first isomerizes to the *trans* form before undergoing photocyclodimerization.

Polyvinyl cinnamate is best when esterification is high.[83] The photosensitivity increases with the molecular weight as fewer photons are needed for crosslinking.[84]

(ii) Cinnamylidene esters

Esters such as polyvinyl cinnamylideneacetate are used,[85,86] i.e.

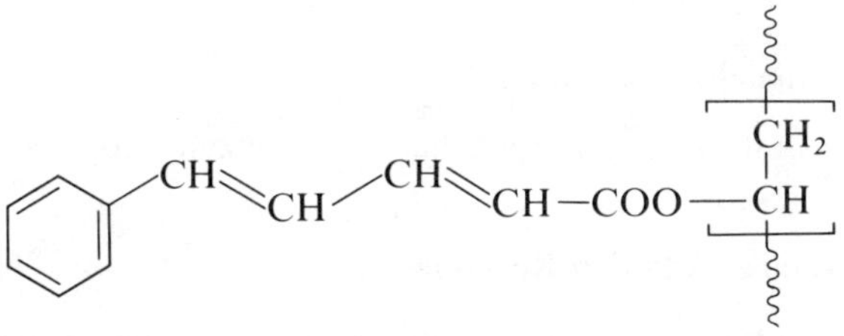

(iii) Cinnamoyl and other unsaturated substituents

These find application such as in chalcones (benzylideneacetophenones)[87] e.g. poly-4-vinyl benzylidene acetophenone

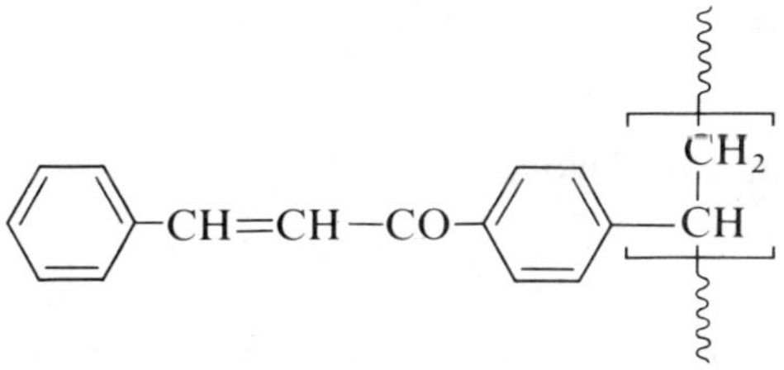

There are many others in this photocycloaddition class such as coumarins, stilbenes, maleimides, and polyacetylenes.[3,88–90]

6.2.2.2 Nitrene Reactions

(i) Bis-azide crosslinking compounds[91–93]

e.g.

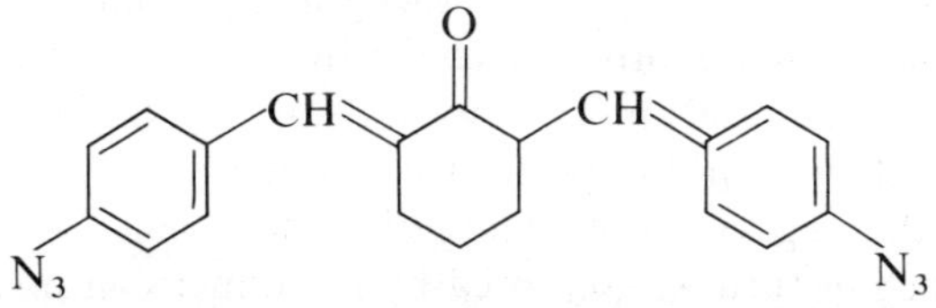

2,6-bis (4′-azidobenzylidene) cyclohexanone.[94]

This photolyses to give nitrenes, the imine radical being created.

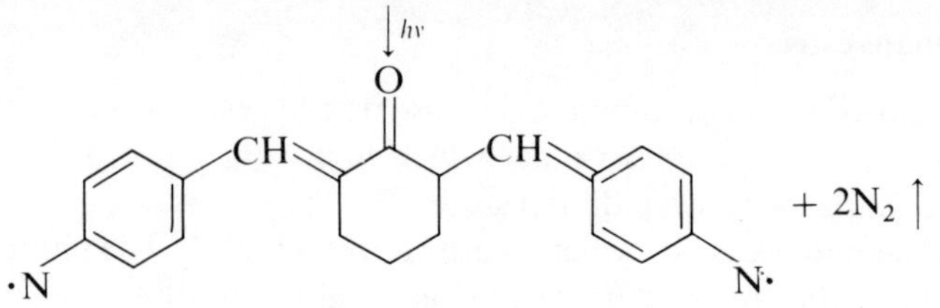

$$·N-C_6H_4-CH=\underset{\underset{O}{\|}}{C}(CH_2)_3C=CH-C_6H_4-N· + 2N_2 \uparrow$$

Covalent azides have an electronic structure which is a resonance hybrid, i.e.

$$R:\overset{+}{\ddot{N}}=N=\overset{-}{\ddot{N}}: \leftrightarrow R:\overset{-}{\ddot{N}}:\overset{+}{N}\equiv N:$$

(ii) Azide functionalized polymers

In these systems an azide group has been introduced into a polymer.[95–98]

6.2.2.3 Free-radical Addition Reactions

These are generally procured by the following main types of system.

(a) Allylic esters such as diallyl phthalate[2,99] and diallyl *iso*phthalate. Generally, these are used in conjunction with a bis-azide.[100]
(b) Thiol-ene system.[76,101–103] These are used largely in W. Grace's letterflex photopolymer plate system.
(c) Acrylic and methacrylic esters.[104–109]

6.2.2.4 Ring-opening Crosslinking

This occurs basically in cationic-type polymerization systems where complex diazonium salts yield Lewis acids on exposure to actinic radiation which can open ring systems such as epoxy resins, with subsequent crosslinking reactions.[110] This type of technology has been described in section 3.4.1.1.

There are many limitations in photoresist technology, and novel systems progressing towards an ideal are currently being researched. Some areas of improvement needed are:

1. Time of cure – printed-circuit manufacturers can only allow a few seconds exposure due to their production schedules.
2. Etch solution resistance – copper photoengraving etching may be performed with fairly mild etching mordants. The creation of silicon and molybdenum image-wise patterns needs more powerful solvents. Baths for electroplating often cause pinholing in thin protective coatings.
3. Fine line width images – the resistance of a photostencil is determined by its thickness which in turn is monitored by the dimension or the definition of image required. Etching or plating solutions are oxidative in character and

thin organic coatings need to withstand attack by them. Thick film layers may be used for coarse images eliminating the problem of etch resistance. Images with line widths of less than 25 μm however can present difficulty.

6.3 APPLICATIONS OF PHOTORESISTS

Photoresists find application for several of the printing processes. The major areas of interest include:[73]

1. Printing plates.
2. Photoengraving.
3. Silkscreen printing.
4. Printed circuits.
5. Collotype.
6. Proofing systems.

6.3.1 Printing Plates

Photopolymerization by actinic radiation is used extensively in the preparation of printing plates.

The kind of plates involved may be categorized into the following groups:

(a) *Relief or raised image*;
(b) *planographic, photolithography*;
(c) *gravure or intaglio-photogravure*;

but photopolymerization is finding use primarily in (a) and (b).

6.3.1.1 Relief or Raised-Image Plates[111]

Photopolymerization can be used for presensitized dry and liquid raised image or relief plates for letterpress, dry offset, and flexography.

Currently, liquid plates have chiefly found application in newspaper printing. The better-quality presensitized plates have extended the life of letterpress printing. Presensitized plates from rubber-type polymers are finding application in the rubber-plate industry.

The traditional copper, zinc, and magnesium plates used by the bulk of letterpress printers and platemakers are now being rivalled by photopolymer plate technology.

Photopolymer platemaking is advantageous as a more rapid and simpler process, and permits faster processing. One operation is needed for developing and relief forming of the image areas, requiring one item only of equipment. This process permits all line and halftone combinations to be relief-formed in a single step, which is the same with metal plates such as magnesium and zinc, but not generally in the case of copper where a two-bath technique is used. This necessitates the cutting of a protective mask, printing down on the halftone areas to protect them as the line and dropout areas are etched to greater depths. A notable difference occurs with a four-colour set skeletone black plate, which on making with photopolymer material, gives no relief-forming problem.

Zinc sometimes presents the problem with powderless etching of 'pimpling' on the non-image areas. This can be eliminated with photopolymer plates and results in less need for 'finishing'.

The photopolymer wash-out process is time stable and cheaper in contrast to the compounds employed in the bath preparation for the powderless etching of copper and zinc. These deteriorate all the time, in particular when the bath is not being used. Magnesium plate powderless etching is the only exception, as it is time stable.

Photopolymer plates seem to have more favourable ink transference with less impression and squash than metal plates in general, and permit a contrast reduction between the main shadow and low middle tones, particularly on monochrome work. A different negative screen range, especially between the main shadow and highlight areas, is a sensible consideration.

Powderless etching has the disadvantage of image loss. Photopolymer plates can be made accurate to the negative. There is always a small loss during etching with the metal plate, especially in the copper process. More control is thus in the plate processing.

Photopolymer plate systems currently do have some problems. Some of the photopolymers used are more expensive than metal. A plate processor is not required for developing but a few systems do need a special drying and post-hardening unit. Fireproofing precautions are necessary for those employing an alcohol solvent for wash-out. Local tone correction is impossible on the actual plate. A main disadvantage to metal reproductions is the critical nature of screen negative making for photopolymer plates, which is generally evident in smooth vignetted areas.

It is required to print down from matt-surfaced negative material for some photopolymer plates, which will increase production costs. Even with non-colour work, sometimes it is imperative to contact the initial negative to positive and then revert to negative. Matt negative film guarantees perfect contact between both film and plate throughout exposure time as oxygen left on the plate surface may cause underexposure or a subsequent image loss, especially in the highlight areas while wash-out is being performed. This disadvantage also renders the need for print-down from a one-piece negative film. An exception is when the image areas of each subject are adequately separated from each other. Some photopolymer plate systems need storage in an atmosphere of carbon dioxide to remove oxygen to activate them for exposure.

Press-life for these photopolymer plate systems is satisfactory, generally up to about half a million impressions, although current systems claim longer.

The blank material can have a shelf-life of the order of 1–2 years.

The principle of relief platemaking[111] is depicted in Figure 75.

Photopolymer relief plates fall into two principal categories: liquid types and solid types. A few examples of each will be described to outline the principles of formulation and mode of action of the system.[111,112]

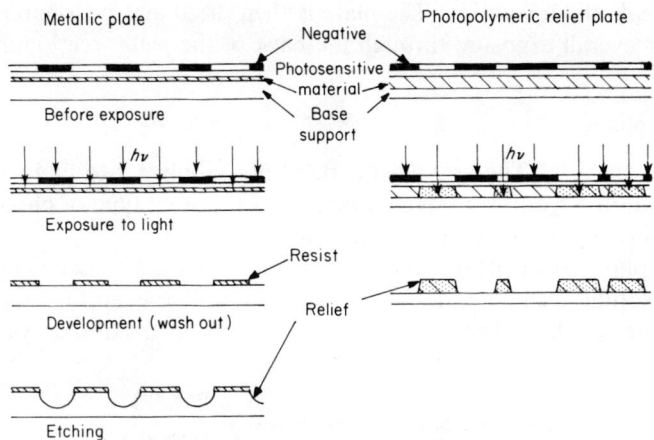

Figure 75 The principle of relief platemaking.[111] (Reproduced by permission of Eiichi Asayama, Plastics Age Co., Ltd.)

(i) Liquid types

The first liquid resin system was the letterflex plate.
(a) *Letterflex*.[112,113] W. R. Grace developed this liquid photopolymer system. The material relies upon the crosslinking reaction of a polyurethan possessing allylic unsaturation and a polythiol.[114]
i.e

$$HS-R'\text{\textasciitilde\textasciitilde\textasciitilde} + CH_2=CH-CH_2-R\text{\textasciitilde\textasciitilde\textasciitilde}$$

$$\downarrow \text{actinic radiation} \; + \text{benzophenone}$$

$$\text{\textasciitilde\textasciitilde\textasciitilde}-R-CH_2-CH_2-CH_2-S-R'\text{\textasciitilde\textasciitilde\textasciitilde}$$

An aluminium support is employed and the liquid photoresponsive coating applied on top. A negative is then used, and exposed to actinic radiation from an Xenon lamp. An ultrasonic washing technique with a solution of surface active agents is used to remove the unexposed areas. Equipment costs are fairly high.

(b) *Asahi photopolymer resin*.[111,112,115] A similar product to letterflex is the Asahi photopolymer resin (APR). In this case the resin is an unsaturated polyester on a polyester backing sheet. The production of the APR plate involves the negative being placed emulsion side up on to the exposure unit and covered by a transparent film under vacuum. The liquid resin is then dispersed over the cover sheet which is in turn backed up with a sheet of the plastic base material. A quick back-exposure to actinic radiation is applied to aid the image area's adhesion to the support carrier and then the subsequent relief exposure from below. The relief is formed after exposure by washing out in a weak alkali solution. The plate is then dried and post-hardened by a further overall exposure through the back of the plate (see Figure 76).

(ii) Solid plates

(a) *Dycril.*[111,112,116] The first relief plate of photopolymer to find commercial application. Dycril is a typical example of the solid type of photopolymer plate invented by Dupont and introduced in 1957.

The photosensitive layer about 0.3–1.0 mm thick is a binder composed of a water-soluble cellulose derivative such as cellulose acetate succinate, a monomer of the divinyl type such as triethylene glycol diacrylate, and a

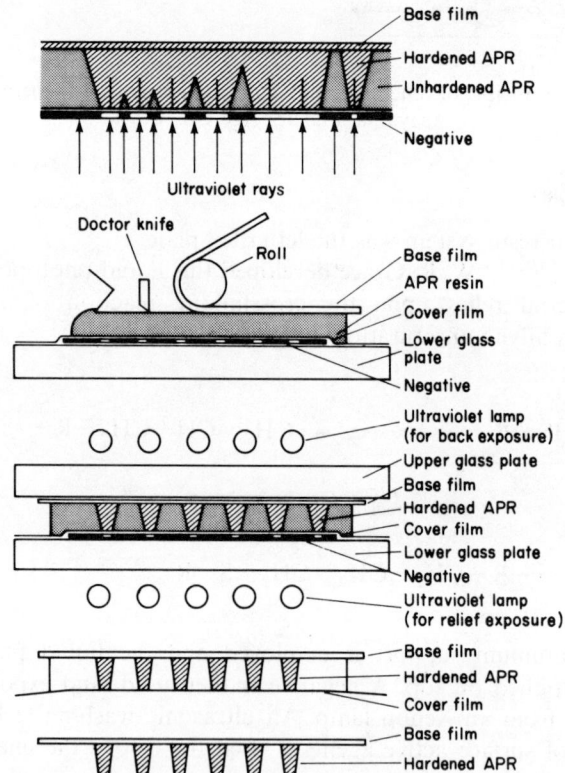

Figure 76 The Asahi photopolymer liquid resin plate system.

photo-initiator such as 2-ethylanthraquinone and a thermal or 'dark-reaction' stabilizer of the *p*-methoxyphenol type. The mixture is placed on an aluminium support or steel sheet with an adhesive sheet and anti-halation sheet between the pair.

The Dycril photo-sensitive layer then has a negative placed over it and this is followed by exposure to actinic radiation from a carbon or mercury arc lamp (usually, placed in a vacuum backing frame having polythene sheet surfaces, so that contact can be improved between the photosensitive coating and the negative film).

Post-exposure treatment, involves washing with 0.2–0.5 per cent caustic soda solution. This removes the unexposed areas alone, a relief remaining on the support. Drying is then performed after washing out, or for some systems the whole plate area is re-exposed to actinic radiation and used as a relief or dry offset plate.

Dycril has a fairly high resolving power and is durable. It is fairly cheap and may be washed out with less hazardous solutions such as alkali.

(b) *Nyloprint.*[111,112,117,118] BASF introduced this solid type of photosensitizer compound coated plate around 1967. The solid photopolymerizing material is a combination of an alcohol-soluble polyamide such as a copolymerized polyamide of hexamethylene diammonium adipate and ε-caprolactam, a vinyl monomer (such as *m*-xylene, bis-acrylamide, hexamethylene bisacrylamide, or triethylene glycol diacrylate, etc.), a photo-initiator such as benzoin methyl ether, and a 'dark reaction' stabilizer (such as hydroquinone and derivatives) and maleic anhydride to function as a compatibility modifier for polyamide with vinyl monomers.

In contrast to Dycril, Nyloprint has a wash-out with alcohol, and its waste disposal may be of concern.

(c) *Dyna-flex.*[111,118] This is a solid photosensitive plate marketed by Dynaflex and has a photosensitive emulsion on an aluminium backing or carrier plate. It is thought to be a system based on polyvinyl alcohol and dichromate chemistry.

It is kept frozen for storage prior to use and after exposure via a negative to actinic radiation is washed out with water and a small amount of defoaming agent, making it a relatively cheap system. It appears to have the disadvantage of some difficulties in the reproduction of images in fine detail.

Photosetting in conjunction with photopolymer plates has a vast potential in three chief areas of printing. These are:

1. Newspaper production.
2. Flatbed cylinder letterpress.
3. Rotary letterpress commercial printing.

Some of the main commercial plate systems available or under trial are:

1. Dycril
2. BASF Nyloprint
3. Letterflex
4. Dyna-flex
5. Flexomer
6. APR
7. Tevista
8. Sonne KPM
9. NAPP
10. Toplon
11. Torelief

The technical aspects of these platemaking systems are shown in Table 14.[112]

Table 14 Technical aspects of various plate systems. Appendix B: from Cannon Paper[112]

	DYCRIL	BASF NYLOPRINT	LETTERFLEX
Name of plate			
Country of origin	USA	BASF Germany	Grace & Co, USA
Type of plate	Solid blank material	Solid blank material	Liquid resin
Material	Polyvinyl chloride	Photosensitive nylon	Polyurethan
Solvent	Sodium hydroxide	Alcohol	Industrial detergent
Thickness gauges	0.43 mm/6.35 mm	0.4 mm/1.75 mm	0.76 mm
Relief depth	0.20 mm/3.81 mm	0.2 mm/0.66 mm	0.51 mm
Pattern plate	Yes	Yes	Yes
Backing material	Steel, aluminium, mylar	Steel or aluminium	Aluminium
Max size of material	Varies with the type of plate	Varies according to type of plate	
Equipment	Manual, semi-automatic	Semi-automatic, fully automatic	Manual, semi-automatic, automatic
Processing time	Average 20 min. Type 40:8 min	15/30 min, 60 per hour	6 min, $2\frac{1}{2}$ min, 1 min
Estimated press life		One million	500 000

	DYNA-FLEX	FLEXOMER	APR
Name of plate			
Country of origin	Dyna-flex Corp, USA	USA	Japan
Type of plate	Solid blank material	Solid blank material	Liquid resin
Material			Unsaturated polyester
Solvent	Water		Weak sodium hydroxide
Thickness gauges	0.508 mm, 0.61 mm	0.25 mm/6.4 mm	0.75 mm
Relief depth	0.406 mm, 0.508 mm	0.20 mm/3.2 mm	0.5 mm
Pattern plate	Being developed		
Backing material	Aluminium	Steel, aluminium, polyester	Polyester
Max size of material	76 cm × 100 cm	147 cm wide and in any length	420 mm × 594 mm
Equipment	Manual, semi-automatic	Similar to Nyloprint and Torelief	Manual, semi-automatic, fully automatic
Processing time	7 min/48 sec	Similar to Nyloprint and Torelief	15/7 min, 8/4 min, 5/1 min*
Estimated press life	One million claimed	500 000	500 000

*The first figure is for the first plate and the second figures for successive plates.

	TEVISTA	SONNE KPM	NAPP
Name of plate	TEVISTA	SONNE KPM	NAPP
Country of origin	Japan	Japan	Kansai Paint Co, Japan
Type of plate	Liquid resin	Solid blank material*	Solid blank material
Material	Unsaturated polyester	Polyurethane	Polyvinyl alcohol
Solvent	Dilute aqueous solution of sodium carbonate	Sodium hydroxide	Water
Thickness gauges	0.8 mm/1.1 mm	0.83 mm, 0.96 mm	0.83 mm, 1.00 mm 0.45 mm, 0.55 mm
Relief depth	0.5 mm/0.8 mm	0.6 mm, 0.64 mm	1.30 mm, 1.63 mm; relief depth of 0.60 mm
Pattern plate	Yes	Standard plate can be moulded at room temperature	
Backing material	Steel	Steel	Steel
Max size of material	400 mm × 500 mm	415 mm × 575 mm	430 mm × 600 mm
Equipment	Manual, automatic	Manual, automatic	Manual, automatic
Processing time	15/10 min, 5/3 min	15 min (manual)	15 min, 10/1 min
Estimated press life	500 000	500 000	400 000

* Liquid resin system also being developed.

	TOPLON*	TORELIEF
Name of plate	TOPLON*	TORELIEF
Country of origin	Tokyo Ohka Kogyo Co, Japan	Toray Chemical Co, Japan
Type of plate	Solid blank material	Solid blank material
Material	Photosensitive nylon	Photosensitive nylon
Solvent	Alcohol	Alcohol
Thickness gauges	1.0 mm, 0.8 mm	0.75 mm/0.95 mm
Relief depth	0.56 mm, 0.6 mm	0.55 mm/0.65 mm
Pattern plate		
Backing material	Steel	Steel, aluminium or polyester
Max size of material	420 mm × 600 mm	420 mm × 594 mm
Equipment	Semi-automatic	Semi-automatic
Processing time	15/20 min	15/20 min
Estimated press life		500 000

* Still field testing

6.3.1.2 Photolithography, Planographic Plates[23]

These are lithographic plates and find application on offset presses. There is no difference in height between the base support material and converted photopolymer. The materials used include diazo monomers, diazo polymers, diazo compounds, in conjunction with additive film formers such as epoxy resins, phenolics, polyvinyl acetals, polyamides, azide compounds, unsaturated chalcone, cinnamate, stilbene, and vinyl polymers and dichromated colloids.[119]

A novel successor to offset planography is the driographic plate which was designed to overcome the need for a water–oil balance. Driography requires that the differential between the oleophilic (oil-receptive) image area and the background polymer that is not grease, oil, or water receptive removes the necessity of a fountain solution. Photopolymerization may be used to remove the development step, the differential between oil receptive and non-receptive occurring by exposure to actinic radiation alone.

There are three kinds of metallic photolithographic plates currently used. These are:

1. The surface plates – negative and positive working plates.[120–122]
2. 'Deep-etch' plates.[123,124]
3. Bimetallic or trimetallic plates.[125]

The latter two may be regarded as 'reversal' positive working processes.

All are prepared differently. For surface plates, direct use is made of the coating material or the resultant compounds of photochemical change as the image bearer. In contrast, in the 'reversal' type method, the coating acts as a hydrophilic stencil and the image itself is composed in essence of the plate surface backed up by the lacquer application.

In general, negative working surface plates are based on water-soluble coatings in which the resultant photochemical products form the image. On the other hand, positive working surface plates use water-insoluble coatings that on exposure to actinic radiation give products that are rendered preferentially soluble to develop the non-printing areas of the plate.

Negative working methods are most often aqueous-developable in water, dilute acids, or an emulsion of the solvent needed to dissolve the original coating substance. Positive working surface plates are developed in appropriate dilute alkali solutions.

(i) The surface plate[23,126]

These are often made of aluminium or zinc. Zinc is oleophilic, in contrast to aluminium which is not. Mechanical graining is performed to render the surface hydrophilic in the non-printing areas. Washing and 'counter-etching' of the plate is then executed with dilute hydrochloric, phosphoric, hydrofluoric, or acetic acid. Some of the etchant is considered to be either absorbed or chemically combined at the plate surface.

Throughout the storage at high relative humidities, there is a danger of corrosion. Zinc plates are protected by application of a dilute solution of sodium or ammonium dichromate containing sulphuric acid (Cronak process). Aluminium plates are treated with dichromate/hydrofluoric acid solution (Brunak process). The plate therefore has a layer of oleophilic chromium oxides on its surface. 'Desensitization', which is the rendering of the plate hydrophilic prior to application of the photosensitive coating, is necessary with sodium alginate, sodium carboxymethyl cellulose, or oxidized cellulose. Absorption of the hydrophilic polymer occurs on the plate surface and must be a compound possessing both hydroxyl and carboxyl group. The photoresponsive material is then applied. This is usually dichromated polyvinyl alcohol or dichromated egg albumin. Exposure of the plate to actinic radiation is then performed under a photographic negative. Post-exposure treatment involves a development ink consisting of a carbon black suspension in a non-drying oil, and then developing the plate with lukewarm water. Unexposed coating dissolves, but in the exposed areas the coating swells minutely and remains attached to the metal. For polyvinyl alcohol usage, a hardening agent such as haematin or catechin is advisable in the developing solution. An absorbed layer of albumin or polyvinyl alcohol stays on the non-image areas after development. This necessitates a further desensitizing process to make them hydrophilic. The plate is then ready for printing after drying out.

(ii) Deep-etch plates[127,23]

These, in contrast to the surface plate, require a positive transparency for their preparation. The plate image areas, corresponding to the black areas of the original, are slightly etched away and then filled with an ink-receptive layer. Graining and counter-etching, and optional Cronaking are performed. Normally dichromated gum arabic or dichromated polyvinyl alcohol is the photosensitive solution then applied. Post-exposure development of a gum arabic coating is carried out with a calcium chloride/lactic acid solution. Polyvinyl alcohol/dichromate coating may be developed with water but will require a further hardening with a dilute chromic acid solution. The acid solution possesses a high level of calcium and ferric chlorides to stop the removal of the stencil during hydrochloric acid etching. Etch depth is as small as 5×10^{-4} cm. Removal of the etching solution is effected by washing with anhydrous ethanol, 99 per cent isopropanol, or ethoxyethanol. The deep-etch layer is applied after drying and must be incompatible with gum arabic or polyvinyl alcohol. A developing ink is then applied as before and the plate scrubbed under warm water to take the stencil off. The non-image areas are then desensitized.

(iii) Bimetallic and trimetallic plates[128,23]

These are produced by the techniques of photolithographic deep-etch platemaking. A bimetallic plate has an oleophilic metal such as copper or zinc in the image area and another one such as chromium or iron in the non-image areas. For trimetallic plates, backing occurs from a third metal for support. Electroplated copper or an aluminium base is the prime constituent of a bimetallic plate (Lithengrave plate) or the base may be stainless steel (Aller plate). Bimetallic plates are generally processed from photographic negatives but the latter can be processed from positives. Ferric nitrate is the etch solution used, dissolving the copper but not attacking the metal underneath. Subsequent to etching the photoresist has to be removed, the plate then inked, and desensitization of the non-image areas carried out.

Processing of trimetallic plates occurs via positives. Their make-up comprises a zinc, steel, or chromium base plated with copper (10^{-3} cm thick) and then chromium (1.25×10^{-4} cm thick). They are processed in a similar manner to deep-etch plates. The polyvinyl alcohol system is favoured for these plates.

The lithographic plate industry uses many systems such as deep-etch coatings, wipe-on plates, negative and positive working presensitized paper, polyester, and metal plates.

Some photosensitive compounds (some of which have been previously described) recorded in the literature[129] are shown in Table 15.

Presensitized plates may be used to avoid the necessity of coating plates with dichromated colloid prior to exposure to actinic radiation. Diazonium salts are used as sensitizers, being soluble in water, and eliminate the azido group after exposure, becoming insoluble.

The most common resin used is the condensation product of 4-diazo-diphenylamine and formaldehyde, eliminating the use of polyvinyl alcohol and related colloids.

Commercial presensitized plates are formulated on two chief groups of photoresponsive compounds. These are:

1. Diazo resins and diazo oxides.

2. Cinnamic esters or epoxy resins.

Typical examples of these which have previously been described are listed[129] in Table 16.

Diazo resins and diazo oxides are mainly applied for presensitized surface plates. Negative or positive working character is dependent upon their solubility parameters and the nature of their photoreaction products. Mechanisms are often obscure but tend to proceed via a decomposition reaction to form complex acids or compounds less basic than the starting compounds.[129]

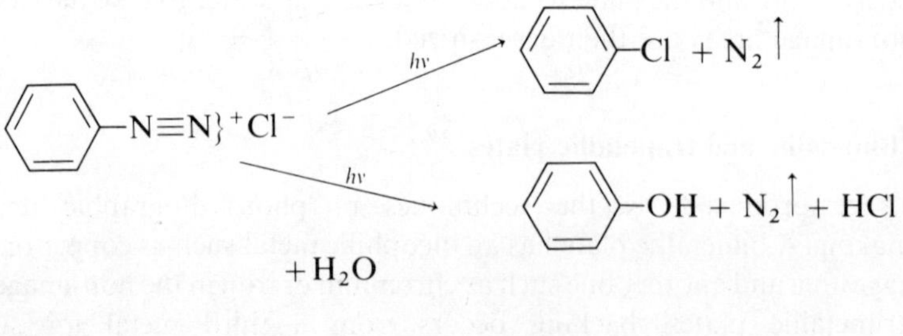

Diazonium salt (cf. diazo resin (I) of Table 16)

o-quinone diazide
(cf. diazo oxide (II) of Table 16)

The scheme above is reproduced with permission of the Institute of Printing.

Decomposition of a water-soluble diazo resin (I) by actinic radiation produces water-insoluble compounds. Compound (I) is water-soluble owing to it possessing ionic diazo groups. Developing the lacquer in water establishes a solubility differential. The products of decomposition have an oleophilic character and present the foundation of a presensitized and wipe-on negative working process.

For the diazo oxide (II), either possibility of a negative or positive working system is afforded by creating a more delicate equilibrium in the difference between the basic nature of the exposed to unexposed regions of the coating. In contrast with the previously discussed diazo resin, these lacquers are insoluble in water, but by appropriate choice of the diazo oxide compound, a solubility differential may be created in either an acid or alkaline solution.

The solubility differential for the negative working process is often based on dilute acid development, using to advantage the fact that the less basic characteristics of the materials formed in the exposed regions of the coating are less acid-soluble than the masked areas of the starting diazo compound. The latter is removed by an appropriate acid solution, and the chemically decomposed material as the image-bearing surface remains. In contrast, for the positive working process, an alkaline solution allows preferential dissolution of the complex acid compounds created by actinic radiation. The original diazo coating is left insoluble as the printing surface on the plate.

A presensitized 'reversal' positive plate exists based on the water soluble diazo resin (I). The technology of this is depicted in Figure 77.

The photosensitive lacquer is composed of the diazo resin (I) in conjunction with a synthetic hydrophilic colloid such as polyacrylamide. Exposure to actinic radiation through a positive film permits a solubility differential to be established by developing the coating in water. For the unexposed areas, dissolving the original water-soluble coating permits development of the underlying ink-receptive surface as the printing image. The coating regions hit by the actinic radiation stay as an insoluble hydrophilic stencil in the non-printing regions of the plate. This is in contrast to the deep-etch or gum-reversal positive process where the hydrophilic stencil is predominantly removed after the image

Table 15 Some synthetic photosensitive coatings used in photolithography. (Reproduced by permission of the Institute of Printing)

Class of chemical compound	Basic chemical structure	Typical coating material	Type of plate or process
DIAZO SALTS condensates with aliphatic or aromatic active carbonyl compounds	$R'{\sim}N{-}Aryl{-}N_2{-}X$ $R''{/}$ where X is an inorganic acid radical	p-diazo diphenylamine-paraformaldehyde resin condensate	(i) negative surface paper, plastic or metal plates (ii) positive reversal metal plates (iii) negative surface wipe-on plates
DIAZO OXIDES sulphonic acid esters of complex alcohols, amides, imidazoles, etc	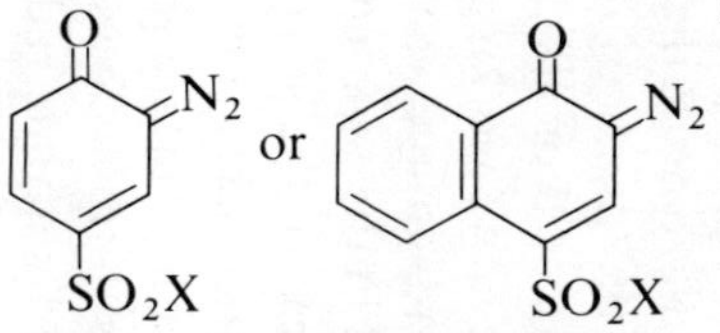 [Benzoquinone diazide structures with N_2 and SO_2X groups] where X is a derivative of a complex aromatic or heterocyclic ring system $-C{=}C{-}CO{-}C{=}C-$	Benzoquinone -1:4-diazide -(4)-2-sulphoanilide	Positive and negative surface plates – paper and metal
UNSATURATED KETONES		α-γ-dicinnamylidene-acetone	Negative surface metal plates (anodized)
NITRO-NAPHTHALENE SULPHONIC ACIDS	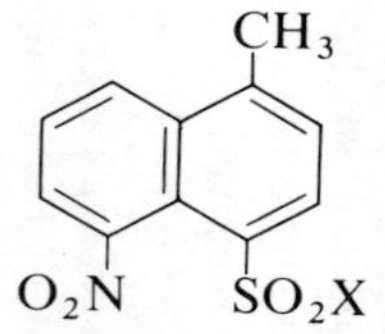 [Naphthalene structure with CH_3, SO_2X, and O_2N groups] where X is an amide ester	1-methyl-5-nitronaphthalene-4-sulphonomethyl amide	Positive surface metal plates

AROMATIC o-NITRO-ALDEHYDES	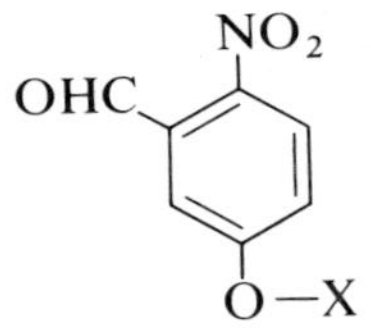 OHC—⟨NO₂, O—X⟩ where X is a complex ester grouping	Bis(formyl-4-nitro-phenyl-phthalate)	Positive surface metal plates
CINNAMATE POLYESTERS	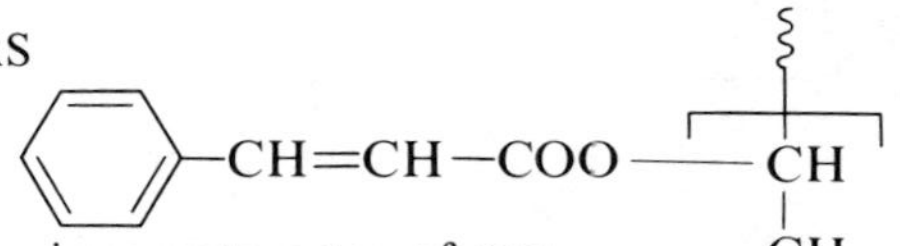 —CH=CH—COO—CH(—⟨Ph⟩)—CH₂— cinnamate ester of PVA	Cinnamate esters of epoxy resins	Negative surface metal plates
CHALCONE POLYMERS	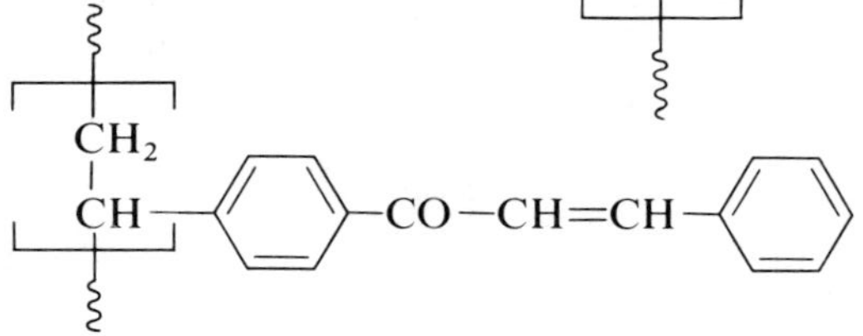 —CH₂—CH(—⟨C₆H₄⟩—CO—CH=CH—⟨Ph⟩)—	Poly (4-vinylbenzalacetophenone)	Negative surface metal plates

Table 16 Some compounds used in presensitized plates. (Reproduced by permission of the Institute of Printing)

Class of chemical compound	Typical compound	Solubility characteristics	Application in photolithography
Diazo-formaldehyde resin (I)	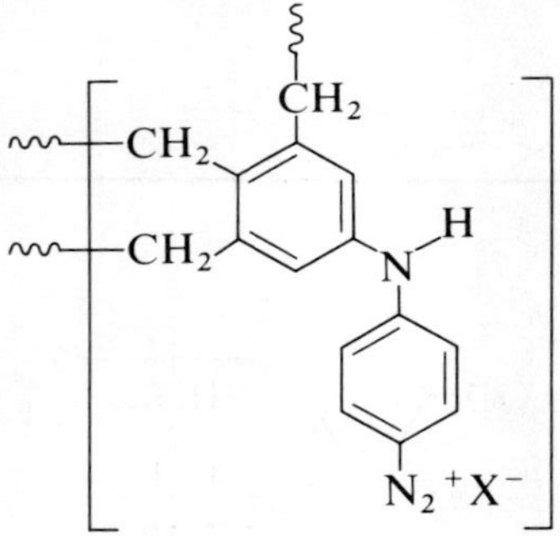 4-diazo-1,1'-diphenylamine formaldehyde condensate	Water soluble	(i) negative working process – presensitized paper, plastic or metal plates, wipe-on plates (ii) presensitized positive working reversal plates
Diazo oxide (II)	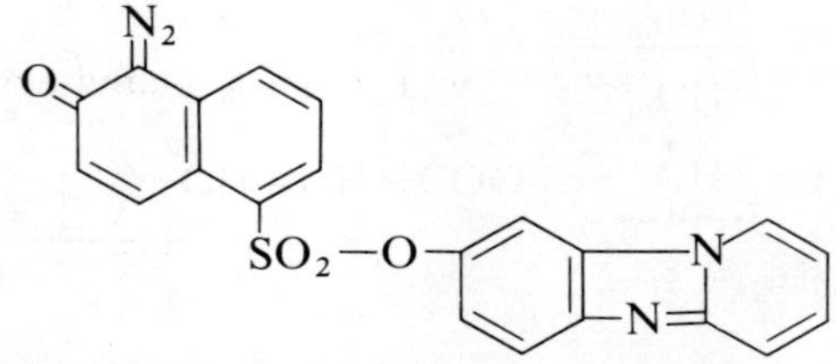 napththoquinone-(1,2)-diazide-(1)-5-sulphonic ester of 6-hydroxy(1',2'-1,2) pyrido-benzimidazol	Solvent soluble, e.g. ethylene glycol monomethyl ether	Negative and positive working presensitized surface plates – paper and metal bases
Cinnamic ester, of epoxy resin (III)	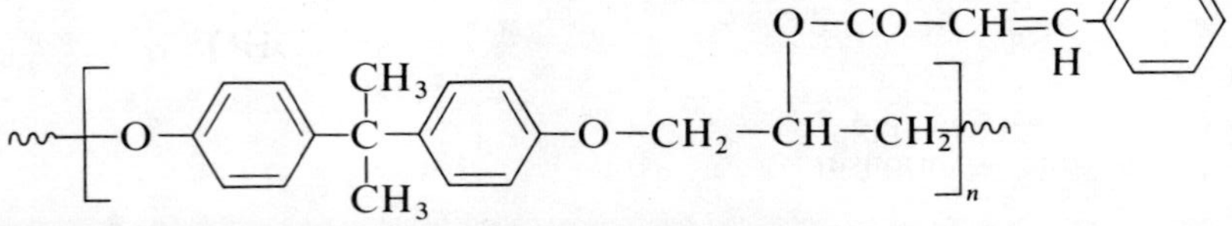	Solvent soluble, e.g. cyclohexanone and toluene	Negative working surface aluminium presensitized plates

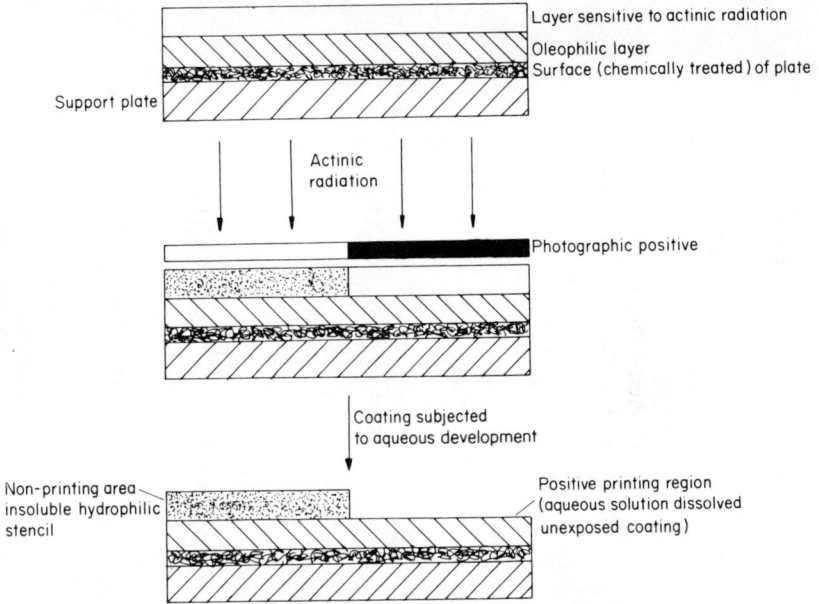

Figure 77 A presensitized 'reversal' positive plate.

development. T. Tsunoda and T. Yamoaka proposed a mechanism for the photochemistry of the reaction, involving the release of aromatic free radicals which crosslink appropriately aligned polyvinyl alcohol chains to create an interlocking structure with less solubility as depicted (see overleaf):[129-131]

The above is adapted and reproduced with permission of the Institute of Printing.

Cinnamate resin compounds, in contrast to the diazo coating compounds, are exclusively employed as negative working coatings, proceeding via a crosslinking mechanism instead of a decomposition process. There are many cinnamate resins, including derivatives of polyvinyl alcohol, cellulose, starch, and epoxy resins. It is the latter type, epoxy resins, that have found most application in lithographic coating materials.

As cinnamate resins are water-insoluble, the solubility differential is achieved by developing the coating in a suitable solvent system. An emulsion of the solvent is generally used dispersed in an aqueous phase of gum arabic and phosphoric acid. The light-exposed coating regions are rendered insoluble and form the printing image.

Polyvinyl cinnamate has a general photoresponse of the order of a tenth of that of dichromated colloids. Its rate can be accelerated by the use of photosensitizers such as:

1. Nitroamines – increase of the order of 100 times.
2. Quinones – increase of the order of 200 times for specific ones.
3. Aromatic amino ketones – increase of the order of 300 times for specific ones.

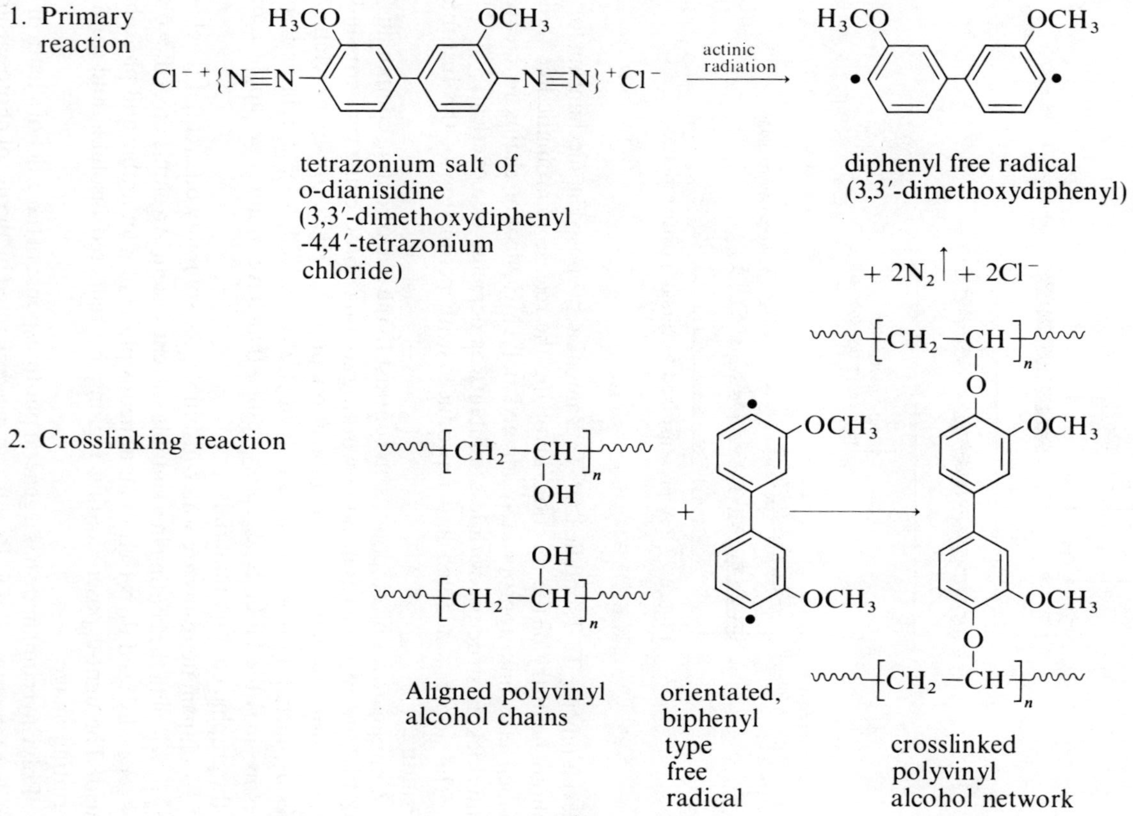

1. Primary reaction

$$\text{Cl}^-\ ^+\{N\equiv N\}\text{-}\underset{\underset{\text{OCH}_3}{|}}{\text{C}_6\text{H}_3}\text{-}\underset{\underset{\text{H}_3\text{CO}}{|}}{\text{C}_6\text{H}_3}\text{-}\{N\equiv N\}^+\ \text{Cl}^- \xrightarrow{\text{actinic radiation}} \text{·}\underset{\underset{\text{OCH}_3}{|}}{\text{C}_6\text{H}_3}\text{-}\underset{\underset{\text{H}_3\text{CO}}{|}}{\text{C}_6\text{H}_3}\text{·} + 2N_2\uparrow + 2\text{Cl}^-$$

tetrazonium salt of o-dianisidine (3,3'-dimethoxydiphenyl-4,4'-tetrazonium chloride)

diphenyl free radical (3,3'-dimethoxydiphenyl)

2. Crosslinking reaction

$$\sim\!\!\!\left[CH_2\text{-}CH\right]_n\!\!\!\sim \quad + \quad \text{·}\underset{\underset{\text{OCH}_3}{|}}{\text{C}_6\text{H}_3}\text{-}\underset{\underset{\text{OCH}_3}{|}}{\text{C}_6\text{H}_3}\text{·} \quad \longrightarrow \quad \sim\!\!\!\left[CH_2\text{-}CH\right]_n\!\!\!\sim$$
$$||$$
$$\text{OH}\text{O-Ar-Ar-O}$$

Aligned polyvinyl alcohol chains

orientated, biphenyl type free radical

crosslinked polyvinyl alcohol network

The Kodak photoresist[132] is based upon the polyvinyl cinnamate system and consequently the preparation of printing plates follows a somewhat different procedure to that used for dichromated colloids. The plates must be thoroughly dried prior to coating owing to the water insolubility of the Kodak resist. Development is carried out with an organic developer such as trichloroethylene. As the solvent-swollen resist is very delicate a vapour degreaser can be used, eliminating the swabbing with a cotton-wool technique.

In order to circumvent the need for using organic solvents as developers it has been attempted to incorporate free carboxyl groups into polyvinyl cinnamates, allowing aqueous alkali to be used instead.

The Kodak photo-resist is particularly suitable for printed-circuit manufacture but also finds application in some invert halftone photogravure processes and photolithographic plates. Superior adhesion to metal has resulted in the preferred use of cinnamic esters of particular epoxy resins for platemaking.

Polyvinyl cinnamate is used to offset some of the disadvantages inherent in the dichromated colloid system, as described in section 6.1.1.1, which include:

1. The sensitivity varies with temperature, age, relative humidity.
2. The 'dark reaction' may insolubilize the film.
3. Irritant/toxic effect of dichromates.

Polyvinyl cinnamate is often prepared by partial esterification of polyvinyl alcohol with cinnamoyl chloride.

Two crystalline forms of cinnamic acid exist, α and β, differing in orientation of the molecules on the crystal lattice. Actinic radiation of certain wavelengths causes dimerization to give α-truxillic or β-truxinic (*isotruxillic acid*) which are acid derivatives of cyclobutane. This system has been described in section 6.2.2.1[129] and is expanded in further detail here.[133–135]

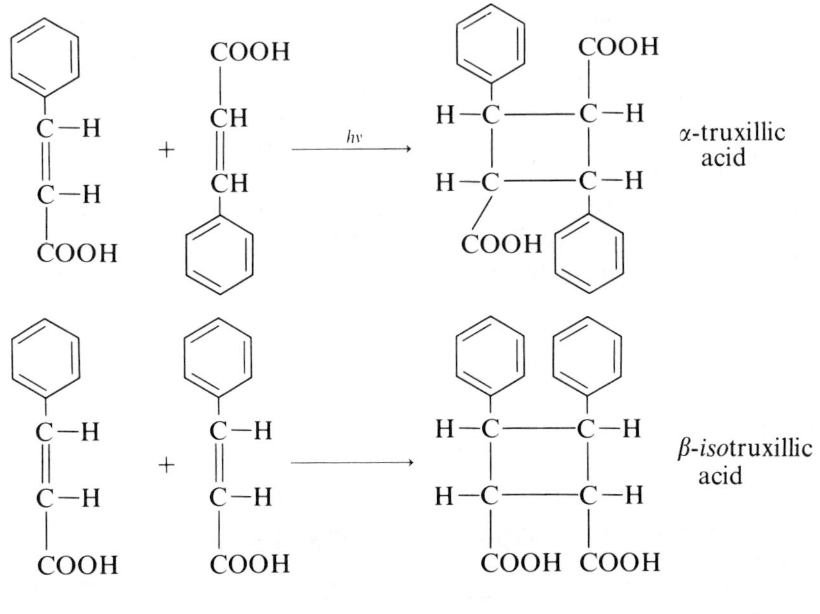

Exposed and developed polyvinyl cinnamate may be hardened by heating to 200 °C; above 250 °C elimination of cinnamoyl groups occurs.

The longest wavelength of light that will photosensitize the reaction is about 483 nm. It has been found by Nakamura, Sakata, and Kikuchi[143] that it needs only about 1.3 cinnamate groups per polymer molecule to react to insolubilize the polymer.

Polyvinyl cinnamate is only weakly absorbent above 320 nm but sensitizers may be used absorbing in the 350 nm region.

6.3.1.3 Photogravure[23,73,136]

In this intaglio process, ink is transferred from recesses in a metal plate or cylinder to the surface to be printed upon. In contrast to the halftone processes previously described, depth of colour variations are produced by varying the recess depth and consequently the ink quantity transferred to the printed surface. Both the depth and the size of the recesses may be varied depending upon the printing tone required.

Preparation of the photoresist is from 'carbon tissue' or 'pigment paper'. Gelatin or polyvinyl alcohol/dichromate may be used. Sensitization of the paper is performed by soaking it in a dichromate solution and then glazing with gelatin in contact with a glass sheet or ferrotype plate. A calcium chloride dessicant is backed to the paper by a blanket to aid drying. The paper detaches on drying and is then subjected to exposure underneath a screen comprising of two sets of transparent parallel lines arranged orthogonally, the enclosed areas being opaque. This creates a mesh-like pattern or hardened gelatin enclosing small squares of unhardened gelatin, corresponding to the cells which will finally constitute the printing areas. A positive transparency is then used to expose the paper so that every square is hardened to a depth dependent on the incident actinic radiation intensity. The paper is then transferred to a copper plate or cylinder with the metal against the gelatin. The image is then aqueously developed after the removal of the backing paper, the water removing the unhardened gelatin. A photoresist remains, consisting of squares of varying depths partitioned by walls of completely hardened gelatin. Ferric chloride is used as an etch to prevent gas production as this would tend to lift the resist. The greatest depth of etch occurs in the least hardened areas and also the converse is true, giving shadow and highlight regions. Prior to printing the resist is removed.

6.3.2 Photoengraving[73,137]

This is the process by which the formation of blocks for illustrations are made. As the quantity of ink transferred from a unit area of printing surface in letterpress work cannot be varied, the illustrative material has to use either:

(a) line drawings – lines of differing thickness or crosshatching represent various tones; or

(b) photographs reproduced as halftone blocks.

For halftone work, the continuous tones of the original photograph are transformed to a uniform pattern of varying sized dots by either:

(i) a 'crossline screen' which is placed a short, critical distance in front of the negative; or
(ii) a 'contact screen' – this has the chessboard pattern of varying optical density across the squares placed in contact with the negative prior to exposure.

Clean copper, zinc, or magnesium plates are used, with the fresh photosensitive coating applied over the surface which on drying is about 1–4 μm thick. The plate is then put under a negative in a vacuum printing frame and exposed to actinic radiation from a carbon arc lamp. The plate is washed after exposure in running water and the unexposed film areas removed with cotton-wool. The acid resistance of the stencil may be augmented by a post-treatment such as fixing with a solution of dilute chromic acid prior to 'burning in'. The reverse side of the plate is coated with an acid-resistant lacquer such as an alcoholic shellac solution. The plate is then electrolytically or chemically etched. Zinc or magnesium plates are chemically etched with dilute nitric acid whereas copper plates are etched with ferric chloride solution. It is necessary to prevent undercutting of the lines and dots by lateral etching. This is achieved by, from time to time, taking the plate out of the etch bath, drying it, and brushing powdered 'dragon's blood' resin onto the sides of the dots and then heating to fuse the resin and then recommence the etch process. The 'Dow-etch' process overcomes this by emulsifying diethylbenzene into the etch solution. The solution is sprayed onto the plate and the hydrocarbon covers the sides of the dots. Electrolytic etching is often carried out with copper plates, where they are rendered as the anode in a cell with an iron cathode. A solution of ammonium and sodium chlorides is used as an electrolyte. The current density may be monitored to control any lateral etching.

6.3.3 Silkscreen Printing[23,73,138,139]

Stencils[140] for screen printing may be made independently of the screen or on the screen itself. Stencils may be produced photographically by impregnating the screen with dichromated gelatin and exposing it to actinic radiation by contact printing under a line or halftone positive, to reproduce fine and complicated subjects. The image is developed by removing the gelatin from the unexposed areas with warm water.

Two chief photosensitizers are used for these types of emulsions:

1. Alkali metal and ammonium dichromates.
2. Diazo compounds.

Early binders involved the use of gelatin but currently, combinations of polyvinyl alcohol and polyvinyl acetate are among those used owing to their superiority in heat resistance compared to gelatin.

There are three chief kinds of photostencil[139] used in practice. These are:
1. *Indirect photostencil.*
2. *Direct photostencil.*
3. *Direct/indirect photostencil.*

6.3.3.1 Indirect Photostencil

This consists of a gelatin or synthetic material coated onto a paper or film base. Presensitization of the emulsion is carried out during manufacture or prior to the stencil production. These photostencils are called indirect as processing is always performed before the film is transferred to the mesh. There are many variations in processing techniques. A typical base film would be a polyester coated with a presensitized gelatin emulsion. Exposure to visible or near-UV light occurs through a positive in intimate contact with the photostencil, and is hardened chemically. Warm water is used to wash away the emulsion from the unexposed areas and the stencil then mounted onto the mesh. The base support is peeled off after drying.

Advantages of the indirect system are:
(a) handling flexibility;
(b) high definition without ink spread as the stencil film adheres to the screens underside thus guaranteeing intimate contact with the printing stock.
(c) Print runs are feasible up to about 5000.

6.3.3.2 Direct Photostencil

For this method, processing is performed directly onto the screen. The light-sensitive liquid emulsion is coated onto both sides of the mesh with a trough. Generally two or more coats are applied to embed the mesh in the photoemulsion. The whole screen is, after drying with a warm air fan or at ambient temperature, exposed to actinic radiation through a positive kept in contact with the screen in a vacuum frame. The unexposed areas of the stencil are then washed out with warm water and the screen then dried.

This method allows durable, long-run stencils to be made but they have the disadvantage of limitation in quality. This is owing to the fact that the emulsion does not readily form a flat surface across the individual mesh filaments (as it would do if perfectly coated) but tends to follow on drying the contour of the mesh, resulting in ink spread owing to inadequate substrate contact.

Sawtoothing is a problem often encountered, owing to the fact that emulsions often cause the stencil to contract and shrink during drying along the edge of the printed image.

6.3.3.3 The Direct/Indirect Photostencil

This is an amalgamation of the two previous systems, retaining some of the best features of both of them. A machine-coated emulsion film on a polyester base support is put into contact with the mesh on the stencil side. A sensitized emulsion is then squeegeed across the inside of the mesh, laminating the film to the screen and at the same time sensitizing it. The base support is peeled off after drying and the screen treated for processing similarly as the direct photostencil.

Excellent contact with the printing stock is guaranteed by the even surface of the film and allows a resolution akin to an indirect stencil. The sandwiching of the mesh between the film and emulsion gives a stencil life of the same older as that of the direct photostencil.

The quality and life of the stencil is determined to a large extent by the nature of the exposure. Some important parameters are adhesion and thickness, and definition.

1. Adhesion of the stencil needs a fine layer of soft (or undercured) emulsion on its surface to promote good contact with the mesh. Film thickness is therefore an important parameter. Adhesion and thickness are influenced by two prime factors:

    (a) Exposure level
       (i) intensity of light source
       (ii) exposure time
       (iii) light source/photostencil distance (inverse square law). Overexposure results in reduced adhesion and brittleness, likely to cause cracking when printing occurs.

    (b) The light source spectrum must match the absorption characteristics of the photostencil material otherwise excessive exposure times needed or no cure occurs.

2. The definition, that is, the positive image of an ideal photostencil, should be reproduced exactly under ideal conditions. The definition is affected by several factors which include:

    (a) Photostencil material resolution. This has a finite resolving power owing to its nature of construction, i.e. pigment colour and particle size, sensitizer, emulsion composition, etc.

    (b) Level of exposure. Overexposure will result in light spread within the photostencil material, causing the 'close-in' of detail.

    (c) Geometry of light source
       (i) size of irradiation source
       (ii) irradiation source/photostencil material distance
       (iii) reflector shape.
    Actinic radiation passing through the positive and dried photoemulsion should be parallel or a change will occur in the magnitude of the reproduced image in the photostencil. It can become either larger or smaller due to this undercutting.

    (d) Contact between positive and photosensitive material – if this is poor, light undercutting can result, giving loss in definition.

6.3.4 Printed Circuits[73,141]

These significantly reduce the time factor necessary to wire electronic apparatus. The circuit design ensures that contacts between the components may be created by a metal coating on an insulating base, covered with a thin copper layer. Application of a photoresist is then either by:

(a) photoengraving; or
(b) screen printing polyvinyl alcohol/dichromate or any usable photosensitive system onto the copper. The resist is hardened and the copper regions not forming part of the circuit are treated by etching, with aqueous ferric chloride or peroxodisulphate solution.

6.3.5 Collotype Printing[23,73,142]

The use of halftones is dispensed with in this process. Collotype approaches continuous tone reproduction more than any of the other printing methods.

Collotype is similar to photolithography as the image areas are made by photocuring dichromated gelatin, with the exception that it has a non-planar printing surface, covered with microscopic reticulations.

6.3.6 Proofing Systems[73]

Photoresist technology now allows an approximation in the graphic arts industry to the final full-colour print to be made with the advent of proofing systems.

Coatings of photoresists on polyester clear film in the three primary colours, yellow, magenta, cyan, and also in black can be contacted with colour-separation negatives. These plastic-supported resists after exposure are developed similar to offset plates and on superimposition upon each other in close registration, often give a good approximation to the final full-colour work, allowing customer approval prior to the press run.

Also, the transfer-type system exists and the Dupont Cromalin TM system, where superimposed photopolymer layers are dusted with various toners. Each layer is exposed through a colour-separation negative. These exposed areas are therefore photoconverted to non-tackiness and the unexposed areas remain tacky. The latter therefore accepts the powdered pigment toner. The final proofs are bright and clear with no adverse effects from the polyester layers. Superimposition of the exposed separations occurs in the transfer system.

6.4 REFERENCES

1. Hepher, M., *The Journal of Photographic Science*, **12**, 181–190, (1964).
2. De Forest, W. S., *Photoresist: Materials and Processes*, McGraw-Hill Book Co., New York, (1975).
3. Delzenne, G. A., 'Photoresists', in *Encyclopaedia of Polymer Science and Technology*, Suppl. 1, Interscience, New York, (1976).

4. Thompson, L. F., and Kerwin, R. E., 'Polymer Resist Systems for Photo and Electron Lithography', *Ann. Rev. Mat. Sci.*, **6**, 267, Annual Reviews Inc., Palo Alto, California, (1976).
5. Bryce-Smith, D., *Photochemistry: A specialists report*, V. **1–6**, The Chemical Society, Burlington House, London, (1970–75).
6. Loening, E. E., *Process Engravers' Monthly*, 266–268, 297–301, (1950).
7. Scot Mungo Ponton, *New Phil. J.*, 169, (1839).
8. Becquerel, E., *Compt. rend*, **10**, 469, (1840).
9. Fox Talbot, *Brit. Pat.*, 565, (1852).
10. Duncalf, B., and Dunn, A. S., *Printing Technology*, 125–133, (Dec. 1970).
11. Shuttleworth, S. G., *J. Soc. Leather Trades Chemists*, **34**, 410, (1950).
12. Eder, J. M., *Ausführliches Handbuch der Photographie.*, IV, **4**, Wilhelm Knapp, Halle, (1929).
13. Lumiere, A. L., and Seyewitz, A., *Bull. Soc. Chim.*, **3**, 1040, (1905).
14. Wensley, G. C., *Process*, **65**, 15, (1958).
15. Howard, H. C., and Wensley, G. C., *Ptg. Tech.*, **4**, 9, (1960).
16. Stiehler, H., *Internat. Bull. for Ptg., and Allied Trades*, No. 73, 12, (1956).
17. Mees, C. E. K., *Theory of the Photographic Process*, (2nd edn), Macmillan, p. 74, (1954).
18. Davies, W. G., and Prue, J. E., *Trans. Farad. Soc.*, **51**, 1045, (1955).
19. Smethurst, P. C., *Proc. Eng. Monthly*, 198, 229, 254, (1946).
20. Kläning, U. K., and Symons, M. C. R., *J. Chem. Soc.*, 977, (1960).
21. Kläning, U. K., *Acta Chem. Scand.*, **12**, 576, 807, (1958).
22. Wiberg, K. B., 'Oxidation by chromic acid and chromyl compounds', in *Oxidation in Organic Chemistry* (Ed. K. B. Wiberg), Academic Press, New York, Ch. 2, (1965).
23. Finch, C. A., *Polyvinyl Alcohol, properties and applications*, John Wiley & Sons Ltd, (1973).
24. Driscoll, G. L., Ph.D. Thesis, University of Delaware, 1968 (*Dissertation Abs.* B, **29**, 1606 (1968)).
25. Hasseberger, F. X., Ph.D. Thesis, University of Delaware, 1969 (*Dissertation Abs.* B, **30**, 3095 (1970)).
26. Plotnikov, I., *Chem. Zt.*, **52**, 669, (1928).
27. Holloway, F., Cohen, M., and Westheimer, F. M., *J. Amer. Chem. Soc.*, **73**, 65, (1951).
28. Schläpfer, K., *Schweiz. Archiv. Angew. Wiss. Tech.*, **31**, 154, (1965).
29. Schläpfer, K., *Adv. Print. Sci. Technol.*, **6**, 1, (1971).
30. Elöd, E., and Schachowrsky, T., *Koll. Zeit.*, **72**, 69, (1935).
31. Trudelle, Y., and Neel, J., *Bull. Soc. Chim. France*, 223, (1969).
32. Maillet, G., *Bull. Soc. Franc. Phot.*, **22**, 202–204, (1935).
33. Dunn, A. S., Coley, R. L., and Duncalf, B., 'Thermal decomposition of Polyvinyl Alcohol', in *Properties and Applications of Polyvinyl Alcohol* (Ed. C. A. Finch), Monograph No. 30, Society of Chemical Industry, London, p. 208, (1968).
34. Loening, E. E., *Process Engravers' Monthly*, 209–210, (1948).
35. Brit. Pat. 737,767.
36. Brit. Pat. 738,958.
37. Hepher, M., and Loening, E. E., *Penrose Annual*, **48**, 139, (1954).
38. Oster, G. K., and Oster, G., *J. Appl. Polymer Sci.*, **48**, 321–327, (1960).
39. Brit. Pat. 764,380.
40. U.S. Pat. 2,100,063.
41. U.S. Pat. 2,099,404.
42. Brit. Pat. 296,008.
43. Brit. Pat. 418,011.
44. Jorgensen, G., *Tech. Ass. Graphic Arts, Proc. of 4th Ann. Meeting*, 97–104, (1952).
45. French Pat. 904,255.
46. Jorgensen, G., and Powers, A. J., *Tech. Ass. Graphic Arts, Proc. 4th Ann. Tech. Meeting*, 103, (1952).

47. Hepher, M., *Penrose Annual*, **49**, 122–123, (1950).
48. U.S. Pat. 1,574,356–359.
49. U.S. Pat. 1,751,908.
50. U.S. Pat. 1,587,269–273.
51. Davis, R., and Pope, C. I., 'Technique for Ruling and Etching Precise Scales on Glass and their Reproduction by Photoetching with a New Light Sensitive Photoresist', *National Bureau of Standards Circular 565*, U.S. Dept. of Commerce, (Aug. 1955).
52. Cohen, M. D., Schmidt, G. M. J., and Sonntag, F. J., *J. Chem. Soc.*, 2000, (1964).
53. Bertram, J., and Kursten, K., *J. Prakt. Chem.*, **51**, 324, (1895).
54. U.S. Pat. 1,965,710.
55. U.S. Pat. 2,544,905.
56. Minsk, L. M. et al., *J. Appl. Polymer Sci.*, **11**(6), 302–11, (1959).
57. Brit. Pat. 965,262.
58. Brit. Pat. 717,709.
59. Robertson, E. M., Van Deusen, W. P., and Minsk, L. M., *J. Appl. Polymer Sci.*, **2**, 308–311, (1959).
60. Brit. Pat. 846,908.
61. Brit. Pat. 822,866.
62. French Pat. 886,716.
63. Brit. Pat. 767,985.
64. Brit. Pat. 886,100.
65. Merrill, Stewart, H., and Unruh, C. C., *J. Appl. Polymer Sci.*, **7**, 273–279, (1963).
66. Brit. Pats. 825, 795, 826, 272, 834, 337, 850, 453, 875, 377–8, 835, 849. U.S. Pats. 2,927, 022-3, 2, 902, 365.
67. Süs, O. et al., *Liebigs, Ann. Chem.*, **593**, 91–126, (1955).
68. Süs, O., *Z. Wiss. Phot.*, **50**, 476–517, (1955).
69. Brit. Pat. 708,834.
70. Belg. Pats. 497, 206, 500, 222.
71. Shaw, J. M., Frisch, M. A., and Dill, F. H., IBM, *J. Res. Develop.*, **21**, 219, (1977).
72. Pacansky, J., and Johnson, D., *J. Electrochem. Soc.*, **124**, 862, (1977).
73. Kosar, J., *Light Sensitive Systems*, J. Wiley & Sons Inc., New York, N.Y., pp. 143–146, (1965).
74. Morgan, C. R., Magnotta, F., and Ketley, A. D., *J. Polym. Sci. Polym. Chem. Ed.*, **15**, 627, (1977).
75. Morgan, C. R., and Ketley, A. D., *ACS Organic Coatings and Plastic Chemistry*, **33**(1), 281, (1973).
76. Saunders, K. J., *Organic Polymer Chemistry*, Chapman & Hall, London, (1973).
77. Flory, P. J., *Principles of Polymer Chemistry*, Ch. XIII-3, Cornell University Press, Ithaca, London, (1953).
78. Y.-O Tu, and Ouano, A. C., IBM, *J. Res. Develop*, **21**, 131, (1977).
79. Feit, E. D., Kammoltt, G. W., and Wurtz, M. E., *J. Vac. Sci. Tech.*, 15, (1978).
80. Fahrenholtz, S. R., Goldrick, M. R., and Hellman, M. Y., *ACS Organic Coatings and Plastics Chemistry*, **35**(2), 306, (1975).
81. Rubner, R., and Kuhn, E., *ACS Organic Coatings and Plastics Chemistry*, **37**(2), 118, (1977).
82. Curme, H. G., Natale, C. C., and Kelly, D. J., *J. Phys. Chem.*, **71**, 767, (1967).
83. Agnihotri, R. K., Hood, F. P., Lesoine, L. G., and Offenbach, J. A., *Photo. Sci. Eng.*, **15**, 141, (1971).
84. Tsuda, M., *J. Polym. Sci.*, **A2**, 2907, (1964).
85. Tanaka, H., Tsuda, M., and Nakanishi, H., *J. Polym. Sci.*, **A1**(10), 1729, (1972).
86. Shankoff, T. A., and Trozzoto, A. M., *Photo. Sci. Eng.*, **19**, 173 (1975).
87. Tsuda, M., *J. Polym. Sci.*, **A2**, 2907, (1964).
88. Stuber, F. A., Ulrich, H., Rao, D. V., and Sayigh, A. A. R., *J. Appl. Polym. Sci.*, **13**, 2217, (1969).
89. De Schryver, F. C., Boens, N., and Smets, G., *J. Polym. Sci.*, **A1**(8), 1939, (1970).

90. Hay, A. S., Bolon, D. A., and Leimer, K. R., *J. Polym. Sci.*, **A1**(8), 1022, (1970).
91. Shimuzu, S., and Bird, G. R., *J. Electrochem. Soc.*, **124**, 1394, (1977).
92. Bokov, Y. S., Korsakov, V. S., Kalyzhnaya, V. G., and Lavrishev, V. P., *Polym., Sci. USSR*, **18**, 74, (1976).
93. Harita, Y., Ichikawa, M., Harada, K., and Tsunoda, T., *Polym. Eng. Sci.*, **17**, 372, (1977).
94. Jinno, K., Matsumoto, Y., and Shinozaki, T., *Photo. Sci. Eng.*, **21**, 290, (1977).
95. Merrill, S. H., and Unruh, C. C., *J. Appl. Polym. Sci.*, **7**, 273, (1963).
96. Takeishi, M., and Okawara, M., *J. Polym. Sci. Polym. Lett.*, **7**, 201, (1969).
97. Ulrich, H., Staber, F. A., Peters, G. M. Jr., and Sayigh, A. A. R., *J. Polym. Sci. Polym. Chem. Ed.*, **14**, 565, (1976).
98. Stuber, F. A., Ulrich, H., Rao, D. V., and Sayigh, A. A. R., *Photo. Sci., Eng.*, **17**, 446, (1973).
99. Schild Knecht, C. E., *Allyl Compounds and their Polymers*, Ch. 11, Wiley-Interscience, New York, N.Y., (1973).
100. Gilano, M. N., and Lipson, M. A., *Tech. Reg. Conf. Soc. Plastic Eng.*, Mid Hudson Section, Society of Plastic Engineers, p. 30–32, (1970).
101. Kehr, C. L., and Wazolek, W. R., *ACS, Organic Coatings and Plastics Chemistry*, **33**(2), 295, (1973).
102. Morgan, C. R., and Ketley, A. D., *ACS Organic Coatings and Plastics Chemistry*, **33**(1), 281, (1973).
103. Morgan, C. R., and Ketley, A. D., *ACS Polymer Preprints*, **17**(2), 500, (1976).
104. Celeste, J. R., U.S. Patent 3,469,982, (1969).
105. Schoenthaler, A. C., U.S. Patent 3,418,295, (1968).
106. Labana, S. S., and McLanghlin, E. O., *J. Elastoplastic*, **2**, 3, (1970).
107. Nordstrom, J. D., and Hinsch, J. E., *ACS Organic Coatings and Plastics Chemistry*, **29**(1), 160, (1969).
108. Burlant, W. J., and Taylor, C. R., U.S. Patent, 3,528,844, (1970).
109. Bartelt, J. L., and Feit, E. D., *J. Electrochem. Soc.*, **122**, 541, (1975).
110. Schlesinger, S. I., *Photo. Sci. Eng.*, **18**, 389, (1971); *Polym. Eng. Sci.*, **14**, 513, (1974).
111. Tatsuo, Warashina, and Tsunetoshi Kai (Asahi Chemical Ind. Co. Ltd.), *Japan Plastics Age*, 19–24, (July 1972).
112. Cannon, R. V., *Printing Technology*, **17**(4), 10–15, (1973).
113. *Printing World*, pp. 166, 180–181, (5 Sept. 1974).
114. Japanese Pat. Announced No. 46-29,525.
115. Japanese Pat. Announced No. 43-19,125.
116. US Pat. 3, 168, 404, 3, 252, 800.
117. Japanese Pat. Announced No. 45-7330.
118. Belg. Pat. 686241.
119. Hartsuch, P. J., *Proc. Ann. Tech. Meet. Tech. Ass. Graphic Arts*, **2**, 140, (1950).
120. Adams, R. A. C., U.S. Pat. 3,073,765, (1963).
121. Algraphy Ltd., and Adams, R. A. C., Brit. Pat. 879,767, Brit. Pat. 879,768, (1961).
122. Wood, W. H., U.S. Pat. 2,760,432, (1956).
123. Wood, W. H., U.S. Pat. 2,297,932, (1942).
124. Wood, W. H., U.S. Pat. 2,270,712, (1942).
125. Mertle, J. S., *National Lithographer*, **54**, July–Dec. 1947; **55**, June–May 1948.
126. Tyrell, A., *Basics of Reprography*, Focal Press, London and New York, (1972).
127. Askew, F. A., *Printing Ink Manual*, W. Heffer & Sons Ltd., Cambridge, (1969).
128. Apps, E. A., *Inks for the Major Printing Processes*, Part II, Leonard Hill, London, (1963).
129. Smith, A. H., *Printing Technology*, **11**(1), (April 1967).
130. Tsunoda, T., and Yamoaka, T., *J. Appl. Polymer, Sci.*, **8**, 1379, (1964).
131. Tsunoda, T., *Graphic Arts, Japan*, **7**, 15, (1965).
132. U.S. Pat. 2,610,120, (1952).
133. Bernstein, H. I., and Quimby, W. C., *J. Amer. Chem. Soc.*, **65**, 1845, (1943).

134. de Jong, A. W. K., *Chem. Ber.*, **55B**, 463, (1922).
135. Schmidt, G. M. J., *Abst. Intern. Congr. Pure and Applied Chem.*, 18th Congr., 86, (1961).
136. Astrua, M., *Manual of Colour Reproduction for Printing of the Graphic Arts*, Fountain Press, England, (1973).
137. Cox, R. S., and Cannon, R. V., *Penrose Ann.*, **44**, 116, (1950).
138. Sericol Ltd., Technical Releases.
139. E. T. Marler Ltd., Technical Releases.
140. *Autotype Stencil Techniques*, Technical Release from Autotype International Ltd., Wantage, Oxfordshire, England, (1977).
141. Byer, M., *Materials & Methods*, **43**, 134–137, (1956).
142. *How Things Work – The Universal Encyclopaedia of Machines*, Paladin, Granada Publishing Ltd., p. 420, (1972).
143. Nakamura, K., Sakata, T., and Kikuchi, S., *Bull. Chem. Soc. Japan*, **41**, 1977, (1968).

7
Potential hazards of ultraviolet systems

7.1 GENERAL AREAS OF HAZARDS

An ultraviolet system has several areas of potential hazards which are listed in the following sections.

7.1.1 Lamps and machinery

These can pose a hazard in three main areas:

(a) Escaping radiation – eyes, skin
(b) Ozone gas formation – respiratory
(c) Electric currents – shock, heart failure.

Superficial eye damage and burning of the skin may arise from even a brief exposure to UV radiation, more serious consequences resulting from prolonged exposure.

The effects can be similar to that produced by overexposure to the sun. An unscreened mercury arc over several metres distance produces on exposure for a few seconds an excruciatingly painful eye condition. The symptoms resemble that of sand in the eyes with accompanying intense pain and intolerance to bright lights, also with eye watering and sometimes temporary vision losses.

The American Medical Association has recommended a maximum exposure limit of UV light intensity to individuals of 0.5 $\mu W/cm^2$ (7 hr day), or 0.1 $\mu W/cm^2$ (24 hr day).

An alternative recommended standard is one supported jointly by the American Council of Governmental Industrial Hygienists (ACGIH) and the National Institute for Occupational Safety and Health (NIOSH). This standard includes skin and eye exposures to UV radiation. These type of standards employ a relative effectiveness curve for the biological effect of UV radiation. This is shown in Figure 78. The exposed radiation's UV spectrum is measured in absolute units of energy such as microwatts per square centimetre ($\mu W/cm^2$) and appropriately weighted to the biological effectiveness curve. The ACGIH and NIOSH standards advocate for the spectral range 315–400 nm a total irradiance of 1 mW/cm² for an exposure time of greater than 1000 seconds. For less time the limit is recommended not to exceed 1000 mW-s/cm². Also, in the range 200–315 nm an exposure of less than 0.003 W-s/cm² of weighted UV energy is proposed.

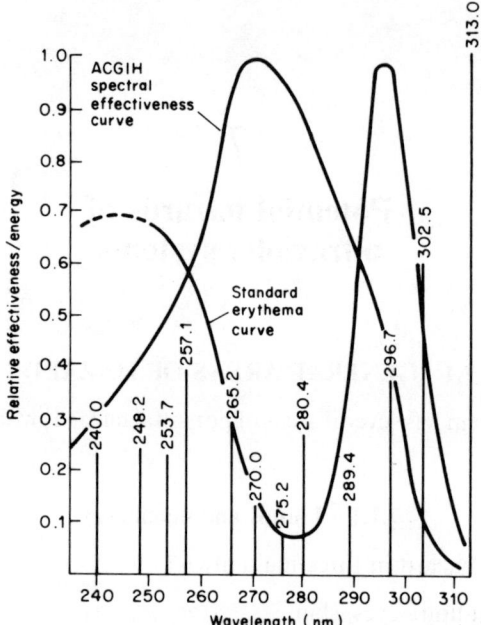

Figure 78 Ultraviolet curing radiation and biological effectiveness (Hg vapour lamp).[29]

Ozone gas has an odour threshold of 0.1 ppm in air by volume. The AIHA (American Industrial Hygiene Association) and OSHA (Department of Labour) advocate an 8-hour time-weighted average for ozone concentrations to be less than 0.1 ppm. Ozone can be fatal to humans but low doses can produce undesirable effects such as tiredness, headaches, dryness of upper respiratory tract, and many other respiratory problems. Ozone must therefore be vented from the work area.

There are many instruments available for measurement of these levels.[1]

The above are relatively simple to remedy with standard precautions. Those below are much more difficult.

7.1.2 Inks and Coatings

The two main areas for problems are:

1. Chemical reactivity – explosion, fire, etc.
2. Effect on the human body:
   (a) Dermatitis   (d) Mutagenic/carcinogenic
   (b) Respiratory   (e) Toxicity
   (c) Sensitization

In the composition of UV curing inks and coatings the chief factors of concern can be the UV photopolymerizable resins and monomers and the photo-initiator/sensitizers.

Care must be taken with some pigments such as lead chromes, which under some conditions may act as photosensitizers in the system but are nevertheless poisonous if taken internally. For photoemulsion work, dichromates must be treated with care as chromium compounds are reported by Ethel Browning[2] as being hazardous. This work emphasizes the following points regarding chromium.

The two most used trivalent chromium compounds are chromic acid, Cr_2O_3, and chromic sulphate, $Cr_2(SO_4)_3$.

Hexavalent (chromates) are the most important industrially and also the most toxic. The maximum allowable concentration is $0.1 \, mg/m^3$ for Cr_2O_3.

Administration of soluble chromium salts orally does not result in retention in the body, as they are readily excreted, unlike many other metals. Subcutaneous, intravenous, and intratracheal administration results in storage of water-soluble salts in the lungs.

Principal toxic effects from industrial usage are:

1. On the skin.
2. To the nasal mucous membrane.
3. To the lungs
4. Lesions of kidneys
5. Potential carcinogen (lung carcinoma).

Ingestion of potassium dichromate results in vomiting owing to its irritant action and absorption. Limited but central nervous disturbances result with convulsions, stupor, and dilated pupils. No evidence of nephritis occurs but possibly an enlarged liver may result.

External application of chromic acid results in poisoning and nephritis.

Dermatitis from chromates occurs commonly on hands, wrists, forearms, eyelids, neck, etc. Evidence has been available of chrome dermatitis as high as 25 per cent in lithographers. Sensitization to chromates is a serious problem in industry. Chrome ulcers are thought to be caused by acid salts of chromic acid but not by alkaline chromates. There is no evidence to suggest that chrome ulcers of the skin cause sensitization to potassium dichromate.

The chief effects from chromic acid dust are:

1. Nasal mucosa.
2. Larynx – carcinogenesis is the chief potential risk.
3. Lung carcinoma is a potential risk.

The toxicity ratings of some important chromium salts are shown below:[3]

Potassium dichromate
 Toxic dose: Oral – Humans   Least lethal dose = $430 \, mg/kg$
       Subcutaneous – Rabbits   Least lethal dose = $12 \, mg/kg$

Lead chromate (chrome yellow pigment)
 Toxic dose: Subcutaneous – Rat   Least lethal dose = $1200 \, mg/kg$
       Intraperitoneally – Guinea pigs $LD_{50} = 400 \, mg/kg$

Chromium itself has the value

Toxic dose: Intraperitoneally – Rat   Least lethal dose = 1.2 mg/kg/6wI

American legislation (Federal Register 1971),[11] gives the following TLV values for industrial air:

chromic acid + chromates (as CrO_3) 0.1 mg/m^3,
chromium, soluble chromic,
chromous salts (as Cr) 0.5 mg/m^3

7.1.2.1 Possible Hazards of Acrylic Ester Monomers

(i) Chemical reactivity

Polymerization can occur by:

1. Heat
2. Light.
3. Impurities – alkaline substances, trace metals, etc.

Polymerization is a strongly exothermic reaction and can cause thermal decomposition and/or rupture of containers. This thermal decomposition can cause the evolution of irritant vapours and gases and could, in some cases, lead to fire.

Material storage should be in a cool, shaded, well-ventilated area, away from heat sources and sunlight, and not allowed near foodstuffs, organic peroxides, or other polymerization initiators.

(ii) Toxicology

Data is very scanty. There is no positive evidence to suggest that acrylates are carcinogenic. They are, however, irritants to the skin, eyes, and respiratory passages via inhalation (misting should be minimized in litho UV curing inks as this can cause ingestion of fine particles).

Prolonged exposure to the skin to either liquid or vapour will cause burns.

Allergic responses in sensitive individuals can occur with repetitive exposure and they must be isolated from any further exposure.

(iii) Handling

All contact with the skin, eyes, clothing, and the respiratory system should be avoided.

The materials should be handled in well-ventilated areas, and protection of the eyes, face, and any other exposed skin areas is necessary. Goggles, suitable gloves, face shields (and masks) should be used. Adequate protective clothing should be worn. Polythene is probably one of the best materials. PVC gloves readily absorb acrylate esters and can, therefore, unknowingly produce a continual exposure.

Dirt contamination of these materials must be avoided, as being organic in nature, it can cause polymerization.

(iv) First aid

For accidental skin contact:

1. Remove any contaminated clothing, discard or ensure thorough laundering before re-use.
2. Wash affected areas with copious quantities of warm soapy water, not solvents, as these will carry the acrylates into the skin.
3. When burns or irritations develop, seek medical attention, where steroids will probably be administered to the wound.

Respiratory problems occurring via inhalation may manifest themselves as: chest pains; dry parched mouth and throat; dizziness; and headaches.

The patient who has inhaled these compounds should be kept warm and oxygen administered if needed. For breathing stoppage, use artificial respiration and refer to a hospital.

For fire, sand, earth, or a chemical extinguisher should be used.

There are several important tests that may be needed prior to the acceptance of UV photopolymerizable materials, which include:

1. Skin irritation.
2. Eye irritation.
3. Skin sensitization.
4. Acute oral toxicity.
5. Ames test for mutagenicity.
6. Inhalation studies.

These type of tests are carried out at several research institutes, which include in the United Kingdom:[6-10]

1. Hazleton Laboratories.
2. Inveresk Research International.
3. Toxicol Laboratories.
4. Huntingdon Research Centre.
5. Consultox.

Typical test procedures (with modification where necessary) carried out at these type of laboratories are outlined in the next section for reference purposes.[8]

7.2 TESTING FOR PHYSIOLOGICAL HAZARDS

7.2.1 Skin Tests

7.2.1.1 Draize Skin Test

Often a modified version of that laid down in the FDA Handbook, *Appraisal of the Safety of Chemicals in Foods, Drugs and Cosmetics*,[4] p. 47 is employed.

The Inveresk Research International,[7] Edinburgh, have kindly provided the following protocol for skin tests, updating many previous tests.

PRIMARY SKIN IRRITATION TEST
(24 hour exposure)

(FDA recommended method)
(21 CFR 191.11)

Introduction

This protocol is designed to determine the primary skin irritancy potential of the test material on intact and abraded skin of rabbits and to compare this, where possible, with the effects on skin of a control material.

Animals and Management

Six New Zealand rabbits (3 males and 3 females) will be used weighing between 2.5–3.0 kg. They will be housed individually in cages with a grid floor beneath which is a peat moss tray. They will be fed on Spratts Rabbit Diet (Spillers) and allowed food and water *ad libitum*. Room temperature will be a target 15 °C $\pm$ 2 °C and humidity will be recorded.

The hair will be clipped from the backs and flanks of all animals and 2 of the 4 test areas on the back of each animal will be abraded.

Test Material

Should the test material be a solid, it will be moistened with physiological saline before application. A control material will be included with which to compare the test material.

TEST METHOD

0.5 ml or 0.5 g of the test substance moistened with physiological saline will be applied under 2.5 × 2.5 cm patches of chromatography paper to intact and abraded skin on each animal.

The patches will be covered with Blenderm tape and the whole trunk will be bound with Sleep occlusive tape which will remain in position for 24 hours. At the end of this period, the patches will be removed and the skin wiped (not washed) to remove any test substance remaining.

Skin reactions will be scored according to the following recommended system, immediately after patch removal and again 48 hours later to give the required 24 hour and 72 hour readings.

SCORING SYSTEM

Erythema and eschar formation	Grade
Very slight erythema (barely perceptible)	1
Well defined erythema	2
Moderate to severe erythema	3
Severe erythema (beet redness) to slight eschar formation (injuries in depth)	4

Oedema formation	Grade
Very slight oedema (barely perceptible)	1
Slight oedema (edges of area well defined by definite raising)	2
Moderate oedema (area raised approximately 1 mm)	3

Severe oedema (raised by more than 1 mm and extending beyond the area of exposure)

The reaction score is the average value of the 6 test animals.

The primary irritation score is obtained in the following way:
The values of erythema and eschar formation for intact and abraded skin at 24 and 72 hours (4 values) are added to the equivalent values for oedema formation for intact and abraded skin at 24 and 72 hours (4 values) and the resulting figure is divided by 4.

A material which has a primary irritation score of 5 or more is a primary irritant under the definitions of the FDA recommended system.

It is intended that this study will be performed to meet the IRI Code of Good Laboratory Practice and the standards laid down by the USA Food and Drug Administration on 22 December 1978.

7.2.1.2 Repeat Insult Skin Test[10]

This test is used to determine the local irritation produced by repeated administration of materials intended for topical application.

Four New Zealand white rabbits are restrained. The hair is removed from their backs by close clipping and a depilatory cream. Twenty-four hours after this treatment, the animals are again restrained and the backs examined to ensure freedom from blemishes likely to affect the test.

A 0.5 ml sample is applied to the shaved back of each of the rabbits, which are then restrained for six hours. This procedure is repeated on five successive days, the assessment of any skin reaction being made 24 hours after each application.

Skin irritancy of UV resins and acrylate monomers is commonly assessed by the 'Draize' test which has been established for the testing of cosmetic goods and related applications. Values of 0–2 are considered mild irritants, 2–5 moderately irritant, and 6–6+ are severe irritants. The test has several possible limitations as it is a single application test and does not relate to conditions of repeated or prolonged exposure, or supply data as to whether the test materials are skin sensitizers. Variations of results occur with biological systems as they are dependent on such factors as strain, weight, age, and sex of the test animals plus environmental factors such as the time of the year. Often a comparative test is included on a material of known effect alongside any new materials. Minor variations in the manufacture or compositions of the materials under test can have definite effects on the values obtained on the 'Draize' test.

Owing to these factors, it is unlikely that Draize numbers have any significant objectivity and it is perhaps better to indicate whether materials fall into the classifications of mild, moderate, or severe irritants. It should also be borne in mind that there is not a sharp transition between these classifications. Variations will occur amongst different batches of material from the same source and amongst the same material from alternate sources.

Some examples of monomers with possible irritant classifications are shown below:

1. *Mild irritants*
   (a) Pentaerythritol tetra-acrylate
   (b) *Iso*-bornyl acrylate

2. *Moderate irritants*
   (a) Trimethylol propane triacrylate
   (b) 1,6 Hexanediol diacrylate
   (c) 2-phenoxyethyl acrylate
   (d) Tripropylene glycol diacrylate
   (e) Diethylene glycol diacrylate
3. *Severe irritants*
   (a) Neopentyl glycol diacrylate
   (b) 1,4-Butanediol diacrylate
   (c) 2-ethyl hexyl acrylate
   (d) Hydroxyethyl acrylate
   (e) Hydroxypropyl acrylate

7.2.2 Eye Tests

7.2.2.1 Draize Eye Test[10]

This test is a modified version of that laid down in the FDA Handbook.[4]

The Inversk Research International,[7] Edinburgh, have kindly provided the following updated protocol for eye tests according to the standards laid down by the USA Food & Drugs Administration on 22 December 1978.

EYE IRRITATION IN RABBITS

(FDA recommended method)
(21 CFR 191.12)

Introduction

This protocol is designed to determine the irritancy potential to rabbit eyes of a test material following one instillation only. No rinsing will be carried out following treatment.

Animals

Six New Zealand White rabbits (3 males and 3 females) weighing 2.5–3 kg will be used. They will be acclimatized for at least one week before the test. They will be housed separately in a room at a target temperature of 15 °C $\pm$ 2 °C and recorded humidity. They will be fed on Spratts Rabbit Diet (Spillers) and allowed food and water *ad libitum*.

Both eyes will be examined to ensure the absence of defects or irritation, 24 hours before testing. Should damage be suspected, this will be confirmed by fluorescein staining. Fluorescein sodium ophthalmic solution (BP) 2 per cent will be applied as one drop directly onto the cornea. After 10 seconds, the excess fluorescein will be washed away with sterile saline (BP) at body temperature; injured areas of the cornea will appear yellow under ultraviolet illumination in a darkened room.

Test Material

The test material will be tested as supplied or as requested by the Client.

The quantity of material applied will be 0.1 ml (liquids) or if powdered, the maximum practical amount that can be instilled into the eye – up to a maximum of 100 mg; the weight of material instilled will be recorded.

The rabbit will be held firmly but gently and the test material placed into one eye by gently pulling the lower eyelid away from the eyeball to form a sac into which the material will be dropped. The lids are then gently held together for 1 or 2 seconds. The other eye, remaining untreated, will serve as a control.

A note will be made of any abnormal behaviour of the rabbits which indicates a sting effect.

The eyes will be examined for irritation using standard illumination. The ocular reaction will be recorded at 1 and 24 hours and at 2, 3 and 7 days after treatment.

Observation of suspected corneal damage will be facilitated by applying fluorescein dye directly to the cornea.

Ocular reactions will be assessed numerically using the scoring system as detailed below.

Scoring System

Cornea	*Grade*
No ulceration or opacity	0
Scattered or diffuse areas of opacity, details of iris clearly visible	(1)*
Easily discernible translucent areas of opacity, details of iris slightly obscured	2
Nacreous areas of opacity, no details of iris visible, size of pupil barely discernible	3
Complete corneal opacity, iris not discernible	4
Ulceration, absence of a gross patch of corneal epithelium	4

Iris	*Grade*
Normal	0
Markedly, deepened folds, congestion, swelling, moderate circumcorneal injection (any of these or combination of any thereof), iris still reacting to light (sluggish reaction is positive)	(1)*
No reaction to light, haemorrhage, gross destruction (any or all of these)	2

Conjunctivae

Redness (refers to palpebral and bulbar conjunctivae excluding cornea and iris)

Vessels normal	0
Some vessels definitely injected	1
Diffuse crimson red, individual vessels not easily discernible	(2)*
Diffuse beefy red	3

Chemosis

No swelling	0
Any swelling above normal (including nictitating membrane)	1
Obvious swelling with partial eversion of lids	(2)*
Swelling with lids about half closed	3
Swelling with lids more than half closed	4
Ulceration or necrosis of palpebral and bulbar conjunctivae or nictitating membrane	4

Discharge	*Grade*
Mild	1
Moderate	2
Severe	3

* Bracketed figures indicate the lowest grades considered positive under section 19-12 of the Federal Hazardous Substances Labelling Act Regulations of the USA.

Federal Hazardous Substances Act Regulations Quote

'An animal shall be considered as exhibiting a positive reaction if the test substance produces at any of the readings ulceration of the cornea (other than fine stippling), or opacity of the cornea (other than slight dulling of the normal lustre), or inflammation of the iris (other than slight deepening of the folds, or slight circumcorneal injection of the blood vessels) or if such substances produce in the conjunctivae (excluding the cornea and iris) an obvious swelling with partial eversion of the lids, or a diffuse crimson red with individual vessels not discernible.'

Interpretation of the Test Results

The test shall be considered positive if 4 or more of the animals in the test groups exhibit a positive reaction as defined above. If only one animal exhibits a positive reaction, the test shall be regarded as negative. If 2 or 3 animals exhibit a positive reaction, the test should be repeated using a different group of 6 animals. The second test shall be considered positive if 3 or more of the animals exhibit a positive reaction. If only one or 2 in the second test exhibit a positive reaction the test shall be repeated with a different group of 6 animals. Should a third test be needed the substance will be regarded as an irritant if any animal exhibits a positive response (21 CFR 191.12).

7.2.2.2 Eye Test – FHSLA (1964)[10]

This test is performed according to the method outlined in the US Federal Register.[12]

7.2.3 Sensitization Tests

7.2.3.1 Draize/Landsteiner Test

This method is detailed in the FDA Handbook *Appraisal of the Safety of Chemicals in Foods, Drugs and Cosmetics*,[4] and is based upon the method devised by Landsteiner and Jacobs.[13]

Six albino male guinea pigs are used for each test material. Hair is removed from the back and flanks by close clipping. A series of ten intradermal injections is made into this area on alternate days and the animals are challenged with a similar injection two weeks after the last priming injection. A comparison of the reactions elicited by the priming exposure to those of the challenge exposure enables an assessment to be made as to whether the material has produced sensitization.

7.2.3.2 Occluded Patch Test[10]

The method is based upon that described by Buehler.[14]

Six albino male guinea pigs are used and prepared as described above. The test material is dissolved or suspended in a non-irritating vehicle at the highest concentration found to be essentially non-irritant. This is then applied on a gauze pad which is further occluded by a plastic overlap held in place with adhesive tape. One patch is placed on each animal and left in place for six hours. Insult schedules of thrice-weekly applications for three weeks are used, with a challenge application made two weeks after the last priming exposure. Assessment of sensitization is made by comparison of reactions elicited by challenge application with reactions elicited throughout the induction period.

7.2.3.3 Magnusson and Kligman Maximization Test[10]

This method is based upon that described by Magnusson and Kligman,[15] and consists of a two-stage induction procedure whereby the material is administered by intradermal injection followed by topical application.

Ten albino male guinea pigs are used for each test material. An area on the shoulder region approximately 4 cm × 6 cm is clipped free from hair. Three intradermal injections, comprising the test material, the test material with Freund's adjuvant, and Freund's adjuvant alone, are made into this region. One week later the same area is shaved and the test material applied to the sites of injection under an occlusive patch which is left in place for 48 hours.

The challenge application is made two weeks later using a sub-irritating concentration of the test material applied under an occlusive patch to a shaved area of the flank of each material. Twenty-four hours after removal of the patch the challenge site is evaluated for evidence of a skin reaction attributable to sensitization.

7.2.4 Toxicity

7.2.4.1 Acute Toxicity – LD_{50} Determination[10]

The purpose of this test is to define the acute toxicity of the test material using a two-stage assay which is usually performed on mice or rats. Initially a rangefinding determination is carried out using small groups of animals to establish the order of acute toxicity. This is followed by either a more comprehensive study using larger numbers of animals in conjunction with a smaller range of dose levels in order to determine the median lethal dose (LD_{50}) more precisely or a gross toxicity test to confirm that the least toxic dose is greater than 5 g or 5 ml/kg (this value is generally considered to be the upper limit used to differentiate toxic and non-toxic substances in these species).

Rangefinder

Animals	Young adult albino males or females of a uniform outbred strain with known characteristics fasted overnight.
Treatment	Pairs of males or females are dosed at a number of levels ranging from 0.5–5.0 g or ml/kg administered by the appropriate route (any of the conventional routes can be used e.g. oral, dermal, intradermal, intravenous, intramuscular, intraperitoneal, subcutaneous, etc.).
Observations	All animals are observed frequently for signs of toxicity and mortality over a period of seven days.

LD_{50} examination

Animals
: Young adult albino males or females of a uniform strain with known characteristics, fasted overnight.

Treatment
: Five males and five females are dosed at five levels (determined by the results from Stage 1) administered by the appropriate route.

Observations
: All animals are daily observed for signs of toxicity and mortality rate over a period of 14 days. Animals will be weighed prior to dosing and then at 3, 7, and 14 days.
Gross post-mortem examinations are made on selected animals dying throughout this period in order to establish the cause of death if possible.

Statistical analysis

Analysis of the figures by the method of Litchfield and Wilcoxon[16] to give the LD_{50} value together with 95 per cent confidence limits.

The oral toxicity of the materials can be rated according to Table 17.[7]

Table 17 Oral toxicity values

Toxicity rating	Commonly used term	LD_{50} single oral dose rats	Possible lethal dose for man (g or ml)
1	Extremely toxic	1 mg or less/kg	<0.07
2	Highly toxic	1–50 mg	0.07–3.5
3	Moderately toxic	50–500 mg	3.5–35
4	Slightly toxic	0.5–5 g	35–350
5	Practically non-toxic	5–15 g	350–1050
6	Relatively harmless	15 g +	>1050

The Inversk Research International,[7] Edinburgh, have an upgraded procedure in the following protocol, according to the standards laid down by the USA Food and Drugs Administration on 22 December 1978.

DETERMINATION OF ACUTE ORAL TOXICITY IN RATS (MALE AND FEMALE)

Introduction

This protocol is designed to determine the oral median lethal dose (LD_{50}) of the test material and to record any clinical signs induced by administration of a single dose to young adult male and female rats. The route of administration will be orally by gavage.

Animals and Management

Male and female rats of the Sprague Dawley strain, in the weight range 150–170 g will be used. Rats for dose ranging will be ordered to arrive at 125–150 g body weight while main study rats will be ordered to arrive at 100–120 g body weight.

The rats will be housed in groups of 5 according to sex in suspended plastic cages in a room at a target temperature of 20 °C ± 2 °C and recorded humidity.

The rats will be fed BP Nutrition expanded Rat and Mouse Maintenance Diet No. 1, (of known analysis) and water (quality known) will be available *ad libitum*.

The rats will be allowed an acclimatization period of one week before dosing.

Test Method

1. Dose Range Study

Five pairs of rats (1 male and 1 female) will be used in the dose range study. They will be deprived of food overnight prior to dosing.

Using freshly prepared dosing solutions and a constant dose volume or constant concentration, the rats will be dosed orally by gavage. Five dose levels will be investigated.

These rats will be observed for 14 days after compound administration. Sacrifice at 14 days will be followed by a gross *post-mortem* examination.

2. Main study*

Five dose levels

Twenty-five male and 25 female rats will be randomized into 5 groups each of 5 male and 5 female.

At least one week will elapse after the dosing procedure of the dose range study before commencement of the main study.

The rats will be deprived of food overnight prior to dosing.

From the results of the dose range study, 5 dose levels will be selected for the main study, one dose level per group of 5 male and 5 female rats.

Each rat will be dosed once by gavage and observed frequently during the day of dosing and once each day thereafter for 14 days during which time any induced clinical signs will be recorded. At death, or at the end of the observation period and sacrifice, each animal will be subjected to a gross post-mortem examination.

* N.B.

One dose level

If data based on testing with at least 5 rats per sex indicates that the LD_{50} is greater than 10 g/kg, no further testing at other dose levels will be carried out.

Calculation of LD_{50} and its 95 per cent Confidence Limits

From the death pattern recorded in the main study, the LD_{50} value of the test material will be calculated by a suitable method.[16,27,28]

7.2.4.2 Gross Toxicity

Animals	Young adult albino males or females of a uniform outbred strain with known characteristics, fasted overnight.
Treatment	Five males and five females are dosed at 5 g or 5 ml/kg administered by the appropriate route.
Observations	All animals are daily observed for signs of toxicity and mortality rate over a period of 14 days. Animals are weighed prior to dosing and then at 3, 7, and 14 days.

Longer-term tests can be carried out for chronic toxicity which involves a three/six-month toxicity study (rats). The purpose of these experiments is to study the effect of daily administration of compounds for three/six months by daily observation of general condition, weekly study of body weight and food consumption, together with periodic blood and urine analysis and comprehensive terminal histopathology.

7.2.5 Carcinogenic and Mutagenic substances

Three main procedures are used: bioassays; the cell transformation test; and mutagenicity tests.

7.2.5.1 Bioassays

Administration to mammalian species is mainly by the following routes:
1. Feeding
   (a) mix the compound in diet;
   (b) dissolve it in the drinking water;
   (c) put directly into the stomach by a special tube.
2. Skin 'painting' with a solution of the compound.
3. Subcutaneous injection of the compound solution.
4. Inhalation of fine suspensions (aerosols) of the substance.
5. Intrachael intubation. The substance is introduced into the trachea by means of a special tube.

After killing the survivors of the test procedure, bioassay yields two important parameters. These are:

(a) The tumour incidence.
(b) The latent period for the observation of the first tumour growth.

A widely used index to specify the carcinogenic potential of a chemical compound is the Iball index:

$$\text{Iball index} = \frac{\% \text{ Tumour incidence}}{\text{No. of days for latent period}} \times 100$$

It should be noted that any bioassay procedure has a statistical limitation and is very time consuming as well as being expensive.

For the last two reasons, two quick test procedures are used, the cell transformation test and the mutagenicity test.

7.2.5.2 Cell Transformation Test

This is based upon the actual transformation of *in vitro* cultures of normal mammalian cells into tumour cells by short exposure to small quantities of carcinogenic agents. On an even surface in the culture medium, normal cells divide and spread by forming a one-cell-thick layer that indicates an ordered pattern under the microscope. In contrast, 'transformed' cells spread in a characteristically disordered manner and accumulated on top of each other. These transformed cells when reinjected into a similar type of animal become full malignant tumours.

7.2.5.3 Mutagenicity Tests

These assume that carcinogenesis is a result of damage to the heriditary material, DNA, of the cell. Mutations are also a result of damage to DNA and hence it is assumed that mutagenic substances are likely to be carcinogenic as well.

The Ames test[17] is most frequently used. This is a bacterial test that uses certain mutants of the bacterium *Salmonella typhimurium*.

These *Salmonella* mutants have lost the ability to synthesize the amino acid histidine, and cannot grow in a culture medium that does not contain it.

On exposure of the bacteria to mutagenic chemicals, additional mutations occur. These cause repair in the original defective DNA, as in the presence of mutagenic substances they are again able to grow without histidine (revertants) being present.

The Ames test provides a useful inexpensive, rapid, qualitative test for mutagenicity, supplementing but not replacing bio assay testing.

Some mutagenicity tests employed are outlined below.

Four mutagen-detection assays are currently employed.

(i) Microbial assay of Ames *et al.* (1973)[10]

This procedure uses the fact that certain strains of *S. typhimurium* can be mutated from histidine-dependent to histidine-independent forms which will grow in the absence of histidine. Briefly, the procedure involves the incubation on minimal agar plates (with very little histidine) of the chemical under test and the tester strain of bacteria in the presence of liver microsomes (usually from rats although they can be human) which are present to allow metabolism of the chemical to occur. The plates are examined after a suitable time interval for the presence of growing (histidine-independent) and therefore mutated colonies.

M. J. Davis and P. N. Green of Ward Blenkinsop & Co. Ltd.[18-20] have applied the Ames test to several compounds with an interesting result in the case of the photoresponsive compound benzil. Benzil was found by them to be mutagenic to *Salmonella typhimurium* strains TA98 and TA1538 in the presence of daylight, and in the presence or absence of S9 homogenate, but not under similar testing in the dark. As benzil is known to be photosensitive, they deduced that one or more of its photodecomposition products are likely to be the cause for the observed mutagenic effects in daylight.

Other compounds that were tested by them and found not to be mutagenic to *Salmonella typhimurium* strains TA98, TA100, TA1535, and TA1538 in either the presence or absence of S9 homogenate, or the presence or absence of daylight, are listed below.

Photoinitiators:
>benzoin *n*-butyl ether
>benzoin *iso*-butyl ether
>benzoin ethyl ether
>benzoin *iso*-propyl ether
>2-chlorothioxanthone
>2-phenyl-2,2-dimethoxyacetophenone
>1-phenyl-1,2-propanedione-2-(*o*-ethoxycarbonyl) oxime
>2-*iso*-propyl thioxanthone.

Synergistic agents (photoactivators):
>2-(dimethylamino) ethyl benzoate
>ethyl *o*-dimethyl amino benzoate
>ethyl *p*-dimethyl amino benzoate

The oral LD_{50}'s (mice) and dermal LD_{50}'s (rats) were found by these workers for the above compounds to be greater than 1 g per kg body weight and 0.5 g per kg body weight respectively.

(ii) Host-mediated assay[21,10]

This assay is also based upon *S. typhimurium* auxotroph, but in this case the metabolism of the test material is *vivo*. The test material is administered to groups of rats or mice and approximately three hours later the indicator organism (*S. typhimurium*) is injected i.p. Three to six hours later the intraperitoneal fluid is withdrawn and the number of bacteria which have mutated is determined by plating out the fluid on a minimal agar medium.

The Inveresk International,[7] Edinburgh, have provided the following protocol for mutagen testing updated according to the standards laid down by the USA Food and Drugs Administration on 22 December 1978.

TESTING FOR MUTAGENIC ACTIVITY WITH SALMONELLA TYPHIMURIUM TA1535, TA100, TA1537, TA1538, AND TA98 AND ESCHERICHIA COLI WP2uvrA(pKM101)

Summary

An initial toxicity test will be performed with *S. typhimurium* TA 100 which will indicate the highest exposure levels to be used in the mutation tests.

Both the toxicity test and the mutation tests will be performed in the presence and absence of a post-mitochondrial supernatant fluid (S-9) from the livers of adult male rats treated with Aroclor 1254 and an NADPH-generating system. This activation system will be checked for: (1) total protein concentration; (2) cytochrome $P450/P_1$-450 series concentration; (3) benzo(a)pyrene hydroxylase activity; (4) N-demethylase activity. It will also be checked for its activating potential using some known mutagens.

The mutation tests will be performed in two parts.

A. Using six bacterial strains:

    S. typhimurium TA1535, TA100, TA1537, TA1538, and TA98; and
    E. coli WP2uvrA(pKM101)

six exposure levels will be used which are spaced at half-log intervals. Concurrent positive and vehicle controls will be run.

Triplicate plates will be poured for each exposure level, bacterial strain and activation system.

B1. Results of A are negative. *The full test will be repeated* after an interval of at least 7 days.

B2. Results of A are positive. Retesting will be restricted to those bacterial strains and activation conditions giving the suspect positive results.

Evaluation

For *S. typhimurium* strains TA1535, TA1537, TA1538, TA98, and *E. coli*, at least a doubling of the concurrent vehicle control mutation frequency is required before mutagenic activity is suspected. For *S. typhimurium* TA100, a 1.5-fold increase over the control value is indicative of a mutagenic effect.

Introduction

Chemicals which react with DNA may cause different types of mutations and many of these can be detected with the use of bacteria.

The specific damage caused by a mutagen may not be detectable using a particular strain of bacteria. This is because the DNA site coding for the feature selected in the test system may not be mutable by the type of agent under examination. It is, therefore, necessary to use a variety of bacterial strains in order to comprehensively test for bacterial mutagens. At the present time, available data suggest that the use of the 6 strains described in this protocol permit the detection of a wide spectrum of mutagens.

It is well recognized, however, that many chemicals which may be reactive in a mammalian cell, following metabolic activation, are quite inactive in bacterial cells. Extracts of mammalian cells are, therefore, combined with the bacterial indicator cells in a tissue-mediated assay to increase the relevance of the test in assessing the mutagenicity of chemicals to man.

Several validation studies have been undertaken in recent years employing the bacteria and general experimental conditions described here. This laboratory has been involved in one of the most detailed of these studies and the lessons learned from this experience have prompted some of the divergencies described in this protocol from what is, basically, that suggested by Ames, McCann and Yamasaki (in *Handbook of Mutagenicity Testing Procedures*, eds. Kilbey *et al.*, Elsevier, Amsterdam, 1977).

Methods

Bacteria

Salmonella typhimurium

Five strains of *Salmonella typhimurium* will be used:

    S. typhimurium   TA1535
    S. typhimurium   TA100
    S. typhimurium   TA1537
    S. typhimurium   TA1538
    S. typhimurium   TA98

All these strains contain mutations in the histidine operon, thereby imposing a requirement for histidine in the growth medium. Three mutations in the histidine operon are involved. Operon denotes a whole block of genes under a common control; i.e. can be switched on or off together; usually a coordinated function:

his G 46 in TA1535 and TA100
his C 3076 in TA1537
his D 3052 in TA1538 and TA98

his G 46 is a mis-sense mutation which is reverted to prototrophy by a variety of mutagens which cause base-pair substitutions.

his C 3076 contains a frameshift which appears to have added a $\frac{-G-}{-C-}$ base-pair resulting in $\frac{-GGGG-}{-CCCC-}$. This mutation is reverted by 9-aminoacridine, ICR-191, and epoxides of polycyclic hydrocarbons.

his D 3052 also contains a frameshift mutation with the sequence $\frac{-CGCGCG-}{-GCGCGC-}$ which is reverted with the deletion of 2 base-pairs, $\frac{-CG-}{-GC-}$. It is readily reverted by aromatic amines and derivatives.

All 5 strains contain the deep rough (rfa) mutation, which deletes the polysaccharide side chain of the lipopolysaccharide coat of the bacterial cell surface.

This deletion increases cell permeability to more hydrophobic substances and, furthermore, greatly decreases the pathogenicity of these organisms.

The second deletion, through uvrB, renders the organisms incapable of excision repair of bulky lesions from DNA and thus is more susceptible to mutagenicity by certain chemicals (e.g., polycyclic hydrocarbons). These 2 deletions include the nitrate reductase (chl) and biotin (bio) genes also. UvrB is the notation given to a gene coding for a repair enzyme which, *inter alia* can repair lesions induced by ultra-violet light.

Differences between TA1535 and TA1538, on one hand, and the corresponding TA100 and TA98 strains on the other hand, are due to a plasmid the latter pair contain. A plasmid, R-Utrecht, was originally shown to increase the sensitivity of the *his* G 46 mutation in *S. typhimurium* to methyl methanesulphonate and trimethyl phosphate. The particular R-factor involved here, pKM101, increases the sensitivity of strains TA1535 and TA1538 to the mutagenicity of certain chemicals. The mechanism by which sensitivity is increased is uncertain, but utilization of an error-prone repair mechanism may be involved.

Escherichia coli

One strain of *Escherichia coli* is used. This is *E. coli* WP2uvrA(pkM101). This strain contains an ochre mutation in the trpE locus and can be mutated to tryptophan independence either by a base-pair reversion of an A–T base-pair in the trpE locus, or more likely, by a base-pair substitution within a number of transfer RNA loci elsewhere in the chromosome. The latter causes the original defect to be suppressed (ochre suppression) and involves only base-pair substitution transitions at G–C base-pairs.

Thus, while the trp$^+$ reversion system can detect mutations resulting from chemical attack at both A–T and G–C base-pairs, it does not detect frameshift mutagens. The uvrA mutation causes the bacteria to be deficient in the excision of bulky lesions from the DNA and so is more readily mutated by certain agents (UV-radiation, polycyclic hydrocarbons). The strain also carries the plasmic pKM101 which is present in the *S. typhimurium* strains TA98 and TA100. The sensitivity spectrum of the *E. coli* strain is, therefore, also broadened.

Animals

Male rats weighing 250–300 g will be injected once i.p. with Aroclor 1254 (diluted in corn oil to a concentration of 200 mg/ml) at a dosage of 500 mg/kg, 5 days before they are killed. The animals will be allowed drinking water continuously, but food will be withheld 16 hours before they are killed.

Preparation of the 9000 g supernatant fluid from liver

Freshly killed animals will be totally immersed in cold 2 per cent Tego (an ampholytic detergent), then excess fluid wiped off. The abdomen will then be opened and the liver removed, taking special care not to cut into the gastro-intestinal tract. The livers from several animals will be collected in a tared beaker containing ice-cold homogenization medium. The medium used will be 0.15 M KCl.

The beaker will be weighed and the collected livers transferred to the homogenization vessel. A volume of ice-cold 0.15 M KCl equivalent to 3 times the weight of the liver will be added to the vessel and the livers chopped using long-handled scissors. The chopped livers will be homogenized by eight strokes of a glass tube vessel while the Teflon pestle (radial clearance 0.14–0.15 mm) is rotating at about 1200 rpm. The homogenate will be transferred to sterile polypropylene centrifuge tubes and spun to give a relative centrifugal force of 9000g for 10 min at 0° to +2°C. The supernatant fluid will be decanted leaving behind a thick pellet of (mainly) whole cells, nuclei and mitochondria.

Post-mitochondrial supernatant fluid will be prepared in sufficient quantity for the experiment and stored, as 10 ml samples in sterile plastic tubes, immersed in liquid nitrogen ($-196\,°C$).

Enzymic properties of 9000 supernatant fluid

Experience within this laboratory (unpublished) and elsewhere (Ashwood-Smith (1980), *Mutation Res.*, **69**, 199–200) indicates that, while storage of S-9 preparations at higher temperatures result in loss of activating potential, storage in liquid nitrogen ($-196\,°C$) preserves the samples without loss of activity for many months. Such losses which do occur, in comparison with totally fresh S-9 samples, are associated with freezing and thawing rather than the duration of the frozen state. Therefore, it is sufficient to perform the enzymic measurements at some convenient time after the S-9 samples have been prepared and frozen.

The measurements which will be made are as follows:

1. Total protein concentration.
2. Cytochrome P450/P_1-450 series.
3. Benzo(a)pyrene hydroxylase activity.
4. N-Dealkylase activity.

Testing with *Salmonella typhimurium* TA1535, TA100, TA1537, TA1538, and TA98, and *Escherichia coli* WP2uvrA(pKM101)

Samples of each strain will be grown up by culturing for 16 hours at 37 °C in nutrient broth (8 g Difco-Bacto nutrient broth, 5 g NaCl/l). Such cultures will be kept for up to 3 days at +4 °C.

Ice-cold 0.05 M phosphate buffer, pH 7.4, will be added to pre-weighed NADP and glucose-6-phosphate, etc., as follows to give a final concentration in the 'S-9 mix' of:

NADP-di-Na-salt	4 mM (= 3.366 mg/ml)
Glucose-6-phosphate-di-Na-salt	5 mM (= 1.521 mg/ml)
$MgCl_2 \cdot 6H_2O$	8 mM (= 1.626 mg/ml)
KCl	33 mM (= 2.460 mg/ml)

This solution will be immediately filter-sterilized by passage through a 0.45 μm Millipore filter and mixed with the liver 9000g supernatant fluid in the following proportion:

co-factor solution 9 parts
liver preparation 1 part

Diluted agar (0.6% Difco-Bacto agar, 0.6% NaCl) will be autoclaved and, just before use, 5 ml of a sterile solution containing 1.0 mM L-histidine.HCl, 1.0 mM biotin, and 0.206 mM L-tryptophan will be added to each 100 ml of soft agar and thoroughly mixed. This molten agar will be kept in a water bath at a temperature not exceeding 45 °C.

In the course of testing a chemical, 2 ml soft agar will be dispensed to a small, plastic, sterile tube. 0.5 ml of S-9 mix or 0.05 M phosphate buffer, pH 7.4, will be added first and this will be followed by 0.1 ml of bacteria and, finally, the test substance solution (up to 0.5 ml buffer or saline or up to 0.1 ml of a suitable organic solvent such as dimethylsulphoxide, methanol, or ethanol). The tube contents (which are continually cooling) will be mixed then poured on to minimal medium plates. These plates contain 20 ml of 1.5 per cent Difco-Gibco agar in Vogel–Bonner Medium E (*J. Biol. Chem.*, **218**, 97) with 2 per cent glucose. When the soft agar has set, the plates will be inverted and incubated at 37 °C for 48 hours. The colonies on the plates are then counted using a New Brunswick Incorporated Biotran II automated counter, set for maximum sensitivity (colonies of 0.1 mm or more in diameter counted). The plates are also examined under a microscope to check for precipitates and the growth of non-mutated his^- bacteria.

Toxicity test

An initial toxicity test in the presence and absence of S-9 mix will be performed to establish suitable exposure levels for the mutation tests. A single strain of bacterium, *S. typhimurium* TA100, will be used and one plate per exposure level of chemical will be poured. In the absence of toxicity information or restrictions on test compound availability the following exposure concentrations will be used: 33 μg, 100 μg, 333 μg, 1.0 mg, 3.3 mg, 10.0 mg per plate. Plates will be incubated for 48 hours. The number of mutant colonies will be noted and the plates carefully examined, microscopically if necessary, for thinning of the background lawn of microcolonies. The result will be assessed as normal, thin lawn (TL), very thin lawn (VTL), or no lawn (NL). The lowest dose resulting in noticeable thinning of the lawn will be chosen as the highest dose for the mutation experiments. Alternatively, if toxicity is not observed, the lowest dose causing precipitation of the test chemical will be taken as the highest dose for the mutation studies.

Mutation test

Six exposure levels will be selected, the highest being chosen on the basis of the toxicity test. The exposure levels will be spaced at half-log intervals (e.g., 10 μg, 33 μg, 100 μg per plate).

Triplicate plates will be poured for each exposure level (6), bacterial strain (6), and activation system (2). The experiment will be repeated after an interval of at least 7 days. *If the first experiment does not indicate any mutagenic potential of the test compound, the second experiment will be a complete replicate, but if any mutagenic potential is indicated then re-testing will be restricted to the particular indicator strain(s) and activation conditions concerned.* It may also be necessary to decrease the exposure level intervals.

The following controls will be run with each experiment:

1. Each *S. typhimerium* and *E. coli* strain (0.1 ml) will be plated (one plate) onto complete medium and tested for ampicillin resistance and crystal violet sensitivity.
2. 0.5 ml S-9 mix will be plated (one plate) onto complete medium and incubated for 48 h as a sterility check.
3. The vehicle used for the test substance will be used as the negative control and plated in triplicate with each strain used, both in the presence and absence of S-9 mix.

4. Positive controls will be used for each strain in order to aid verification of the strains and demonstrate activity of the S-9 mix.

   (a) *With S-9 mix*
   Pyrene, 400 μg/plate with *S. typhimurium* TA1537.
   4-Acetylaminofluorene, 1 mg/plate with *S. typhimurium* TA1538 and TA98.
   (This non-carcinogen requires S-9 mix prepared from Aroclor 1254-induced liver. Non-induced liver is ineffective).

   (b) *Without S-9 mix*
   Sodium azide, 1 μg/plate with *S. typhimurium* TA1535.
   Methyl methanesulphonate (MMS), 100 μg/plate with *S. typhimurium* TA100 and *E. coli*.
   9-Aminoacridine, 50 μg/plate with *S. typhimurium* TA1537.

The pattern of results to be expected with these controls is as follows:

		S-9 mix	\multicolumn{5}{c}{*S. typhimurium*}	*E. coli* WP2uvrA (pKM101)
			1535	1537	1538	98	100	
A	S-9 mix sterility test	Ster						
B1	Ampicillin resistance		Sens	Sens	Sens	R	R	R
2	Crystal violet sensitivity		Sens	Sens	Sens	Sens	Sens	±Sens
3	Pyrene with S-9 mix			+				
4	4-AAF with S-9 mix				+	+		
5	Sodium azide		+					
6	MMS						+	+
7	9-Aminoacridine			+				

Evaluation

For *S. typhimurium* strains TA1535, TA1537, TA1538, and TA98 and *E. coli* at least a doubling of the concurrent vehicle control mutation frequency is required before mutagenic activity is suspected. For *S. typhimurium* TA100, a 1.5-fold increase over the control value is indicative of a mutagenic effect. If a dose response is found over 3 dose levels (with, possibly, a reduction in the number of colonies per plate at high dose levels) then the result is scored positive. If a positive result or an indeterminate result is obtained, a second experiment will be performed, if necessary over a narrower dose range. This test would be performed only in those strains and activation conditions which had produced the initial response. If no response is obtained, the full experiment will be repeated after an interval of at least seven days. A full assessment of the mutagenic potential will be included in the report. This evaluation assumes that all associated data are satisfactory, e.g., lack of contamination, vehicle and positive control results acceptable.

Reporting

Details of the methods used will be given. Any special problems encountered will be described.
  Enzyme activity results will be summarized and all original and transformed data from the mutagenesis tests tabulated and evaluated.

Quantity of compound required: 1.5 g
Time required for testing: 3 weeks.

(iii) Micronucleus test

This test is based on the principle that chromosomal breakage will result in the production of 'chromosomal debri' which does not migrate to the poles of the dividing cells during anaphase and is not therefore incorporated into the telophase nucleus. These fragments form micronuclei in the cytoplasm of the daughter cells. The method was described by Boller and Schmid.[10]

The test uses any small laboratory animal species (usually mice) and the highest tolerated dose of the chemical is administered twice with an interval of 24 hours between treatments. The animals are killed six hours after the last dose, and the marrow is removed from the femur. A suspension of the marrow is prepared and a film made which is stained with May Grunwald/Giemsa stain. The film is then examined and the incidence of polychromatic erythrocytes containing micronuclei determined.

(iv) Dominant lethal assay in male mice[10]

Groups of ten male mice are treated with the test material at three dose levels (the high dose being the maximum tolerated dose) on each of five consecutive days. In addition to the negative control group, a further group will be included in which male mice are treated with a material known to produce dominant lethal mutations in the particular strain of mouse used.

Each male is then mated with four untreated females and transferred each week to four new females. This procedure lasts for eight consecutive weeks, thus ensuring that all stages of spermatogenesis have been examined.

Pregnancy is interrupted on day 12 or 13 of gestation. The ovaries and uteri removed and the following data recorded.

1. Number of corpora lutea.
2. Total number of implantations which are subdivided into:
    (a) Early deaths (deciduomata)
    (b) Late deaths (evidence of placenta and probably foetal tissue)
    (c) Live foetuses.

7.2.5.4 Carcinogenicity Tests

A typical carcinogenicity study (mice) could be for eighteen months, under the following conditions.

Animal groups — Random-bred C.D. mice are used and kept under isolation conditions.

Group no.	No. of animals Male	Female	Dose
1	50	50	Zero
2	50	50	Low
3	50	50	Intermediate
4	50	50	High

The animals are housed in a temperature- and convection-controlled environment. The test material is normally incorporated into the diet in the appropriate proportions, although alternative dose routes can be employed if required.

Observations Body weights are measured weekly for the first month, twice-weekly thereafter. Food consumption is measured weekly.

Laboratory Investigations A peripheral blood film is examined and haemoglobin, packed cell volume, and red blood counts measured on five males and five females from each group at 1, 3, 6, 12, and 18 months. Further examinations are made only if deemed necessary and after consultation with the client.

Moribund animals Moribund animals are killed and a full post-mortem carried out. Animals dying during the experiment are examined and tissues submitted for histopathological studies as far as practical, depending on the degree of autolysis present.

Post-mortem studies All animals are killed and examined macroscopically. The control and high-dose groups are examined as follows: liver, kidney, adrenals, thyroid, uterus/prostate, and gonads are weighed and organ:body weight ratios calculated.

The following tissues as well as all tumours seen at post-mortem are examined histopathologically:

Adrenals	Kidneys	Large and small intestine
Bone marrow	Liver	Spleen
Brain	Lungs	Pituitary gland
Gonads	Thyroid	Lymph node
Uterus or prostate	Pancreas	Urinary bladder
Heart	Stomach	
	Salivary glands	

All tumours are classified according to origin, type, and grade and their incidence in groups statistically analysed. Any tissues in which tumours are present significantly more than in controls are examined in the medium and, if necessary, in the low group.

Report A full report of all methods and findings includes analyses of all data obtained from the study and a final assessment by the pathologist of the carcinogenic potential of the test substance.

Of the photo-initiators used for UV curing it would appear that one in particular has come under scrutiny for possible carcinogenic properties. This is Michler's ketone. There appears to be no reports of it having this activity in man, but data obtained with rats would indicate that it is carcinogenic by ingestion.

The report stating this is shown below (*Abstracts of Papers*, Society of Toxicology Incorporated, Fourteenth Annual Meeting, Williamsburg, Virginia, 9–13 March 1975, p. 101.)

125. CHEMICALLY INDUCED HEPATIC NEOPLASMS IN THE FISCHER RAT (F.344).
F. M. Garner and B. V. Cockrell, Litton Bionetics, Kensington, Maryland. (E. R. Hart)

A spectrum of lesions including hyperplasia, hepatic cell adenoma, and hepatocellular carcinoma was induced in rats by administration in the diet of NCI Compound No. CO2006. Malignant liver tumours were observed in female rats receiving 1000 ppm in the diet as early as 12 mo. By the end of the study of this feeding group of 50 animals, 28 % had hepatic cell adenomas and 60 % hepatocellular carcinomas. Few survived to the end of this 2 yr study. A similar feeding group of 50 males and 50 females fed 500 ppm had an equivalent tumour incidence. Approximately 70 % of both sexes had benign liver tumours and 20 % of each sex had hepatocellular carcinomas. Nearly all animals survived to the end of the experiment. Fifty males fed 250 ppm had a much lower incidence of benign and malignant liver tumours, 28 % had benign tumours while only 2 % developed hepatocellular carcinoma. Nodular hyperplasia was found in one control female rat. No other hepatic lesions were observed in any of the controls. Clear cut distinction between hyperplasia and neoplasia could not always be determined with certainty, however, nearly all of the tumours classified as malignant metastasized to other sites, usually the lungs. No other compound-related gross or microscopic lesions were observed in the study. *This compound is obviously carcinogenic* and the induced neoplasia is dose related. (Supported by Tracor Jitco, Inc., Subcontract No. 74-25-106002). (Reproduced by permission of Drs Garner and Cockrell.)

Japanese reports[22,23] originally indicated tumours occurring on oral dosing of rats as far back as 1936. The limit appears to be administration of 5 g of Michler's ketone per rat over 1 year. Other reports have subsequently repeated this finding.[24]

7.2.6 Inhalation Tests[10]

Inhaled toxic materials can cause injury to the lungs, and can also be transported from the lungs to other parts of the body, interfering with normal function there. Adverse response may be immediate or delayed. The investigation of the toxicity of airborne substances in animals has therefore to consider the variety of consequences that would result from their contact with the respiratory tract.

(i) Classification into a toxicity class[25]

LC_{50} determination (rats)[10]

The concentration of the test material in air likely to cause death in 50 per cent of the exposed animals is estimated.

Animals	Albino rats: one control and up to five test groups, each of five male and five female animals.
Exposure	Single continuous whole-body exposure of each test group to one of a series of concentrations of the dispersed test material. Depending on animal response, the exposure may continue for up to six hours; analysis of test atmospheres, particle size distribution.
Observations	During exposure and for 14 days after exposure: mortality, body weight, clinical signs, food and water consumption, macroscopic pathology of the major internal organs.
Analysis of data	The LC_{50}, including standard error, is calculated. The material is allocated to a toxicity class, according to the Combined Tabulation of Toxicity Classes.[25]

Federal Register inhalation test (rats)[10]

The allocation to a toxicity class is based on the proportion of deaths following a single exposure to a high concentration of the test material in air.

Animals	Albino rats: one control and one test group, each of five male and five female rats.
Exposure	Single continuous whole-body exposure of the test groups for one hour to 2 mg per litre or 200 ppm of the test material; analysis of the test atmosphere, particle size distribution.
Observations	During exposure and for 14 days after exposure, mortality, body weights, clinical signs, food and water consumption, macroscopic pathology of the major internal organs.
Analysis of data	The substance is defined as either highly toxic or toxic according to the definitions made in the regulations under the Federal Hazardous Substances Labelling Act, published in the Federal Register of the USA.

(ii) Thresholds of response and mortality[10]

Westinghouse specifications (rats or mice)

The test gas should exhibit no greater toxicity than nitrogen.

Animals	Albino rats or mice: one test group either of two male and two female rats or of five male and five female mice.
Exposure	Single, continuous whole-body exposure for 16 hours to a mixture of 75 to 80 per cent test gas and 20 to 25 per cent oxygen; analysis of the test atmosphere.

Observations	During exposure and for 24 hours following exposure: mortality, body weights, clinical signs, macroscopic pathology of the major internal organs.
Analysis of data	The specification is fulfilled in the absence of adverse effects at the termination of the observation period.

ASTM tentative specifications (guinea pigs)[10]

The test gas should exhibit no greater toxicity than Group VI gases of the Underwriters Laboratories classification.

Animals	Albino guinea pigs: one test group of two male and two female animals.
Exposure	Single, continuous whole-body exposure for two hours to the test gas at a concentration of 20 per cent by volume in air.
Observations	During exposure and for 24 hours following exposure, mortality, body weights, clinical signs, macroscopic pathology of the major internal organs.
Analysis of data	The specification if fulfilled in the absence of adverse effects at the termination of the observation period.

FDA acute inhalation test (rats, guinea pigs, or rabbits)[10]

Animal response to an aerosolized preparation is observed. The method is essentially that recommended by J. H. Draize in the *Appraisal of the Safety of Chemicals in Foods, Drugs and Cosmetics.*[4]

Animals	Albino rats, guinea pigs, or rabbits: one control and one test group, each of either five male and five female rats or guinea pigs or two male and two female rabbits.
Exposure	15 minute exposure to 30 seconds continuous spray release repeated 10 times over a five-hour period; either head only or whole body exposure; analysis of O_2 concentration, estimation of aerosol dose.
Observations	During exposure and for 14 days after exposure: mortality, body weights, clinical signs, food and water consumption, macroscopic pathology of the major internal organs.
Analysis of data	Animal response is compared with response elicited by exposures to similar aerosol preparations.

(iii) Subacute and chronic studies

FDA subacute inhalation test (rats, guinea pigs, or rabbits)[10]

Repeated daily exposures to an aerosolized preparation are made over an extended period of time and animal response is observed.

Animals	Albino rats, guinea pigs, or rabbits: one control and one test group, each of either ten male and ten female rats or guinea pigs, or four male and four female rabbits.
Exposure	15 minute exposure to 30 seconds continuous spray release, twice daily for 90 consecutive days. Analysis of O_2 concentration, estimation of aerosol dose.
Observations	Mortality, body weight, clinical signs, food and water consumption, haematology, lung radiography, macroscopic pathology, organ weights, histopathology.
Analysis of data	Animal response is compared with response elicited by exposures to similar aerosol preparations.

Subacute inhalation toxicity tests in rats and dogs[10]

Animals	Albino rats of the Wistar strain. Purpose-bred beagles.
Groups	Five groups, each of ten male and ten female rats and four male and four female dogs.
Exposure	Up to one hour, twice daily, head-only exposures on 30 consecutive days. Control, vehicle control, low, medium and high dose groups. Particle size distribution; monitoring of O_2 and aerosol concentration.

Observations:

Mortality	The time and mode of any death is recorded. Autopsy is made and macroscopic appearance of the major internal organs is noted.
Signs of response	Animals are observed throughout the exposure periods and during periods between exposures. The onset, duration and intensity of any signs are recorded.
Body weights	Animals are weighed twice each week during a pre-exposure acclimatization period and during treatment.
Food consumption	The food consumed is recorded and food utilization ratios are calculated.
Water consumption	The water consumed is recorded when abnormal consumption is noted.
Ophthalmoscopy	The eyes of all animals are examined before treatment and after the last exposure.

Haematology	Red cell count, haemoglobin, packed cell volume, red cell indices (mean corpuscular volume, mean corpuscular haemoglobin, mean corpuscular haemoglobin concentration). White cell count (total and differential).
Blood chemistry	Plasma urea. Plasma glucose. Total serum protein, alb/glob ratio, serum alkaline phospatase, serum glutamic-pyruvic transaminase, serum glutamic oxaloacetic transaminase, electrolytes Na^+ and K^+. Analysis of the drug and propellant concentration.
Urinalysis	Volume, S.G., pH, proteins, glucose, ketones, bile pigments, blood pigments, microscopy of spun deposits.
Terminal studies	The ciliary activity of the respiratory epithelium is measured in four rats and four dogs per group.
Macroscopic pathology	Moribund animals and those surviving until the end of treatment are killed and autopsied. The macroscopic appearance of the major internal organs is recorded.
Organ weights	The following organs are weighed: brain, adrenals, lungs, heart, liver, kidneys, spleen, and gonads. Statistical evaluation of absolute and relative organ weight is made.
Histopathology	Organ samples retained.

Chronic inhalation toxicity test in rats.[10]

Animals	Albino rats of the Wistar strain. Initial weight approximately 150 g. One hundred male and one hundred female rats are kept in a holding room three to a cage. Animals are acclimatized for ten days before treatment: they have free access to Dixon's Diet 41B and tap water throughout the experiment except during exposure periods.
Treatment	Group 1: Restrainer control 20 male + 20 female rats. Group 2: Propellant aerosol control 20 male + 20 female rats. Group 3: Low-dose test aerosol 20 male + 20 female rats. Group 4: Medium-dose test aerosol 20 male + 20 female rats. Group 5: High-dose test aerosol 20 male + 20 female rats. Each group is kept in restrainers for up to one hour, twice daily, during five-day weeks, for 26 consecutive weeks.

Group 1 animals are exposed to fresh air only during restraint.

Test groups are exposed, head only, for up to one hour, twice daily, respective aerosols at the required concentrations.

Particle size distribution; monitoring of O_2 concentrations.

Observations:

Mortality	The time and mode of any death is recorded. Autopsy is made and the macroscopic appearance of the major organs is noted.
Signs of response	Animals are observed throughout the exposure periods and during periods between exposures. The onset, duration, and intensity of any signs are recorded.
Body weights	Animals are weighed weekly during the pre-exposure acclimatization period and during treatment.
Food consumption	The food consumed is recorded and food utilization ratios are calculated.
Water consumption	The water consumed is checked by visual inspection. The volume consumed is recorded when abnormal consumption is noted.
Haematology	Red cell count, haemoglobin, packed cell volume, red cell indices (MCH, MCV, MCHC). White cell count (total and differential), prothrombin time.
Blood chemistry	Plasma urea, plasma glucose, total serum protein, Alb/glob. ratio, SAP, SGPT, SGOT, electrolytes Na^+, K^+, and Cl^-, cortisol/corticosterone. Analysis of the drug and propellant concentration.
Urinalysis	Volume, S.G., pH, proteins, glucose, ketones, bile pigments, blood pigments, microscopy of spun deposits.
Interim kill	Five male and five female rats per group are killed after 13 weeks' exposure.
Terminal studies	The cillary activity of the respiratory epithelium is measured after six months' treatment in four rats per group.
Macroscopic pathology	Moribund animals and those surviving until the end of treatment are killed and autopsied. The macroscopic appearance of the major internal organs is recorded.
Organ weights	The following organs are weighed: brain, adrenals, lungs, heart, liver, kidneys, spleen, gonads. Statistical evaluation of absolute and relative organ weights is made.

Histopathology	Organ samples retained.

Chronic inhalation toxicity test in dogs[10]

Animals	Purpose-bred beagle dogs. Initial weight approximately 5 kg. Twenty-five male and twenty-five female dogs are kept in pens, three to a pen. Animals are premedicated against infections and parasites during a four-week acclimitization period. Each dog receives daily 500 g Spratt's Dog Diet and *ad libitum* tap water.
Treatment	Group 1: Restrainer control five male + five female dogs. Group 2: Propellant aerosol control five male + five female dogs. Group 3: Low-dose test aerosol five male + five female dogs. Group 4: Medium-dose test aerosol five male + five female dogs. Group 5: High-dose test aerosol five male + five female dogs.
	Each group is kept in restrainers for up to one hour, twice daily, during five-day weeks for 26 consecutive weeks. Group 1 animals are exposed to fresh air only during restraint.
	Test groups are exposed, head only, for up to one hour to the respective aerosols at the required concentrations.
	Particle size distribution: monitoring of O_2 concentrations.

Observations:

Mortality	The time and mode of any death is recorded. Autopsy is made and the macroscopic appearance of the major internal organs is noted.
Signs of response	Animals are observed throughout the exposure periods and during periods between exposures. The onset, duration, and intensity of any signs are recorded.
Body weights	Animals are weighed weekly during the pre-exposure acclimatization period and during treatment.
Food consumption	The food consumption is recorded and food utilization ratios are calculated.
Haematology	Samples of blood are taken during the acclimatization period from five male and five female dogs selected at random, and from two male and two female dogs per group after one, three, and six months' exposure. The

	following parameters are examined: red cell count, haemoglobin, packed cell volume, red cell indices (mean corpuscular volume, mean corpuscular haemoglobin, mean corpuscular haemoglobin concentration). White cell count (total and differential). Prothrombin time.
Blood chemistry	Plasma urea, plasma glucose. Total serum protein, Alb/glob. ratio, serum alkaline phosphatase, serum glutamic-pyruvic transaminase, serum glutamic oxaloacetic transaminase, electrolytes Na^+, K^+, and Cl^-, cortisol/corticosterone. Analysis of the drug and propellant concentrations.
Urinalysis	Volume, S.G., pH, proteins, glucose, ketones, bile pigments, blood pigments, microscopy of spun deposits.
Electro-cardiography	During the acclimatization period, as well as after three and six months' exposure.
Interim kill	Two male and two female dogs per group are killed after 13 weeks' exposure.
Terminal studies	The ciliary activity of the respiratory epithelium is measured after six months' treatment in four dogs per group.
Macroscopic pathology	Moribund animals and those surviving until the end of treatment are killed and autopsied. The macroscopic appearance of the major internal organs is recorded.
Organ weights	The following organs are weighed: brain, adrenals, lungs, heart, liver, kidneys, spleen, gonads. Statistical evaluation of absolute and relative organ weights is made.
Histopathology	Organ samples retained.
Analysis of data:	At the termination of the experiment the results are collated and their toxicological significance assessed.

(iv) Effect of inhaled materials on reproduction

Test animals are exposed to aerosols:

1. Before mating for fertility studies.

2. During embryogenesis for teratogenic evaluation.

3. During lactation for post-natal studies.

The observations performed in these studies are identical to observations made during routine reproduction studies.

7.3 SOME CHEMICAL ASPECTS OF PHOTOBIOLOGY

Coyle[26] has presented a review concerning the interaction of light and biological systems and some relevant aspects of this are summarised below.

7.3.1 Eye Effects

Photokeratitis is the name given to the excruciatingly painful inflammation of the cornea due to incident UV radiation on the eye. Damage to nucleoprotein is a probable consequence of the photochemical reaction.

Lens damage involves the photo-oxidation of tryptophan residues in the protein structure.

Aromatic residues and disulphide links in proteins do not transmit much UV radiation above 290 nm. Of the natural amino acids, tryptophan possesses in this region the highest extinction coefficient as well as the lowest triplet energy state of the amino acids. The importance of this is that tryptophan degradation occurs after energy transfer, as a consequence of light absorbed by other residues.

Tryptophan's main reaction in the photo-excited state is by electron transfer to an acceptor molecule, eventually undergoing conversion by bond cleavage to N-formylkynurenine. This is depicted below:

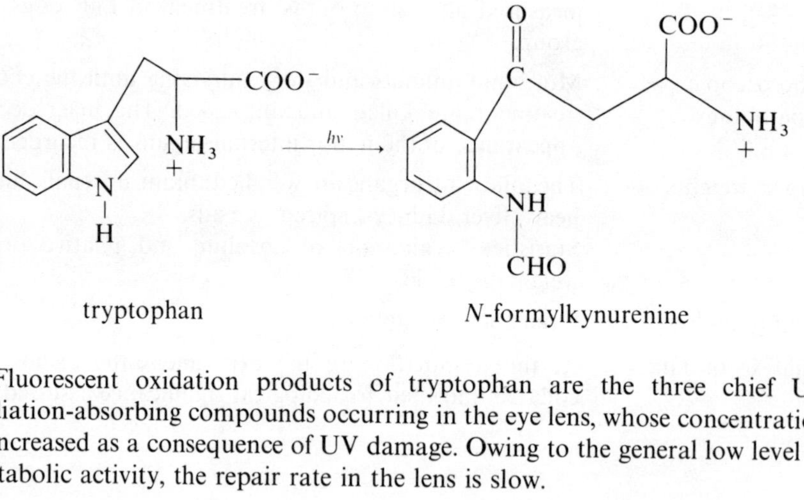

tryptophan $\xrightarrow{h\nu}$ N-formylkynurenine

Fluorescent oxidation products of tryptophan are the three chief UV radiation-absorbing compounds occurring in the eye lens, whose concentration is increased as a consequence of UV damage. Owing to the general low level of metabolic activity, the repair rate in the lens is slow.

7.3.2 Skin Effects

Ultraviolet radiation has several actions on skin. It may cause inactivation of enzymes as a consequence of the photoreactions of proteins involving tryptophan oxidation or cystine S–S bond cleavage. The prime damaging effect is, however, a consequence of photochemical reactions of nucleic acids and their nucleoproteins.

Cell mutation or death can result from photodamage to deoxyribonucleic acid. The pyrimidine bases are the chief reaction centres, with the sugar residues involved in a more minor role. The purine bases act mainly as energy transfer agents.

Two major types of photoreaction can occur. These are photohydration and dimerization.

(i) Photohydration

Water is added across the ring double bond and the hydrate thus formed can thermally dehydrate, but the uracil product is not so reactive as other hydrates of pyrimidine bases. The coding parameters of ribonucleic acid (RNA) are affected by the formation of hydrates and may well be significant in mutation. Compounds other than water may photochemically add across the bond.

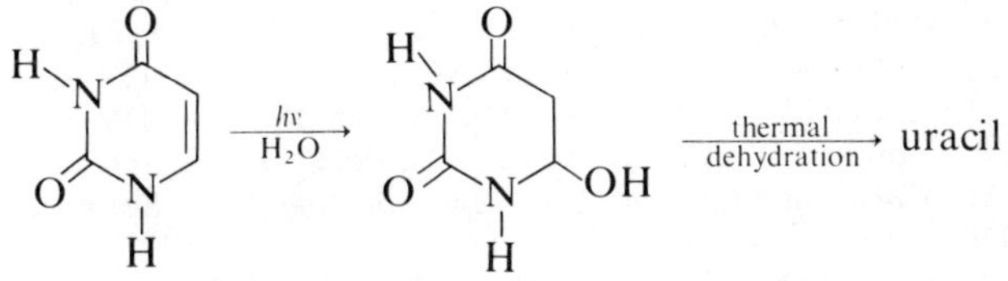

uracil $\xrightarrow{h\nu / H_2O}$ [hydrate] $\xrightarrow{\text{thermal dehydration}}$ uracil

(ii) Dimerization

Cyclobutane dimers are created by (2 + 2) cycloaddition with thymine in particular, by direct irradiation or after triplet sensitizer energy transfer.

In DNA, the pyrimidine dimers formed are the chief lesion sites. Of paramount importance from the photodimerization reaction is the possibility of reversal to regenerate pyrimidine bases. Thymine dimers can be photocleaved by changing the wavelength. Dimerization is the preferred reaction at 280 nm whereas cleavage is at 240 nm.

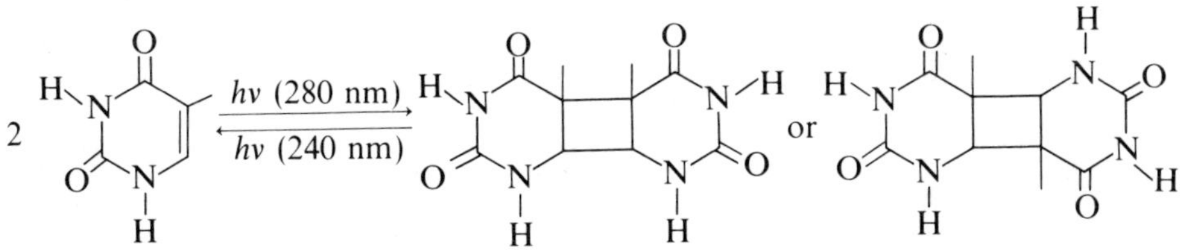

$\xrightleftharpoons[h\nu\ (240\ nm)]{h\nu\ (280\ nm)}$

The skin's sensitivity to UV basically depends upon the ability to repair DNA damage on irradiation. Mild effects on the skin are known as erythema and are a slight initial reddening which is due mainly to blood-vessel dilation. Chronic effects are wrinkling and formation of cancers due to tissue degeneration.

7.4 REFERENCES

1. Andrews, H. L., *Rev. Sci. Instrum.* **16**, 253, (1945).
2. Browning, E., *Toxicity of Industrial Metals*, Ch. 12, Butterworth, (1961).

3. *Toxic Substances List*, N.I.O.S.H. (National Institute of Occupational Safety and Health) (1973).
4. *F.D.A. Handbook, Appraisal of the Safety of Chemicals in Foods, Drugs and Cosmetics*, F.D.A., p. 47, (1959).
5. *Journal Officiel de la Republique Francaise*, 21 April (1971).
6. Hazleton Laboratories – *Literature Release*.
7. Inveresk Research International – *Literature Releases*.
8. Toxicol Laboratories – *Literature Release*.
9. Huntingdon Research Centre – *Literature Release*.
10. Consultox Laboratories – *Literature Release*.
11. *Federal Register*, Vol. 36, No. 105, 29 May, (1971).
12. *U.S. Federal Register*, 29 FR 13009, 17 September (1964).
13. Landsteiner and Jacobs, *J. Exp. Med.* **61**, 643, (1935).
14. Beuhler, *Arch Dermatol.*, **91**, 171–175, (1965).
15. Magnusson, B., and Kligman, A. J., *J. Invest. Dermatol.* **52**(3), 268–276, (1969).
16. Litchfield and Wilcoxon, *J. Pharm. Exp. Therap.* **96**, 99–108, (1949).
17. Ames, B. N., McCann, J., and Yamasaki, E., *Mutation Research*, **31**, 347–364, (1975).
18. Davis, M. J., and Green, P. N., *Paint Manufacture*, **48**(1), 38, (1978).
19. Davis, M. J., and Green, P. N., *Paint Manufacture*, **48**(1), 39, (1978).
20. Green, P. N., Young, J. R. A., and Davis, M. J., *Paint Manufacture*, 32, April, (1978).
21. Legator, M., *Chemical Mutagenesis in Mammals and Man*, Springer-Verlag, New York, (1970).
22. Harada, M., *Journal Osaka Iga Khai Zassi*, **35**, 1916, (1936).
23. Kinosita, R., *Trans. Jap. Path. Soc.*, **27**, 665–727, (1937).
24. *Survey of Compounds which have been examined for Carcinogenic Activity*, U.S. Dept. Health, Education and Welfare, Public Health Service, Publication No. 149.
25. Hodge, H. C., and Sterner, J. H., *Amer. Ind. Hyg. Assoc. Q.*, **10**, 4, 93, (1949).
26. Coyle, J., *Chemistry in Britain*, **16**(9), 460–464, Sept., (1980).
27. Weil, *Biometrics*, **8**, 249, (1952).
28. Thompson, W. R., *Bact. Reviews*, **11**, 115, (1949).
29. *Criteria for a Recommended Standard – Occupational Exposure to Ultra-violet Radiation*. U.S. Dept. Health, Education and Welfare, HSM 73-11009, (1972).

Index

Abrasion resistance of films, 173
Absorption effects, 16–27, 178
Acetate,
  butyl, 169
  butyl carbitol, 169
  ethyl, 220
  polyvinyl, 204, 268
  polyvinylcinnamylidene, 277
  vinyl, 164, 169
Acetone,
  dicinnamal, 272
  α-γ-dicinnamalyidene, 290
Acetophenone, 73
  benzylidene, 277
  dichloro, 85
  diethoxy, 83, 84, 183–184
  2,2 dimethoxy 2-phenyl, 82, 83, 184, 203, 320
  ω-ethoxy, 84
  poly-4-vinylbenzal, 291
  poly-4-vinyl benzylidene, 277
  Trichloro, 85
Acetophenone derivatives, 79–85
  chlorinated, 79, 85
  dialkoxy, 79, 83–85
Acetylacetone, 124
Acetyl peroxide, 29
Acid,
  acenitic, 150
  adipic, 155
  benzene sulphonic, 152
  chromic, 307
  cinnamic, dimerization of, 272–273
  4,4′-diazidostilbene, 274
  fumaric, 146, 149–150
  itaconic, 149–150
  lewis, 70, 76, 278
  mesaconic, 150
  nitronaphthalene sulphonic, 290
  α-truxillic, 272–273, 295
  β-truxinic, 272–273, 295
Acridine, 140

9-amino, 322, 325
Acrylamide, 274
  methylene bis, 274
  poly, 268, 274
Acrylate,
  butoxyethyl, 186
  3-butoxy-2-hydroxypropyl, 165
  butyl, 164, 169
  cyclohexyl, 165, 169
  ethyl, 156
  2-ethyl hexyl, 165, 169, 174
  2-hydroxyethyl, 156, 165, 174
  2-hydroxypropyl, 165, 174
  iso-bornyl, 165, 169
  iso-decyl, 165, 169
  phenoxyethyl, 165
  tetrahydrofuryl, 165
Acrylate resin, 192–201
  functionality of monomers and oligomers, 164–168, 181
  mono-, 164–165
  di-, 166, 168
  tri-, 168
  tetra-, 167
  penta-, 167
  UV absorption spectrum of, 194
Acrylic groups (occluded), 185
Actinic radiation,
  scatter of, 178–179
Addition reaction, 146, 278
Adhesion, 76, 176–178, 243
Adipate,
  hexamethylene diammonium, polyamide of, 283
Air inhibition, 183–186
Albumin, 265
Alcohol,
  allyl, 154
  polyvinyl, 204
Aldehydes,
  aromatic *ortho*-nitro, 291
Aluminium, 236–242

339

Ames test, 91, 319, 322
Amide,
  1-methyl-5-nitronaphthalene-4-sulphonomethyl, 290
Amines,
  1-diazo-4, 4'-diphenyl, 271
  4-diazo-1, 1'-diphenyl, 292
  synergistic effect, 88, 184
  tertiary, 183
Aminoplasts, 145
Ammonia, 220
Ampicillin, 325
Angular momentum, 2
Anhydride,
  citraconic, 149
  maleic, 146, 149–151
  phthalic, 150, 151
Annealing,
  batch, 237
  continuous, 237
Anthracene, 140
Anthraquinone, 99–100
  2-ethyl, 100, 282
  sulphonate, 100
  tertiary butyl, 100
Antimony trioxide, 110–113
  UV spectrum of, 111
Area of non-image, 280
Aroclor, 320, 323, 325
Arrhenius model, 33
Aryldiazonium compounds, 70, 75–76
Atmosphere,
  effect on cure, 183–186
Atomic term symbol, 7
Atoms,
  properties of, 1–12
Aurate,
  sodium chloro-, 123
Azeotrope, 152, 156
Azides, 263, 272–273
  bis, crosslinking compounds, 277, 288
Azido groups,
  photolysis of, 142
Azo-bis- (*iso*-butyronitrile), 126
Azo compounds, 126–127

Basecoat, 177, 186
  opaque, photopolymerizable, for metal decoration, 243
Beading, 172
Beer's Law, 17
Beer–Lambert,
  absorption coefficient, 53
  law, 17, 187–188
Benzaldehyde,
  p–N, N'-dimethyl amino, 184
Benzanthrene, 140
Benzene,
  Tetra cyano, 69, 75
Benzoate,
  2-(*n*-butoxy) ethyl *para* dimethylamino, 93, 97
  2-(*n*-butoxy) ethyl *para* dimethylamino, UV spectrum of, 97
  2-(dimethyl amino) ethyl, 320
  2-(dimethylamino) ethyl, UV spectrum of, 96
  ethyl *ortho* dimethyl amino, 93, 96, 320
  ethyl *ortho* dimethylamino, UV spectrum of, 96
  ethyl *para*dimethyl amino, 93, 96, 183–184, 320
  ethyl paradimethyl amino, UV spectrum of, 96
Benzil, 204
  mutagenicity of, 319
  UV spectrum of, 83
  metals, 79, 82, 83
Benzohydrophenone, 86–87
Benzoin, 79–81, 83, 274
UV absorption spectrum of, 83
Benzoin derivatives, 79–85
  alkyl ethers, 79, 80–82, 83
  alkyl ethers, UV spectrum of, 83
  n-butyl ether, 320
  n-butyl ether, UV spectrum of, 193
  iso-butyl ether, 320
  ethers, 144
  ethyl ether, 82, 320
  methyl ether, 82–83
  methyl ether, UV spectrum of, 83
  iso-propyl ether, 320
Benzophenone, 85–88, 158–159, 161, 169, 203, 281
  energy level diagram of, 127
  UV spectrum of, 87
  reaction with oxygen, 87
  synergism with amines, 130
  thio, 71
Benzopinacol, 85–86, 158
Benzoquinone,
  -1: 4-diazide-(4)-2-sulphoanilide, 290
  para, 101
Benzothiazole
  2-mercapto, 119
Bioassays, 318
Biphenyls, 271
Bisacrylamide, 283
  hexamethylene, 283
Bisphenol 'A', 265

Bitumen, 264–265, 269, 272
Black plate, 236–237
Blade,
 doctor or scraper, 209–210, 212, 252
Blanket cylinder, 211–212
Blistering, 172
Blocking, 216
Bohr,
 frequency condition, 14
 theory of the atom, 1–3
Boltzmann,
 distribution law, 52
 factor, 52
Bond energy, 44–45
Born–Oppenheimer approximation, 9
Brønsted acids, 75, 77
Brunak process, 287
Bubbling, 272
Budium, 240
Burning in, 276, 297
Butane,
 cyclo, 295, 337
iso-Butylene, 75
Butyral,
 polyvinyl, 268
Butyrate,
 sucrose acetate, iso-, 169

Cage effect, 27–33, 63, 117–118, 124, 185
Cans,
 beer and beverage, 240
 deep drawn processed, 239–240
 general line, 238
 open top, 238–239
Capillary action, 215–216
ε-Caprolactam, 283
Caps,
 screw and lugged, 241
Carbamate,
 N′, N′, dimethyl dithio, 122
 dithio, 121
 methyl diethyl, 122
 methyl dithio, 140
Carbazole,
 N-vinyl, 71, 75, 128
Carbohydrates, 265
Carbon tetrabromide, 129
Carbonyl compounds,
 aromatic, 71–74
Carbonyl group, 25
 transitions found and orbital shapes, 13
Carcinogenic substances, 318–328
Carcinogenic tests, 326–328
Casein, 265
Cationic,
 cured epoxy systems, 145
 photoinduced polymerization, 74–79
Cell transformation test, 318–319
Cellsolve, 204
 butyl, 169
Cellulose,
 acetate, 106
 derivatives, 282
 sodium carboxymethyl, 287
Chain reaction, 26
Chalcones, 277
Chalking, 112–117
Charge transfer, 125
 complex, 128
 state, 73
Chemical resistance, 175
Chemosis, 313
Chloramide,
 sodium p-toluene, 120
Chloride,
 2-naphthalene sulphonyl, 120
 sulphonyl, 120
Chrome ulcers, 307
Chromium salts,
 chromic chromate, 267
 chromium dioxide, 267
 toxicity of, 306–308
Cinnamate,
 polyvinyl, 264–267, 272–273, 277, 295–296
Circuits,
 printed, 295, 300
Coatings,
 external, 240
 enamel, 234–243
 flexible for plastics, textiles and metals, 234–243
 internal, 234–243
 roller, 234–235
 size basecoat, 234
 UV curing, 231–244
 vinyl film, 234
 vinyl flooring, 234
 wood, 232–234
Cobalt salts, 123
Cockling, 226, 256
Coefficient of friction,
 for films, 173
Collisional deactivation,
 of electronically excited molecules, 13
Condensation reaction, 146
Conjuctivae, 313
Continuum, 42, 43
Convergence limit, 20, 26, 42

Copolymers,
  block, 139–140
  graft, 139–140
  types of, 37
Cornea, 312–314, 336
Coulombic,
  attraction, 104
  effect, 7
  force, 1, 18
Coumarin, 143, 277
Cratering, 172
Crawling, 172
Cronak process, 286–287
Crosslinking, 173, 176, 185–186
  density, 173, 175, 181
  ring opening, 278–279
Crowns, 241
Cromalin TM,
  dupont system, 300
Cure,
  bulk, 43, 181–183, 186–201
  dependence upon,
    coating thickness, 179–180, 187–188
    colourant, 179–180
    intensity, 179–182
    photo initiator/sensitizer concentration, 182–183, 200
    substrate, 186–188, 200–201
    temperature, 186–187
    UV source spectrum and wavelength, 187–188, 198–200
    levels, assessment of, 181
  response of coatings, theoretical treatise of parameters, 187–201
  response of photopolymerizable systems, 180–201
  surface, 43, 69, 144, 181–183, 186–201, 246
Curing,
  aerobic, 175
  anaerobic, 175
  electron beam, advantages and disadvantages of, 229–230
  UV, 230–259
Cyanurate,
  tri-allyl, 168
Cycloaddition,
  intramolecular photochemical, 185–186
Cyclohexanone,
  2,6-bis (4'-azidobenzylidene), 277–288
Cyclopropane, 126
Cyclopropene, 126

Dahlgren dampening, 246
Dark (thermal) reaction, 13, 16, 283, 295

N-Dealkylase activity, 320, 323
De Broglie,
  theory of the wave nature of particles, 3
Deoxyribonucleic acid (DNA), 319–322, 336–337
Dermatitis, 306–307
Dextran, 269
Diacetyl, 274
Diacrylate,
  1,4-butanediol, 166–169
  o-dianosidine tetrazonium salt of (3,3'-dimethoxy diphenyl-4,4' tetrazonium chloride), 294
  diethylene glycol, 165–169, 203
  1,6-hexanediol, 166–169, 184–186
  neopentylglycol, 166–169, 203
  triethylene glycol, 282–283
Diaryliodonium compounds, 75–78
Diazo compounds, 269, 288–292
Diazo oxides, 290–292
Dichromate, 265–269, 296–297
  photolysis of, 124
  potassium, 307
  trimethyl lauryl ammonium, 125
Dicyclopentadienyl titanium dichloride, 75
Diffusion,
  of medium, 216
Diluents, 168–169
  allylic, 168–169
  monofunctional, 163–165
  plasticizing, 169–170
Dimerization, 337
2-(dimethylamino) ethyl benzoate, 93, 96
  UV spectrum of, 96
Dipentene, 169
Display tablets, 240
Dissociation, 18–21, 24–25
  of electronically excited molecules, 13
Disulphides,
  acyl, 119
  alkyl, 119
  aralkyl, 119
  aroyl, 119
  aryl, 119
  cycloalkyl, 119
  diacetyl, 119
  dialkyl xanthogene, 121
  diaroyl, 120
  dibenzyl, 119
  di-benzoyl, 119
  dibornyl, 119
  di-n-butyl, 119
  dimethyl, 119
  diphenyl, 119
  tetra-ethylthiuram, 140

Doctor blade, 209–210, 212, 252
Dominant lethal assay, 326
Doubling, 244
Driers, 217
Dry offset, 279
Dryback, 244
Drying,
  inter colour, 244–245
Dyes,
  binder, effect on photoresponse, 106–107
  chemical constitution, effect on photoresponse, 106
  fading of, 105–110
  logwood, 269
  particle size, distribution and photoresponse of, 106
  sublimatic, 234
  triphenylmethane (basic) types, 105

Einstein Equivalence Law, 27
Electronically excited state,
  fate of, 12–13
  molecular rearrangement of, 13
Electronic transitions, 18–21
  n → π*, 17, 72–73
  n → σ*, 72
  π → π*, 16–17, 72–73
  σ → σ*, 16–17
  permitted and forbidden, 7–8
  stable ground state to stable excited state, 19
  stable ground state to unstable state (anti-bonding orbital), 19
Electron,
  donor/acceptor complex, 69–70
  dual nature of, 4
  probability of finding the position of, 4
  properties of, 1–8
  transfer, 67, 104–105, 123, 336
Electromagnetic radiation, 12
  theory of, 1
Electromagnetic spectrum, 41, 263
  visible region, energy levels of for mercury transitions, 43
Elongation of films, 174
Energies of orbitals for non-conjugated systems, 17
Energy of a molecule, 15–16
Engraving,
  copper, 264, 278
  photo, 264, 267, 296–297
Eosin, 140, 184–185
  mechanism of photo-sensitization, 102–103
Epichlorohydrin, 146

Epoxides, 139
Epoxy resin, 145–149
  acrylated, 146, 148–149
  preparation of, 146–147
Erythema, 310–311, 337
Eschar, 310–311
Escherichia coli, 320–325
Esters,
  acrylated epoxy, 145–149
  α-acyloxime, 90–91
  O-alkyl xanthate, 120
  allylic, 278
  cinnamylidene, 277
Etch,
  dow process, 297
  resistance, 267–268
  solutions, 178
  powderless, 280
Ethanolamine,
  N,N'-dimethyl, 203
Ether,
  benzoin, *iso*-butyl, 184
  benzoin, *n*-butyl, 192–201
  benzoin, butyl, spectral absorption coefficients of, 197
  benzoin, *n*-butyl, UV absorption spectrum of, 193
  benzoin, methyl, 283
  trimethylol propane tri allyl, 169
Exciplex, 88–89, 97–98
  formation of, 70–71
Excited states,
  fates of, 21–26
  metastable, 23
  quenching of, 117–118
  triplet, 186
Exciton, 104–105, 114–116
Exposure limit for UV, 305
Extender, 203
Extinction coefficient, 182
Extruded tubes, 240–241
Eye damage, 305
Eye tests, 312–314
  Draize, 312
  FHSLA, 314

Fading,
  effect of humidity on dyes, 108
  effect of treatments on dyes, 107–108
Filler sealer, 232
Film hardness, 130
Finishing,
  dry, 232
  wet, 232
Fish glue, 268

Flash drying, 58–65
Flexibility of films, 174
Flexographic inks, 217–220
  UV curing, 202
Flexography, 211, 279
Flow,
  agent, 202–203
  of films, 179
Fluorene,
  4-acetylamino, 325
Fluorenone, 73, 92, 95, 97–99
  UV spectrum of, 95
Fluorescence, 21–23
  of dyes, 108–109
  resonance, 22
  sensitized, 23
Folding cartons, 244
Formaldehyde, 25, 271, 292
Formulation, 170–204
  clear, coating varnish, UV, 202–203
  paste ink, offset litho, sheet fed, carton, UV, 204
  paste ink, offset litho, white, metal decorating, UV, 203–204
  silk screen ink, UV, 203
  typical photoemulsion coating for screen application, 204
  typical conventional ink, 201–202
  typical UV curing printing ink, 202
N-formyl kynurenine, 336
Fountain solution, 286
Fragmentation reactions, 68
Frank–Condon principle, 20–21
Fraunhofer bands, 128
Free radical, 26
Free radical polymerization,
  auto-acceleration of, 36–77
  inhibition and retardation reactions of, 36
  initiation reaction of, 34
  methods of, 37
  propagation and chain transfer reactions of, 34–35
  termination reactions of, 35–36
Functionality, 163, 174
Furan,
  tetrahydro, 70

Gelatin, 263–265, 268–269, 274, 296–298, 300
Gerade, 8–12
Ghosting, 244
Glass transition, 186–187
  temperature, 173–176
Gloss, 175–176, 179, 216

Glues,
  animal, 265
Glycol,
  ethylene, 150
  1,2 propylene, 150–151
Gold,
  halides, 75
  salts, 123
Gravure, 212–213
  inks, 217–220
  UV curing inks, 202
Gripper bar, 248
Ground state of atoms and molecules, 1
Grotrian diagram, 43–44
Grotthus–Draper law, 46, 106
Growing chain radicals,
  inter reaction of, 64
  oxygen reaction with, 64
Gum Arabic, 265

Halftone, 279–280
  blocks, 296–297
  printing, 186
  reproduction, 211, 213
Hardness of films, 172–174
Hazards of UV systems, 305–337
  inks and coatings, 306–309
  lamps and machinery, 305–306
  testing for physiological, 309–335
Heliograph, 264
Hexadecanol, 169
Hexanediol, 155
1,2,6-Hexanetriol, 156
Hi-top, 236–240
Histidine, 319, 322, 324
Host-mediated assay, 320
Hund's Rules, 7, 12, 16
Hybridization, 12
Hydrogen abstraction, 68–69
Hydroquinone, 152, 158, 283
Hydrogen peroxide, 104, 109, 112–119
  photochemistry of, 29–31
  thermal and photochemical energy required for bond rupture, 26
Hydroperoxide,
  cumene, 118
Hydroxylase,
  benzo(a)pyrene, activity of, 320, 323

Iball index, 318
Image area, 211–212
Imide, poly-, 276
Imine radical, 278
Impact resistance of films, 174
Impression cylinder, 212, 214

Impulse drying, 58–65
    kinetics of, 61–65
Inert blanketing, 129, 150, 175
Inhalation,
    effect on reproduction, 335
Inhalation tests, 328–336
    ASTM, 330
    chromic, 332–335
    Federal Register, 329
    FDA, acute, 330
    FDA, subacute, 331–335
    LC50 determination, 328–329
    toxicity class, 328–329
    Westinghouse specifications, 329–330
Initiation, 63
Inks and coatings,
    affect on the human body, 306
    chemical reactivity of, 306
Inks,
    metal decorating UV, 203–204
    water-based flexographic, 255–256
Inner filter factor, 191
Interdeck curing, 257
Interfacial contact, 276–178
Internal reflectance, 179
Intersystem crossing, 24–25, 73, 157
Iodine, 25
    vapour, 21–22
Iodoform, 272
Ionic initiation, 69–71
Iris, 313–314
Irradiation intensity, 52–57
Irritant,
    mild, 311–312
    moderate, 311–312
    severe, 311–312
Isocyanates,
    acrylated, 153–156

Jablonski diagram, 24–25
Judea asphalt, 264

Ketone,
    aromatic, 38–39
    hydrogen abstraction by, 69
    aromatic/amine combinations, 85–88, 97–98, 130
    methyl ethyl, 175, 181, 220
    Michlers', (4,4'-dimethylaminobenzophenone), 73, 88–89, 203–204
    Michlers', carcinogenicity of, 328
    unsaturated, 290
Kodak photoresist, 295
Kubelka-Munck Equation, 178

Lambert's Law, 17
Laminate plastic, 268
Lamination, 268
Lamps,
    control units of, 50
    cooling systems for, 51
    flash, 58–65
    future types, 56–65
    mercury arc, high pressure, 41, 43
    mercury arc, low pressure, 41, 43
    mercury arc, medium pressure, 41–46
    metal halide, 46
    pulsed xenon, 58–61
    shielding of, 48–49
    xenon, 28
Larynx, carcinogenesis of, 307
Lead,
    chrome, 306
    iodide, 272
Letterflex, 157
Letterpress, 210–211, 220
    printing presses for, 210
Levelling of films, 170–172
Lewis acid, 70, 76, 278
Light absorption in a partially transparent film, 188–192
Line drawings, 213
Linear combination of atomic orbitals, 4, 8
Linseed oil, 217
Lithographic,
    inks, UV curing, 203–204
    plates, presensitized, 271
    printing, sheet fed, 51–53
    printing, web fed, 51–53
Lithography, 220
    offset, 211–212
Los Angeles (1966) Pollution Act, 230
Loss factor, 225–227
Lung,
    carcinoma of, 307

Magnesium,
    iodide, 46
    oxide, UV spectrum of, 111
Magnetron, 226
Maleic anhydride, 70
Manganese carbonyl, 124, 139
Matt systems, 175
Melamines,
    acrylated, 145
Mercaptans, 119
Mercaptides,
    metal, 120
Mechanical properties of films, 172–174

Mercury,
  arc, high intensity, medium pressure, 186
  spectrum of, 192–193
  vapour, 23–24
Metal,
  carbonyl compounds, 124, 139–140
  closures, 241–243
  compounds and ions, 122–124
  containers, 238–241
  decorating, 177–178, 245–246
  decorating, IR application to, 223
  halides, 124
Metastable states, 42
Methacrylate, 168, 174
  methyl, 169
Methane,
  sp^3 hybridization of, 12, 15
Methylene blue, 269
Micronucleus test, 326
Molecular structure, 1–12
Molybdenum carbonyl, 139
Monomers, 162–169
  acrylic esters, chemical reactivity of, 308
  acrylic esters, first aid for, 309
  acrylic esters, handling of, 308–309
  acrylic esters, hazards of, 308–309
  acrylic esters, toxicology of, 308
  irritancy of, 311–312
Morse curve, 17–21
Multiplicity, 7, 16
Mutagenic,
  substances, 318–328
  tests, 319–326

Naphthalene,
  energy level diagram of, 127
Nasal mucosa, 307
Nephritis, 307
News inks,
  drying of by penetration, 215
Newtonian liquid, 170–172
Nitrenes, 127, 277–278
Nitrile,
  α-bis-1-cyclohexane carbo, 127
Nitrobenzene, 152
Norrish reaction, 80–85, 90
  of aromatic ketones, 73–74
  Type 1, 74
  Type 2, 74
  Type 3, 74

Octane,
  1-aza-5-hydroxymethyl 1-3, 7-dioxabicyclo (3.3.0), 167
Oedema, 310–311

Offset printing,
  sheet fed, 211–212
  web, 212
Oils,
  palm, 236, 246
  rolling, 246
Operon, 323
Orbital, 1–17
  bonding and anti-bonding, 8–12
  definition of, 4–7
  hybrid,
    sp, 12, 14
    sp^2, 12
    sp^3, 12
  molecular, 4
  molecular, energy content of, 72
  non-bonding, 17
  order of energy for filling with electrons, 8
  overlapping of, 8–12
  shapes of, 8–12
  splitting of, 8–12
  type, 16–17
    d, 4–6
    f, 4–6
    p, 4–6, 8–12
    s, 4–6, 8–12
    π, 16–17
    σ, 8, 16–17
Organic,
  peroxides, 117–119
  phosphines, 125
  phosphites, 125
Oven,
  box, 235–236
  conveyor or tunnel, 235–236
  metal decorating, 235–236
  wicket, 246
Overprint varnish, 177
Oxetane, 185
Oxide,
  polyethylene, 155
  propylene, 156
Oxime,
  1-phenyl-1,2-propanedione-2-(o-ethoxycarbonyl), 90–91, 320
  UV spectrum of, 90
Oxygen,
  energy level, diagram of, 127
  inhibition, 183
  photochemical reactions of, 127–130
  triplet state, 127–130
  singlet state, 127–129
Oxyplex, 128–129, 184
Ozone gas, 44–51, 306

Paper,
　diazo or blueprint, 269
Pasteurization, 241
Pauli Exclusion Principle, 7
Penta-acrylate,
　dipentaerythritol (monohydroxy), 167
Pentaerythritol tetrakis (thioglycolate), 161
Peroxide,
　diacetyl, 118
　dibenzoyl, 118
　di-tert-butyl, 118–119
Peroxo,
　disulphate ion, photochemistry of, 31–33, 119
　group, photochemical properties of, 117–119
Peroxymonosulphate ion,
　photochemistry of, 33
Perylene, 78
Phenazine, 140
Phenol,
　p-methoxy, 203, 282
　thio, 119
Phenothiazine, 152
1-phenyl-1, 2-propanedione-2-(o-ethoxy carbonyl oxime), UV spectrum and properties of, 90–91
Phosphate,
　tri-butyl, 169
Phosphine,
　organic types of, 125
　tri*ortho*tolyl, 125
　triphenyl, 125
Phosphites,
　organic types of, 125
Phosphorescence, 23
　α, (delayed fluorescence), 24
　β, 24
　resonance, 42
Phosphorus compounds,
　organic, 125
Photobiology,
　chemical aspects of, 336–337
　eye effects, 336
　skin effects, 336–337
Photocationic,
　olefinic polymerization, 75
　ring opening polymerization, 75
Photochemical reactions,
　kinetics of, 33–37
　thermodynamics of, 37–39
Photochemistry, 12–39
　definition of, 12
　solid state, 104–117

Photoconductive effect, 104–117
Photo-crosslinking reactions, 71
　of saturated polymers, 141
Photocycloaddition reactions, 277
Photodimerization, 142–143
Photoemulsion, 307
　polymerization, 32–33
Photo engraving, 125, 268
Photohydration, 337
Photo-initiation, 231
Photoinitiators,
　classes of, 74–127
　definition of, 67
　general mechanisms of action, 68–72
Photo-ionic polymerizing compounds, 74–79
Photo-keratitis, 336
Photolysis, 110, 112
Photopolymerization, 137–144
　air inhibition of, 127–130
　of saturated polymers, 139–141
　of vinyl unsaturated compounds, 138–139
Photoresists,
　acid/alkali, properties of, 265
　applications of, 263–264, 279–300
　classes of systems, 263–264
　definition of, 263
　dichromate/polyvinyl alcohol system, 265–274
　double layer, 269–270
　Kodak, 295
　negative working, 274, 276–279
　positive working, 270, 274–276
　reversal, 274
　types of, 274–279
Photosensitive functional groups, 141–142
Photosensitization,
　by cinnamoyl and related groups, 142–144
　by dyes, 78, 102–104
　　aerobically, 103–104
　　anaerobically, 102–103
　by ferric salts, 274
　by pigments, 110–117
Photosensitizers, 24
　classes of, 74–127
　definition of, 67–68
　dye, 102–110
　general mechanisms of action, 68–72
Photostencil, 278
　direct, 213, 298
　direct/indirect, 213, 298–299
　indirect, 213, 298
Phototropy, 110–113

Phthalate,
  di-allyl, 169, 278
  bis(formyl-4-nitro-phenyl), 291
Pigment,
  calcium 4B magenta, 204
  extender, 175, 178–179
Pigmentation of photopolymerizable systems, 178–180
Pimpling, 280
Pinacol, 157
Pine-oil, 169
Planck's constant, 2, 15
Plastisol, 234, 255–256
Plasticization,
  external, 170
  internal, 170
Plasticizers, 169–170, 173, 186, 203
  primary, 169–170
  secondary, 170
Plates,
  bimetallic, 247
  deep etch, 247
  photopolymer, 247
  trimetallic, 247
Polyamide,
  unsaturated, 138
Polyesters,
  acrylated, 144, 152–153
  cinnamate, 290
  unsaturated, 144, 149–153
Polyethers,
  acrylated, 145, 156–157
Polymerization,
  addition, 137–138
  condensation, 137–138
  vinyl, 139
Polymers,
  acidic, 276
  azide functionalized, 278
  chalcone, 291
  containing cyclic groups, 140–141
  halogenated, 139–140
  incorporating dye groups, 140
  keto, 140
  latex type, water based, 219–220
  network, 185–186
  possessing, 2,2 dimethyl 1,3 dioxolane groups, saturated, purpose modified, 139–141
  sulphur containing, 140
  thermoplastic, 159
  thermosetting, 160
  unsaturated, 138–139, 141–143
  visco-elastic properties of, 174
  water soluble, 219–220
Polystyrene, 139
Polysulphide,
  dibenzoyl, 121
  dibenzothiazoyl, 121
  Dibenzyl, 121
  Dithiobenzoyl, 121
Polyvinyl,
  acetate, 204, 268
  acetal, 138
  alcohol, unsaturated derivatives of, 138
  cinnamate, 142–143
  esters, 138
  trichloroacetate, 139
Positive hole, 104–105, 114
Post cure treatment, 176–177
Pot life (stability), 179–219
Potential energy curve, 17–21
Powdering, 216
Power law fluids, 171–172
Predissociation, 13, 19, 25
  fluorescence, 13
Prepolymers,
  acid crosslinkable, 219
  reactive at elevated temperature, 219
Pre-treatments, 177–178
Press stability, 217–218
Presses,
  flat bed cylinder, 210
  platen, 210
  rotary, 210
Primary,
  irritation index, 310–314
  photochemical process, 14, 21, 26–34
  radical formation, 63
  radical reaction, 63, 64
  recombination, 63
Printed,
  circuit, 213
  paper, recycling of, 219–221
Printing inks, 213–214
  drying of by absorption, 215–216
  drying of by evaporation, 217–218
  drying of by oxidation and polymerization, 217
  drying of by precipitation, 218
  film thickness of, 214
  glycol based, 218, 247
  'hardening' of, 214
  heatset, 218–219
  letterpress, moisture set type, 218
  lithographic sheet fed offset, IR absorption properties of, 222–224
  methods of drying of, 214–259
  pigmentation levels of, 214

quicksetting type, 216, 220
radiation curing, 221–259
'setting' of, 214
solvent based methods of drying, 214–218
solventless methods of drying, 218–259
solventless, oil based, 220–221
thermally catalysed, 219
UV curing of, 244–259
viscosities of, 214
water based, 219–220

Printing plates, 211, 212, 279–296
Asahi photopolymer resin (APR), 281–285
bimetallic, 287–296
deep etch, 287
driographic, 286
Dycril, 281–285
Dynaflex, 283–284
flexomer, 283–284
letterflex, 281–285
NAPP, 283, 285
Nyloprint, 283–285
photogravure, 296
photolithographic,
 bimetallic or trimetallic, 286–288
 deep etch, 286–288
 planographic, 286–296
 'reversal' positive working, 286–288
 surface negative working, 286–288
 surface positive working, 286–288
relief, 157, 274, 279–285
 liquid type, 281–282
 solid type, 282–285
 Sonne KPM, 283, 285
 surface, 286–287
 Tevista, 283, 285
 Toplon, 283, 285
 Torelief, 283, 285
 trimetallic, 287–296

Printing problems (and possible remedies),
greasing (dot spread), 249
ink not drying adequately, 250
linting/picking, 249
misting or flying, 249
piling, 248–249
poor rub and scratch resistance, 250
poor trapping, 250
scumming/tinting, 248
UV inks, 248–251

Printing processes,
collotype, 300
dry offset application, UV curing of, 245–247
flat sheet metal, lithographic, UV curing of, 245–246
flexographic, UV curing of, 252–254
general, 209–213
gravure, 245
gravure, UV curing of, 251–252
Intaglio, 212–213
letterpress, flat bed, 283
letterpress, rotary, 283
lithographic, 269
newspaper, 279
perfecting, 212
planographic, 211–212
plastics, UV curing for dry offset, 247
relief, 210–211
screen, UV curing of, 254–256
sheetfed, lithographic UV curing of, 244–245
silkscreen, stencils for, 297–299
web offset, UV curing of, 245
Proofing systems, 300
Propagation, 64
Psi, ψ, wavefunction, 3–5, 8–12
Pyrazoles, 126
Pyrazoline, 126
 3-acetoxy-3,5,5-trimethyl-1-, 126
Pyrene, 325
Pyrrolidene,
 N-vinyl, 164
 polyvinyl, 268

Quantum mechanics, 3
Quantum number,
 azimuthal, l, 4, 5, 7
 magnetic, m, 4, 7
 principal, p, 2, 4
 spin, s, 4, 7
 total and resultant, L, S, J, 7, 9
Quantum yield,
 definition of, 26–27
Quenching,
 of dyes, 104–105
 of excited states, 21–22
Quinone(s), 99–101, 204, 275, 293
 p-benzo, 204
 diazides, 275

Radiant absorption by photoinitiators, 187–188
Radiation,
 alpha, α-, 227, 229–230
 beta, β-, 227, 229–230
 black body, 221–225
 electromagnetic, 221–259
 gamma, γ-, 227, 229–230

X-, 230
drying, 221–259
  electron beam, 227, 229–230
  free radical forming, 227–259
  infra-red, 172, 221–225
  infra-red, effect of substrate on coatings cure, 224–225
  path length of, in films, 179
  quantum of, 12
  radio frequency, 225–227
  thermal effect, 221–227
Radicals,
  alkoxyalkyl, 84
  alkoxy benzyl, 81–82
  benzoyl, 81–82
  dimethylamino, 126
  diphenyl, 294
  hydroperoxide, 87
  hydroxyl, 104, 114–117
  hydroxymethyl, 31
  peroxy, 183
  semiquinone, 101
  thiocyanate, 120
  thiyl, 157
Recombination,
  diffusive secondary, 27–28
  primary, 27–28
Reflector units, 46–48, 54–56
  elliptical types, 47–48, 54–56
  non-focussing type, 46–47
  parabolic type, 48
Register, 220, 226
Resins, 144–161
  acrylated epoxy, 203–204
  acrylated polyester, 202
  acrylated polyurethan, 202–203
  acrylic, 276
  bakelite, 269
  cinnamate compounds, 291–296
  cinnamate esters of epoxy, 291
  diazo, 271–272
    p-diazo diphenylamine para formaldehyde, 286, 290
  epoxy, 145–149
    derivatives for cans, 238–241
    modification of, 146–149
    preparation of, 146–149
    properties of, 145
  infra-red absorption properties of, 224
  phenol formaldehyde, 271–272
  polyester,
    alkyd, 219
    modification of, 152–153
    properties of, 152
    with styrene, UV curing systems for
      wood finishing, 232–234
  polyethers, preparation, 156–157
  polyurethan, 153–156, 281
  melamine formaldehyde, 219, 238, 243
  melamine sulphonamide formaldehyde, 106
  nitrocellulose, 269
  novalak, 276
  urea formaldehyde, 219
  urethan, modification of, 154–156
    preparation of, 154
Retorting, 238, 241
Rheology, 170–172, 175
Rhodamine, B, 106
Ribonucleic acid (RNA), 322, 337
Roller,
  application, 209–210
  coater, 170
  coating, 209–210
  feed, 209
  pressure, 210
  transfer, 209
Rolling oils, 177
Roll-on closures, 242–243
Rubber,
  Buna, 247
  nitrile, 247
Russell–Saunders coupling, 7, 42
Rutherford Theory of the atom, 1–3

Safranine, 140
Salmonella typhimurium, 319–325
Saw-toothing, 298
Schiemann Reaction, 70, 76
Schrödinger,
  mathematical description of the atom, 3–5, 8
  wave equation, 4, 8
Schumann-range region, 128
β-Scission, 29
Screen,
  inks, UV curing, 203
  printing, 213
Secondary photochemical process, 14, 21
Selection rules, 24, 42
Semi-conductors (type), 104–105
Sensitization tests, 314–315
  Draize/Landsteiner, 314
  Magnusson and Kligman, maximization, 315
  occluded patch, 314–315
Sensitizers, photo-optical, 67
Set-off, 214, 244, 246
Shear,
  rate, 171–172

stress, 171–172
Shellac, 268
Shielding, 48–49
Shortages,
  energy, 218
  raw materials, 218
Show through, 216
Silane,
  chloro, 127
  trimethylchloro, 126
Silica, 179, 203–204
Silver perchlorate, 75
Singlet,
  oxygen, 101, 108–109
  state, 7, 16
Size coat, 242
Skin irritancy, 174
Skinning, 244
Skin test(s), 309–312
  Draize, 309–311
  repeat insult, 311–312
Sodium,
  alginate, 287
  azide, 325
Solvent(s)
  burning of, 219
  recovery of, 218
  release of, 186
Spectral,
  data, significance to cure rate, 192–195
  lines, intensity of, 52–57
  lines, width of, 53–54
Spectroscopic nomenclature, 8–12
Spin-correlation effect, 7
Spray powder, 244, 248
Squash, 280
Starch, 265
Stark–Einstein Law, 14
Statistical weight effects, 53
Stefan–Boltzmann Law, 221
Stilbene, 277
Strike through, 216
Stoving, 234–236
Styrene, 70, 75, 128, 149–153, 164–169, 172–173, 187
  butyl-, 153
  chloro-, 153
  methyl-, 69, 75, 164
Suberone,
  dibenzo, 92, 95
  UV spectrum of, 95
Substrate,
  cold, 186
  metal, 186, 236–238
  moist, 186

porous, 200–201
reflecting, 179, 186
Sulphate, chromic, 307
Sulphenate(s), 120
  ethyl trichloromethane, 120
  2,4,5, trichloro-phenyltrichloromethane, 120
Sulphide(s),
  β-keto, 120
  tetramethylthiuram, mono, 122
Sulphonamides, chloro, 120
Sulphonate, methyl methane, 325
Sulphonic acid,
  toluene-, 219
Sulphur compounds,
  organic, 119–123
Surface,
  cure, 130
  defects, 170–172
  energy, 176–177
  tension, 171–172, 214–216
Symmetry, 8–12
Synergistic agents, 73, 125, 130, 320

Tack, 179–182, 214, 217, 220
Tacticity, 35
Talc, 179
Tallate,
  iso-octyl, epoxy ester of, 169
Tanning,
  chrome, 269
Tensile strength of films, 174
Term symbol, 7–12
Termination, 231
  processes, 64–65
Terneplate, 236–237
Tertiary amines, 71
Tetra-acrylate,
  pentaerythritol, 167
Tetracyanobenzene, 69, 75
Tetrahydrofuran, 70
Tetrazene,
  tetramethyl, 126
Textile transfer, 234
Thallium vapour, 23
Thermal assist, 204
Thermoplasticity, 173, 241
Thiepin-11-one,
  2-chloro-6,11-dihydrodibenzo-, 92–95
    UV spectrum of, 95
  6-11-dihydrodibenzo-, 92, 95
    UV spectrum of, 95
Thiols, 119
Thiol/ene systems, 145, 157–161

Thiosulphate,
  sodium benzyl, 121
Thioxanthone (UV spectrum), 92, 95
  2-acetyl, 92, 94
    UV spectrum of, 94
  2-benzyl, 92, 94
    UV spectrum of, 94
  2-chloro, 71, 91–93, 95, 183–184, 320
    UV spectrum of, 95
  2-cyclohexyl, 92, 94
    UV spectrum of, 94
  2-ethyl, 92, 95
    UV spectrum of, 95
  2-*iso*-propyl, 92–94, 320
    UV spectrum of, 94
  4-*iso*-propyl, 92, 94
    UV spectrum of, 94
  2-methyl, 92, 95
    UV spectrum of, 95
  2-phenyl, 92, 94
    UV spectrum of, 94
  2-tert-butyl, 92, 94
    UV spectrum of, 94
Thioxanthone derivatives, 91–95
  UV spectra of, 94–95
Thiuram derivatives, 122
Thixotropic system, 172
Through cure, 130
Thymine, 337
Tinplate,
  differential, 236–237
  electrolytic, 236–241
  hot-dipped, 236–241
Titanate, *iso*-propyl, 156
Titanium dioxide, 110–117, 203, 243
  UV spectrum of, 111, 113
Toluene, 220
  di-isocyanate, 154–156
  vinyl, 153, 164
Topcoat, 177, 186, 232
Topography, 177
Toughness,
  of films, 172, 174
Toxicity, 174, 243, 315–318
  acute LD50 determination, 315–317
  gross, 317–318
  oral, rating of materials, 317
Transition,
  intrinsic probability of, 53
  type $n-\pi^*$, 192–194
  type $n-\sigma^*$, 117
  type $\pi-\pi^*$, 192–194
Trapping,
  wet on dry, 244–245
Triacrylate,
  pentaerythritol, 168
  trimethylolpropane, 168, 203–204
Triaryl,
  selenonium compounds, 75, 79
  sulphonium compounds, 75, 78–79
Triazines,
  acrylated, 145
Trimethylol propanetris,
  (β-mercaptopropionate), 161
Triple bond,
  infra-red absorption of, 185
Triplet state, 7, 16, 24–25
  energy transfer reactions of, 71–72
Tromsdorff effect, 63
Tryptophan, 322, 324, 336

Ultraviolet coatings, 231–244
Ultraviolet curing, 230–231
  advantages and disadvantages of, 257–259
  interdeck, 244, 246
Ultraviolet drying systems,
  installation of, 51–52
Ultraviolet inks, 202–204, 244–259
  in the field, 247–251
    equipment compatibility, 247–248
    fountain etch, 247
    ink rollers and blankets, 247
    plates, 247
    press conditioning, 247
  wash-up of, 248
Ultraviolet radiation,
  hazards of, 48–49
  reflectance of, 49
Uncertainty Principle (Heisenberg), 3, 54, 70
Undercoat (basecoat), 232
Ungerade, 8–12
Uracil, 337
Uranyl salts, 123

Van der Waals' forces, 159, 170, 173
Varnishes,
  clear overprint, photopolymerizable for paper and board, 244
  finishing, 238–239, 243
Vinyl,
  acetate, 168, 170
  ethers, 75

Waiter trays, 240
Wave,
  -function, normalization of, 4
  -length maxima for infra-red radiators, 223–224

mechanics, 3–4
Wax compound, 204
Wetting, 172, 176
   agents, 162
Wien's Displacement Law, 222
Wigner Spin Conservation Law, 129
Wolff rearrangement (photo-induced), 275
Wood-finishing,
   acid-curing systems for, 232
   aqueous systems, 232
   nitrocellulose systems, 232
   polyurethan systems, 232
Wrinkling, 178–182

Xanthone, 73
   2-chloro, 92, 95
      UV spectrum of, 95
Xenon lamps, 28, 58–61

Zero point vibrational energy, 18
Zinc oxide, 110–112
   UV spectrum of, 111

/547.84R719P>C1/